AF682643

Automotive Sensory Systems

Road Vehicle Automation Series

Edited by Chris Nwagboso

Vehicle Systems Research Centre
School of Engineering
Bolton Institute
Bolton
England

The increasing level of innovations in the fields of information technology, telecommunications, electronics and materials technology in the automotive industry worldwide requires a framework to allow all those involved in these developments to exchange their knowledge and expertise. This need will be met by a series of books on road vehicle automation.

The series focuses on all aspects of vehicle engineering encompassing advanced transport informatic and telematic technology; the aim is to publish high quality books on all aspects of road vehicle automation. It will bring together the expertise of international specialists, researchers, consultants, engineers and managers whose work and efforts are shaping the present and future of road vehicles. These books will provide a truly unique interface between the industries, research establishments and academic institutions working in this fast evolving and exciting area of technology.

As series editor, I ask prospective authors and editors to submit proposals to me in detailed form, incorporating the following: a statement of the objectives and scope of the book; a definition of the intended readership; an indication of its readership; details of its relationship to existing texts; a note of the intended length; a contents list and, if available, two draft chapters. Chapman & Hall will collaborate with each author/editor to ensure that their proposal is evaluated in a fully constructive manner, and that their books are edited, manufactured, promoted and distributed worldwide in a way likely to maximize sales.

1. Automotive Sensory Systems
Edited by Christopher O. Nwagboso

Automotive Sensory Systems

Edited by

Christopher O. Nwagboso

Vehicle Systems Research Centre,
School of Engineering,
Bolton Institute, Bolton, UK

CHAPMAN & HALL

London · Glasgow · New York · Tokyo · Melbourne · Madras

Published by Chapman & Hall, 2–6 Boundary Row, London SE1 8HN

Chapman & Hall, 2–6 Boundary Row, London SE1 8HN, UK

Blackie Academic & Professional, Wester Cleddens Road, Bishopbriggs, Glasgow G64 2NZ, UK

Chapman & Hall Inc., 29 West 35th Street, New York NY10001, USA

Chapman & Hall Japan, Thomson Publishing Japan, Hirakawacho Nemoto Building, 6F, 1–7–11 Hirakawa-cho, Chiyoda-ku, Tokyo 102, Japan

Chapman & Hall Australia, Thomas Nelson Australia, 102 Dodds Street, South Melbourne, Victoria 3205, Australia

Chapman & Hall India, R. Seshadri, 32 Second Main Road, CIT East, Madras 600 035, India

---

First edition 1993

© 1993 Chapman & Hall

Typeset in 10/12 Times by Thomson Press (India) Ltd., New Delhi, India
Printed in Great Britain at the University Press, Cambridge

ISBN 0 412 45880 2

Apart from any fair dealing for the purposes of research or private study, or criticism or review, as permitted under the UK Copyright Designs and Patents Act, 1988, this publication may not be reproduced, stored, or transmitted, in any form or by any means, without the prior permission in writing of the publishers, or in the case of reprographic reproduction only in accordance with the terms of the licences issued by the Copyright Licensing Agency in the UK, or in accordance with the terms of licences issued by the appropriate Reproduction Rights Organization outside the UK. Enquiries concerning reproduction outside the terms stated here should be sent to the publishers at the London address printed on this page.

The publisher makes no representation, express or implied, with regard to the accuracy of the information contained in this book and cannot accept any legal responsibility or liability for any errors or omissions that may be made.

A catalogue record for this book is available from the British Library

Library of Congress Cataloging-in-Publication

Automotive Sensory Systems / edited by Christopher Nwagboso.–1st ed.
p. cm.–(Road vehicle automation series)
Includes index.
ISBN 0-412-45880-2 (acid-free paper)
1. Automotive sensors. I. Nwagboso, Christopher. II. Series.
TL272.5.A988 1993
621.25′49–dc20 92-47248
CIP

This book is dedicated to
Martha, Louise, Lloyd, Hollie
and Naomi

Contents

List of contributors

C.O. Nwagboso, Vehicle Systems Research Centre (VSRC), School of Engineering, Bolton Institute, Bolton, UK.

E.N. Goodyear, SIRA (UK) Ltd, Chislehurst, UK.

S. Re Fiorentin, Electronic Systems, Fiat Research Centre, Orbassano, Italy.

J.A. Foerster, Electronic Instrumentation Laboratory, Faculty of Electrical Engineering, Delft University of Technology, Delft, The Netherlands.

R.F. Wolffenbuttel, Electronic Instrumentation Laboratory, Faculty of Electrical Engineering, Delft University of Technology, Delft, The Netherlands.

G. Rizzoni, Centre for Automotive Research, Department of Mechanical Engineering, Ohio State University, Columbus, Ohio, USA.

W.B. Ribbens, Vehicular Electronic Laboratory, University of Michigan, College of Engineering, Ann Arbor, Michigan, USA.

P. Tiedema, Deltec Fuel Systems BV, Rotterdam, The Netherlands,

J. van der Weide, TNO Road Vehicles Research Institute, Delft, The Netherlands.

R.E. Bicking, Micro Switch Division, Honeywell Inc., Freeport, Illinois, USA.

M. Bouamra, Photonics Systems Laboratory ENSPS/ULP, Strasbourg, France.

P. Meyrueis, Photonics Systems Laboratory ENSPS/ULP, Strasbourg, France.

A. Chakari, Photonics Systems Laboratory ENSPS/ULP, Strasbourg, France.

M. Mutoh, Electronic Engineering Department, Toyota Motor Corp., Kariya, Aichi, Japan.

M. Masegi, Air Bag Engineering Department, Nippondenso Co. Ltd., Kariya, Aichi, Japan.

Z.S. Younan, Program Management Asian-Pacific Operations, TRW Technar Inc., Irwindale, California, USA.

L. Pastor, Faculty of Informatics, Polytechnic University of Madrid, Madrid, Spain.

P. de Miguel, Faculty of Informatics, Polytechnic University of Madrid, Madrid, Spain.

A. Pérez, Faculty of Informatics, Polytechnic University of Madrid, Madrid, Spain.

F. Rosales, Faculty of Informatics, Polytechnic University of Madrid, Madrid, Spain.

S. Rodriguez, Faculty of Informatics, Polytechnic University of Madrid, Madrid, Spain

A. Cabañas, Faculty of Informatics, Polytechnic University of Madrid, Madrid, Spain.

A. Rodriguez, Faculty of Informatics, Polytechnic University of Madrid, Madrid, Spain.

Z.W. Porada, Kraków Polytechnic, Institute of Electrotechnics and Electronics, Kraków, Poland.

J.H. Beyer, The Environmental Research Institute of Michigan, Ann Arbor, Michigan, USA.

W.F. Pont, Jr, The Environmental Research Institute of Michigan, Ann Arbor, Michigan, USA.

K.D. Kuhnert, TZN, Forschungs- und Entwicklungszentrum, 3104 Unterlüß, Neuensothrletherstr. 20, Postfach 27, Germany.

S. Baranowski, Electronics Department, Lille University, Villeneuve D'ASCQ Cédex, France.

M. Lienard, Electronics Department, Lille University, Villeneuve D'ASCQ Cédex, France.

P. Degauque, Electronics Department, Lille University, Villeneuve D'ASCQ Cédex, France.

K.D. Gill, Rover Group PLC, Longbridge Operations, Birmingham, UK.

E. Zabler, Robert Bosch, GmbH, Automotive Equipment Division, Ettlingen, Germany.

F. Heintz, Robert Bosch, GmbH, Automotive Equipment Division, Ettlingen, Germany.

R. Dietz, Robert Bosch, GmbH, Automotive Equipment Division, Ettlingen, Germany.

O. Ina, IC Engineering Department, Nippondenso Co. Ltd., Kariya, Aichi, Japan.

Y. Yoshino, IC Engineering Department, Nippondenso Co. Ltd., Kariya, Aichi, Japan.

M. Iida, IC Engineering Department, Nippondenso Co. Ltd., Kariya, Aichi, Japan.

Preface

The rapidly growing need for mobility has brought with it a major challenge for improvement in the operation and utilization of automotive systems. The economical, environmental and safety constraints imposed by the increase in the number of road vehicles and subsequent government policies also require substantial product development through the application of information technology. This involves the enhancement of vehicle informatics and telematic systems with additional sensors and systems.

The advance in the design and development of automotive sensory systems is so rapid that there is urgent need for the experts involved in the technology to work together to provide a reference book for the engineer of today and tomorrow. This motivated me to spend two years researching the topics and the basis on which such a book should be written. The result is the present compilation of the work of international experts on the state-of-the-art in the field of automotive sensory systems. Thus, a unique collection has been created for the reference of all those concerned with, or interested in, the design and development of modern, safe and intelligent vehicles. Although this book is intended for engineers, managers, scientists, academicians and policy makers, students should also find it valuable. To meet the requirements of students the basics are explained in simple terms; however, it is hoped that others will appreciate this approach, since most of us are well aware that gaps remain in our knowledge of the elements of our profession. These basics have been kept as short as possible so that experienced readers will almost be able to ignore them and to turn their attention immediately to the more advanced concepts which constitute by far the greater proportion of the content.

This book is divided into eight parts, each of which is devoted to different types of automotive sensory system applied to different aspects of vehicle systems. The individual parts contain one or more chapters, the contents of which are outlined below.

Part One consists of two chapters. Chapter 1 gives an introduction to automotive sensory systems, with their definition, background information and present and future trends in their applications. Chapter 2 presents an overview of motor vehicle sensing requirements. In this chapter, reference is made to the more advanced developments in automotive sensory systems, in particular to the PROMETHEUS programme (PROgraMme for a European Traffic with a Highest Efficiency and Unprecedented Safety).

Part Two, which consists of three chapters, discusses different types of sensor used for power plant and transmission. Chapter 3 presents the development of non-contacting capactive torque sensors for power trains (engine and transmission), and shows the advantages of using such sensors instead of strain gauges. In Chapter 4 a variety of techniques are introduced for measuring crankshaft angular position and velocity in relation to engine performance. Chapter 5 describes the application of modern electronic engine management systems to gaseous fuelled engines. This management is assisted by integrated sensors and control systems which enable gas engines to keep pace with the rapid development in petrol and diesel engines in terms of higher performance and reduced exhaust emissions. The hardware and software used in such systems are also discussed.

Part Three of this book contains two chapters dealing with vehicle ride and comfort. Chapter 6 presents basic approaches and technologies for sensing acceleration. The requirements for accelerometers used for vehicle ride and comfort are explained, and technologies currently under development are described and the best suited are discussed. Chapter 7 describes a model for demonstrating the feasibility of using a multimode fibre optic gyroscope for improving vehicle ride. This chapter presents various applications of optical fibre sensors in vehicle technology and discusses results of experiments using the gyrometer model.

Part Four is devoted to security and safety systems. Chapter 8 describes the sensory systems and types of sensor used in developing the Toyota airbag system. Two types of airbag system are discussed: mechanical and electrical. With the mechanical type, both the inflator and the sensors are integrated compactly into the steering wheel pad. With the electrical type, several sensors are positioned in different parts of the vehicle and only the inflator is built into the steering wheel pad.

Driver behaviour modelling and information systems are presented in Chapters 9 and 10 of Part Five. In Chapter 9, there is a description of a modification made on a standard road vehicle in order for it to be used in driver behaviour studies. The driver behaviour study system consists of two subsystems, one mobile and the other laboratory based. The mobile subsystem includes the modified vehicle with an integrated sensory system used for monitoring and measuring driver parameters and action. The laboratory subsystem includes the computer hardware for data capture and recovery for use in reconstructions and analysis of driving sessions. Particular attention is paid in this chapter to two types of sensory system based on image processing techniques. System integration and exploitation aspects are also discussed. Driver and occupant information systems are presented in Chapter 10, where the development of a dashboard information display system is described. A new concept for an electroluminescent dashboard information display system is introduced in which DC and AC power supply voltages are applied simultaneously; this allows the DC voltage to be reduced to about 25 V. This new display system was produced by vacuum methods and is in the form of a tri-electrode structure on a glass substrate.

Part Six is concerned with vehicle navigation and communication. It contains three chapters (Chapters 11–13), which discuss various aspects of sensory systems used in guidance and communication. Chapter 11 presents an imaging sensory system for autonomous vehicle guidance developed at ERIM, which uses laser range imagery techniques. The data and information processing method is also discussed. Chapter 12 describes both the function of sensory systems for real-time vision techniques for high speed vehicle guidance, and the process of designing such systems. Examples of the performance of a typical system are also given. The final chapter in Part Six describes a communication system using electromagnetic characteristics for a road-to-vehicle link in the 1–10 GHz frequency. Emphasis is placed on the performance of the link and on problems in the system (such as attenuation), and a study is made of multipath and masking effects.

Part Seven deals with the use of sensors in automotive manufacture. Chapter 14 discusses the use of machine vision systems for vehicle manufacture in the Rover Group assembly plant in the United Kingdom. It presents various optical (vision) sensors used at different stages of component manufacturing and car assembly and uses various case studies to describe the application of sensory systems.

The concluding section of this work, Part Eight, covers the topic of smart sensors and systems. Chapter 15 deals with a new trend in sensor development called mechatronics, in which mechanical and electronic parts are integrated in an independent system of intelligent (or smart) sensors. The structure of smart sensors with a localized signal processing capability, their application and the benefits of using this mechatronics are discussed. Finally, Chapter 16 addresses the development of intelligent sensor technology in Japan. New smart sensors, which are in fact integrated single or plural sensors and amplifying circuits laid on a single chip using LSI manufacturing technology, are described and their advantages are discussed.

Christopher Nwagboso

Acknowledgements

My deepest gratitude and sincere appreciation are due to national and international colleagues in the automotive industry whose efforts made my task bearable.

I am deeply grateful to Professor Erik Sandewell, who through our discussion in Brussels provided me with certain inspiration for this book. Appreciation is due to Dr Steve Prosser, manager of the Sensors and Microelectronics Department of Lucas Automotive Ltd, for his kind and positive comments on the book proposal and also to Dr J.W. Dunn of Birmingham University, UK.

Thanks are due to Roy Healey, who encouraged me and provided a conducive working environment and also to colleagues at the Vehicle Systems Research Centre, Bolton, UK.

To those who contributed to this book, directly and indirectly; thank you for taking part in the shaping of the future of road transport.

PART ONE

Automotive Sensors

CHAPTER 1

Introduction to automotive sensory systems

1.1 INTRODUCTION

Within the automotive industry, it is known that a quiet technological revolution is progressively taking place in the area of vehicle system performance and control. This revolution is undoubtedly fuelled by the fact that at the present, automotive component and vehicle manufacturers are confirmed with ever-increasing demands from their customers and government legislation for high-quality, reliable, safe and environmentally less damaging products. Compounded with these demands, there is continuing competition within the industry for the available market share, which is normally influenced by the national and international economic cycle.

The increase in road vehicle automation is through the application of sensory systems for control of vehicle components and subsystems. As part of the sensory systems, the application of information technology and electronics in road vehicles has meant that control, monitoring and information display functions normally carried out mechanically or manually are now performed with the enabling sensors and computer processing systems. Some of these functions include: engine control, chassis control, vehicle safety, vehicle–infrastructure interactions, systems diagnostics, emission control information systems and communications. In all these functions integrated sensory systems play a very important role.

Currently, there is a continued drive towards the development of the so-called 'smart' or intelligent vehicles. If the vehicle of the future is to be made more intelligent with some elements of autonomous control, integrated sensory systems will form the most important and central aspects of the vehicle.

This chapter will define automotive sensory systems in terms of an integrated part of a vehicle. Current applications of sensors and systems are discussed. Future development in vehicle sensory systems is discussed.

1.2 WHAT ARE AUTOMOTIVE SENSORY SYSTEMS?

To define automotive sensory systems, a general definition of some of the devices which make up a basic sensory system will be useful. These devices can be classified into one distinct category, namely measurement or sensory system. The measurement system consists of devices which can obtain a quantitative comparison between a predefined standard and a measured. The word measured is used in this instance to designate the particular physical parameter being sensed and quantified within an automotive system. Transducers, sensors and detectors are some of the names used to describe a measurement system.

Transducers are devices capable of converting one form of energy or physical quantity to another. Often this energy or stimulus determines the quantity of the signal obtainable.

Sensors are devices used for the purpose of detecting, measuring or recording physical phenomena and subsequently responding by transmitting information, initiating changes or effecting systems control.

Detectors are devices capable of sensing the presence of something such as physical phenomena, heat, radiation, etc. within the automative system's environment.

It is apparent from this definition that any difference between these devices will be difficult to distinguish. Therefore for the purpose of defining automotive sensory systems, these devices can generally be classified as sensory systems. Typically, sensory systems will comprise a number of elements integrated into component blocks. Most basic sensory systems will contain the sensor, a signal conditioning element, an analysis element and a recorder or display system. Within the sensory system's structure, a feedback sensory signal can be provided in order to optimize the behaviour of the system.

As part of the evolving application of computers, electronics devices and communication systems in vehicles, automotive sensory systems can be difficult to define completely, since many views are possible within this developing technology. To capture the most important aspects of automotive sensors and systems, an automotive sensory system is defined as the integrated part of the vehicle sensors and systems with the functions of detecting, transmitting, analysing and displaying vehicle performance information from a given stimulus within the internal and external environment of the vehicle. This definition will in future include the capacity of the sensory system to memorize and logically store vehicle performance information for the purpose of performance diagnostics and prognostics with inbuilt intelligence and self-calibration. It is possible that a full definition of automotive sensory systems will also include the ability of the sensory systems to form part of a specialized distributed and multi-tasking capability within a networked multiprocessing system. As shown in Figure 1.1, at a low and normal driving task levels, the driver forms part of the sensory system. An integrated sensory system takes account of the ability of the driver to make some intelligent

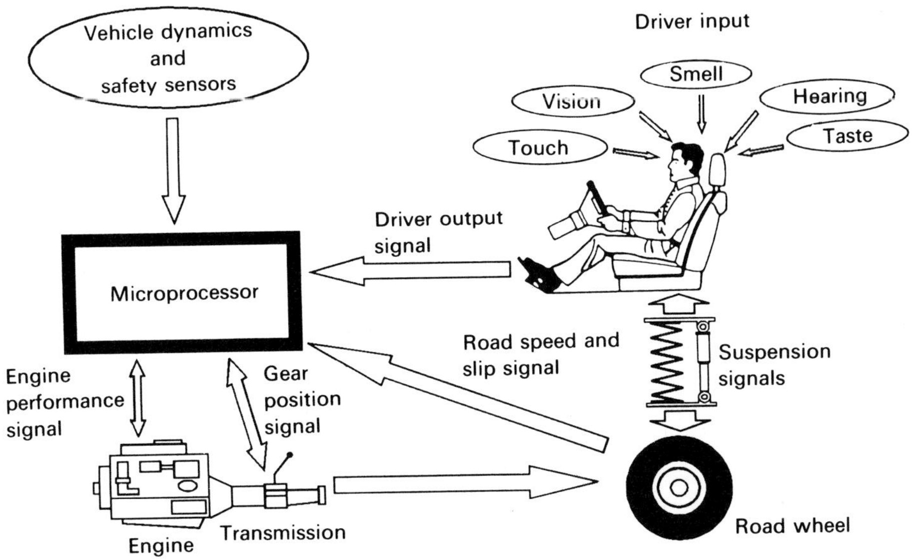

Figure 1.1 Driver as part of the sensory system.

decisions on the progress and state of the road vehicle during driving and parking conditions. The responsibility of the driver in this respect will obviously be reduced as the sensory systems and the vehicle are made more intelligent.

1.3 THE BACKGROUND

The use of sensors and systems in road vehicles was, a decade ago, restricted to the acquisition and display of fairly inaccurate information on vehicle and engine speed. This was extended to simple sensing and display of water temperature, fuel level in the tank and lubrication system failure. With this basic level of sensor technology the demand put on the driver to make up for the deficiency was too high. As can be testified by most drivers under normal driving conditions, this can result in pressures and stress on the driver in busy traffic conditions and during long-distance travel. The stresses and pressures can be even higher on people who drive as part of their job (taxi drivers, bus and heavy goods vehicle drivers).

Today, however, an increasingly competitive market coupled with

- stringent environmental and safety legislation;
- expected high-level driveability and performance;
- moves towards a vehicle with self-management and diagnostics;
- improvements in luxury and comfort;
- the availability of low-cost electronic devices

has encouraged manufacturers to integrate into their vehicles sophisticated sensory systems capable of enhancing the performance and safety of the vehicle. Basic mechanical and some manual operations are now controlled and monitored by these sensory devices. The use of these devices has increased rapidly since early 1970 [1, 2]. This growth has been accentuated by various technical papers [3–6]. The high-level materials discussed in this book indicate that this growth will continue at a rapid rate in the future.

1.4 SOME CURRENT SYSTEMS

Various sensory systems are currently used in road vehicles. The majority of the sensors are used for monitoring and control of powerplants and transmissions. There is also a rapid growth in the use of sensory systems for vehicle ride and chassis control. Sensory systems for passive and active driver and passenger safety continue to be applied as an integrated part of recent road vehicles. Some of the current systems used on road vehicles will now be discussed.

1.4.1 Powerplant and transmission

The types of sensors and systems used for powerplant and transmission control depend on the parameters which will be monitored. Some of these parameters and the sensor systems used are as follows.

Inlet manifold absolute pressure

This is an engine parameter which can be sensed for speed/density and control of fuel mixture. Inlet manifold vacuum pressure allows for ignition control incorporating the appropriate timing sensor. It allows for engine load sensing because as the engine load is increased, the engine slows down and the inlet manifold vacuum pressure tends towards atmospheric pressure; this is accentuated by depressing the accelerator pedal, and thus increasing the opening of the throttle.

The mechanical method of manifold vacuum measurement has generally been in existence since the early days of automotive powerplant (engine) design. The current method uses a piezoresistive silicon diaphragm or other type of electronic pressure sensor connected to the manifold with a system which allows for a frequency or voltage input that represent engine rotational speed.

Crankshaft angle position

As the crankshaft rotates the ignition timing or the injection timing indexes, this yields angular velocity or frequency which is used for ignition and fuel control. Several sensors and systems are available for monitoring the rotational position of the crankshaft. For example, Hall-effect sensors, Wiegand-effect sensors, optical sensors and variable-inductance sensors are used for crankshaft angle position. The sensor signal forms part of the ignition, fuel and powertrain control.

Other powerplant parameters

Apart from the manifold pressure and crankshaft positions, there are other parameters which can be sensed and used to control the performance of a powerplant. The mass airflow into the engine manifold can be sensed using a hot-wire anemometer, a vortex-shedding flowmeter, an ion-drift flowmeter, etc. The output signal is used to control the air/fuel demand of the engine. Throttle position can be monitored with a sensor which helps to control the amount of fuel in the carburettor. The exhaust gas emitted from the engine can be monitored to detect the oxygen content in the gas. This is used as a means of controlling the emission characteristic of the engine. This type of sensory system incorporates an exhaust-gas-oxygen sensor, known as the λ sensor (Figure 1.2(a, b, c)). The sensor voltage is monitored, and, as shown in Figure 1.3, the voltage output can be related to the air/fuel ratio.

The temperature of the engine can be sensed and used as a variable for an engine control system. There are some well-established sensors, such as wirewound resistor temperature sensors and thermistors. These are used in various control systems.

To optimize engine performance and fuel economy, ignition timing and control must be adjusted so that combustion will occur during a specific number of crankshaft degrees, beginning at top dead centre of the power stroke. If it occurs any sooner, knock or detonation will occur. If it occurs any later, less pressure will be produced in the cylinder, and hence there will be a loss of engine efficiency. Even when an ignition system is fitted with electronic advance control, and is used to optimize the engine for best performance and economy, it can, under some conditions, be set sufficiently advanced to cause 'knocking' or 'detonation'. Knock can result in physical damage to the combustion chamber. In order to be able to provide an optimized engine performance without the effect of knock, knock sensors and systems are used. Most of the knock sensor (Figure 1.4) contains a piezoelectric crystal, which acts as an accelerometer for sensing the vibrational response of a knock. Another method of sensing the knock conditions is by measuring the ionization current across the spark plug after a normal firing in the engine [7].

The most recent development is in the in-process combustion monitoring. This is by use of optical fibres and a high-speed camera [8]. There are also other methods, such as the use of combustion chamber pressure sensors [9] and ionization flame front sensor [10].

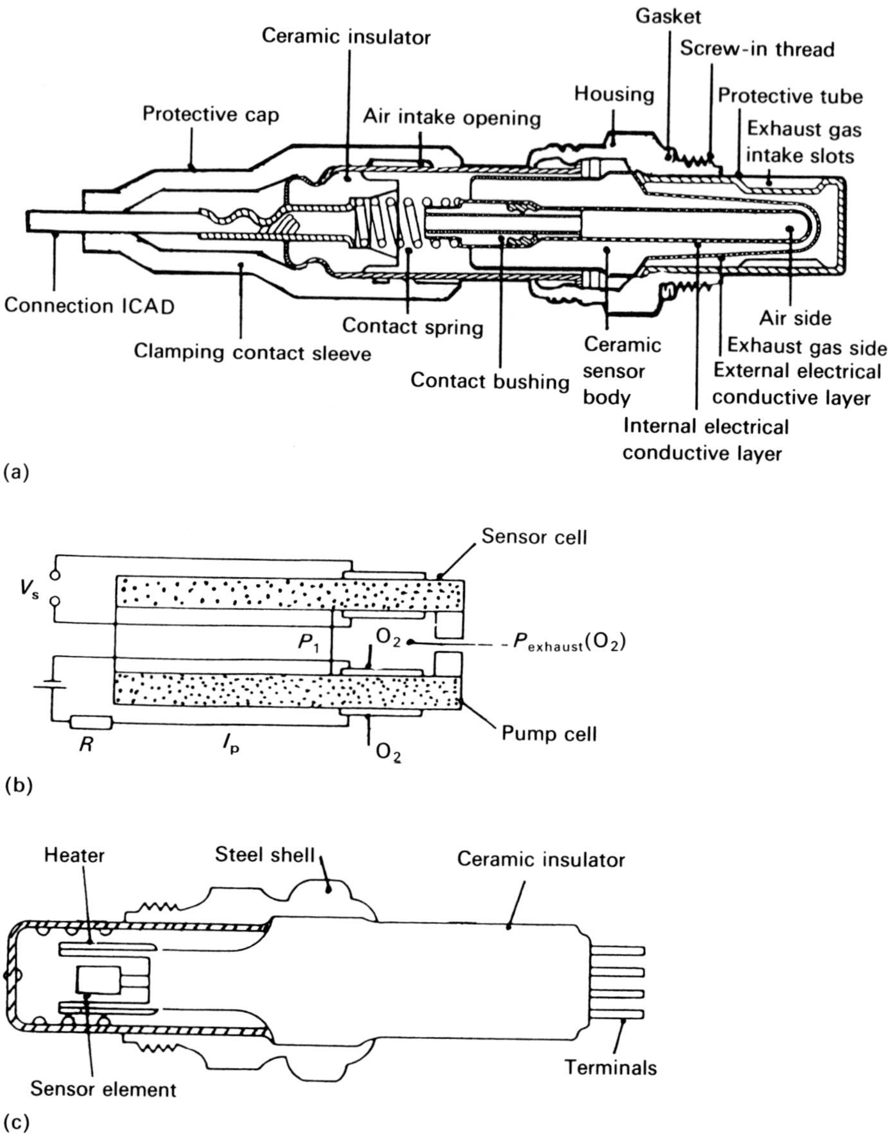

Figure 1.2 Exhaust-gas-oxygen sensor.

Transmission control

The transmission system provides the interface between the powerplant and the road wheels. As shown in Figure 1.5, it acts as a power match system device which is at the disposal of the driver who has the function of engaging or disengaging the power source (the engine) for the purpose of vehicle

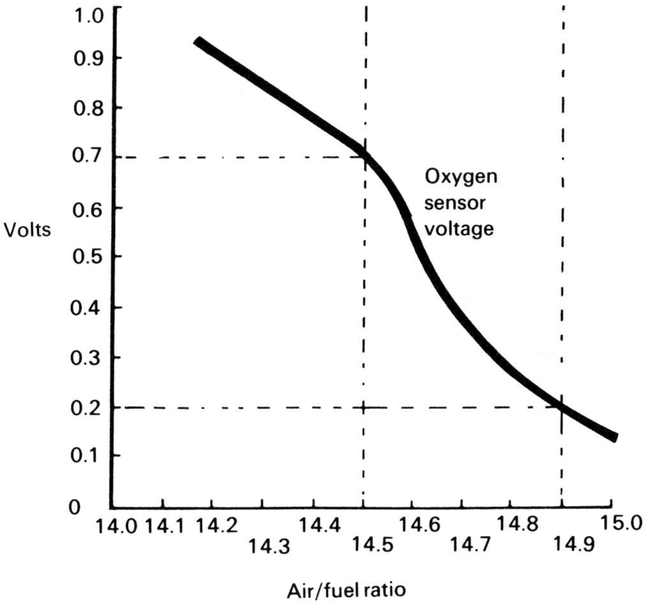

Figure 1.3 Approximate relationship of oxygen sensor voltage to air/fuel ratio.

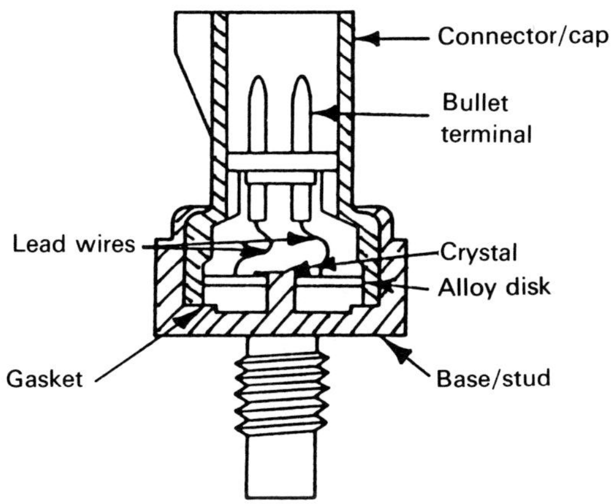

Figure 1.4 Sensor for ignition knock monitoring.

control. For a manual transmission system, the driver acts as part of the feedback loop of an otherwise open loop system, by sensing the engine speed and road wheel speed, and thus changing the gear ratio to match the required road torque (load). This process can be represented as shown in Figure 1.6, where N_{max} is the maximum engine speed at a given gear ratio and N_L is

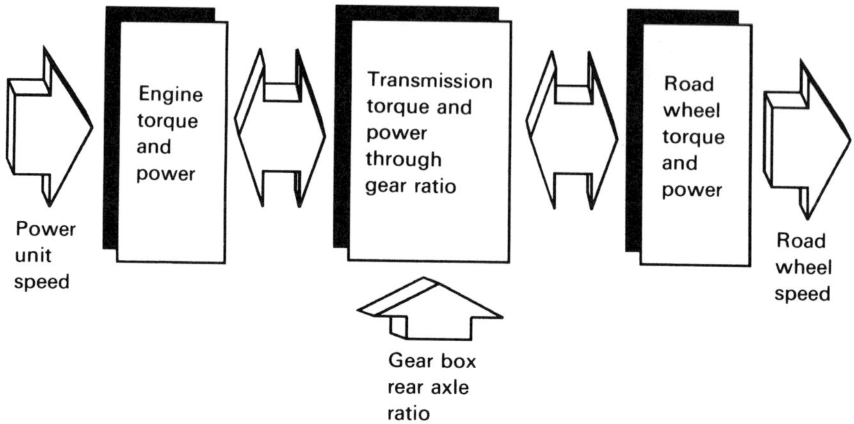

Figure 1.5 Powerplant–transmission interface.

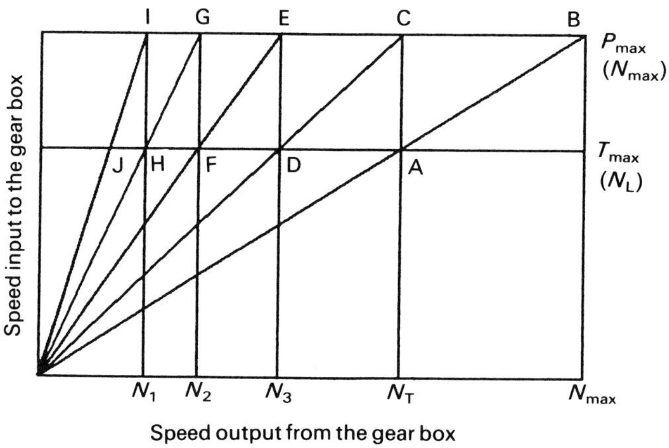

Figure 1.6 Intermediate gear ratio chart.

the load speed when the driver changes from one ratio to the other. Engine speed in the form of the pitch and noise level is one of the parameters intuitively sensed by the driver in order to provide the necessary acceleration and driveability. This is not satisfactory as a means of providing the optimum performance of the vehicle. In changing the gear ratio, the driver has the extra task of ensuring that the right gear ratio is selected.

Against this background, it is now possible to look at various sensing requirements for the transmission control. Powerplant speed information, which can be obtained from the crankshaft, gear box output shaft sensor and road wheel speed sensor, will be required. If it is a manual gear selection system, sensors confirming the gear selector position in relation to the actual gear ratio required can be used [11]. There will be a need to have a sensory system (e.g. with optical sensor) which will monitor the wear conditions of the clutch plate. Some of these sensors can also be used even when operating electronically controlled automatic transmission systems designed to give a better overall performance.

1.4.2 Vehicle ride and control

The sensory systems required for vehicle ride, comfort and handling are defined as those sensors and systems aimed at vehicle chassis management and control systems [12,13]. These include vehicle suspension systems, advanced braking systems and traction control systems. The majority of the sensors used in chassis management and control systems assist the vehicle ride, which is subsequently affected by the vehicle suspension system. The suspension system will provide the focus for the different types of sensors which will be discussed.

The primary function of the suspension system is to control the attitude of the vehicle body with respect to the road surface. This includes the monitoring and control of parameters such as pitch, roll, bounce and yaw motions (Figure 1.7) of the vehicle. There are several types of suspension systems used (e.g. passive, semi-active, active) and they are well defined in

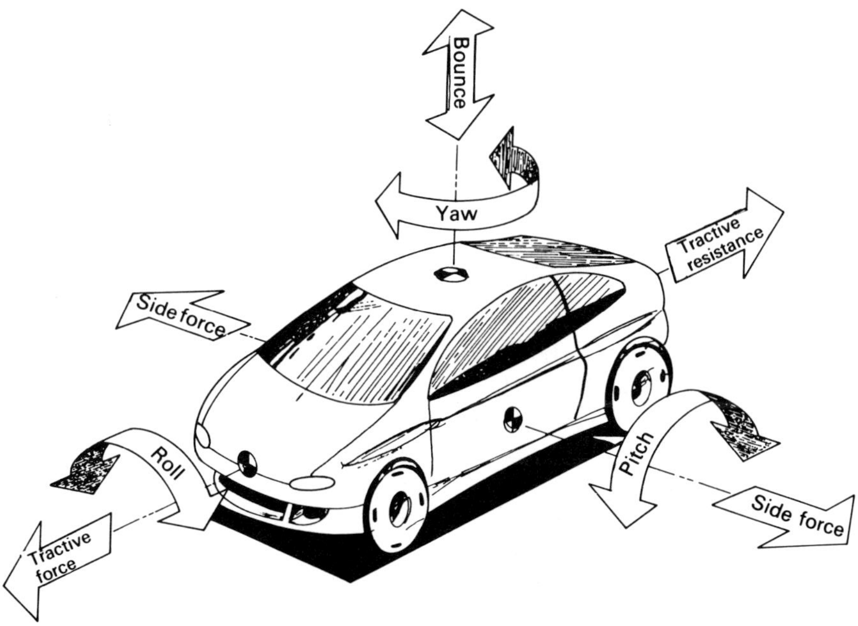

Figure 1.7 Controllable motion of a vehicle.

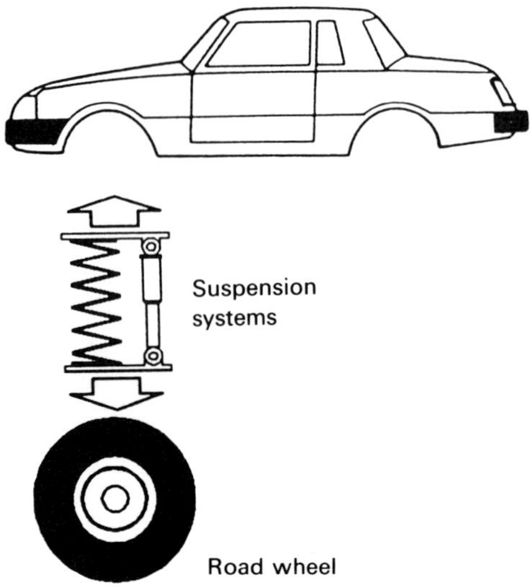

Figure 1.8 Passive suspension.

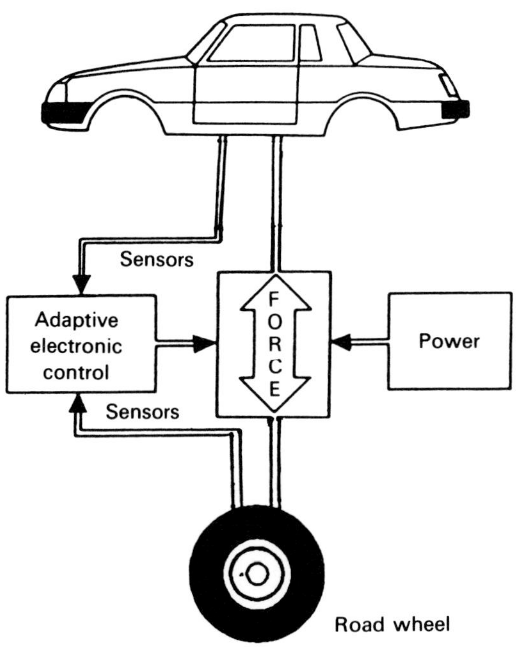

Figure 1.9 Active suspension.

ref. 14. Although contemporary vehicle suspension systems predominantly contain passive systems which include springs and dampers (Figure 1.8), there are several advances in adaptive and active systems (Figure 1.9).

In suspension control, the main objective is to provide a suspension system which will give a suitable and comfortable ride and handling for all conditions. This requires the use of appropriate sensors and systems to monitor the necessary suspension parameters (Figure 1.7) with the necessary electronic control strategy. A control strategy is required in order to minimize the disturbance to the sprung mass and the passengers. It is also required to control the motion of the unsprung mass in order to maximize adhesion of the tyre to the ground surface during handling and ride control (e.g. traction control).

There are several controlled suspension systems available [13, 15–22], but the majority of the sensors and systems used are similar; some examples are described in ref. 22. For a typical controlled suspension system, there are sensors for monitoring the vertical displacement (bounce) of the road wheels relative to the vehicle body. This type of sensor can range from a simple potentiometer to more complex sensors such as linear variable-inductance position sensors and optical sensors which employ the principles of optical

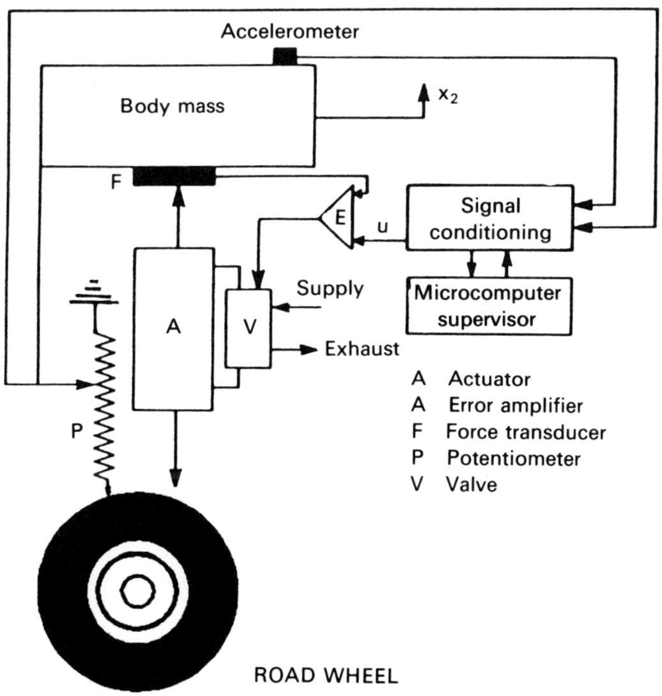

Figure 1.10 Instrumented active suspension system.

grating/Moire fringe technique for linear displacement monitoring. In some cases, sonar sensors can be used to monitor the relative vehicle body to ground vertical displacements. An accelerometer sensor and system is used to monitor vehicle body motion during cornering. As shown in Figure 1.10, the signal from the accelerometer and displacement sensors can be used to empower and control the suspension actuator with an error signal. Accelerometer sensors can also be used for advanced braking systems incorporating antilock braking (ABS)/traction control, steering systems and safety systems (e.g. airbag systems).

For the suspension control, the combination of these sensors is used to monitor and control the bounce, roll and pitch motions of the vehicle, but yaw motion can be monitored by the use of a yaw rate sensor (e.g. optical fibre gyroscope, solid-state piezoelectric devices). An integrated control strategy will include steering wheel rotation sensor (e.g. angular optical sensors–encoders), vehicle velocity sensor (e.g. road speed sensors – Hall-effect sensor) and throttle position sensors (e.g. optical sensory systems or potentiometric sensors – Figure 1.11).

1.4.3 Vehicle safety sensors

It was in 1769 that Nicholas Joseph Cugnot in France made history by constructing the world's first road vehicle. This vehicle, which was a steam-powered tricycle, sped along for 20 min at approximately 3.2 km/h. He also made history by having the world's first road vehicle accident. Since then, safety has increasingly become a very important factor to be considered during vehicle design and development. In recent times, safety requirements have necessitated the development of various sensory techniques for enchancement of vehicle safety.

Sensory systems can be used for passive and active safety systems. An example of the passive system is the supplemental inflatable restraint, which is the airbag system. This is well covered in Chapter 8. Examples of active systems are antilock brakes, traction control systems and, in some measure, tyre performance control.

Antilock braking system

Antilock brakes are an accident avoidance system which, by incorporating relevant sensory systems such as accelerometers, pressure sensors etc., can prevent wheel locking during hard braking in an emergency, especially when the road is slippery. Vehicles lose steering when the front wheels lock. An antilock braking system uses sensors at each wheel to monitor deceleration when the brakes are applied. If any of the wheels begin to lock, the system will modulate the brake pressure, thus pumping the brakes at a rate faster than the average driver could. This will allow the wheel to continue rotating, so preventing locking, and the wheel will continue to respond to the steering

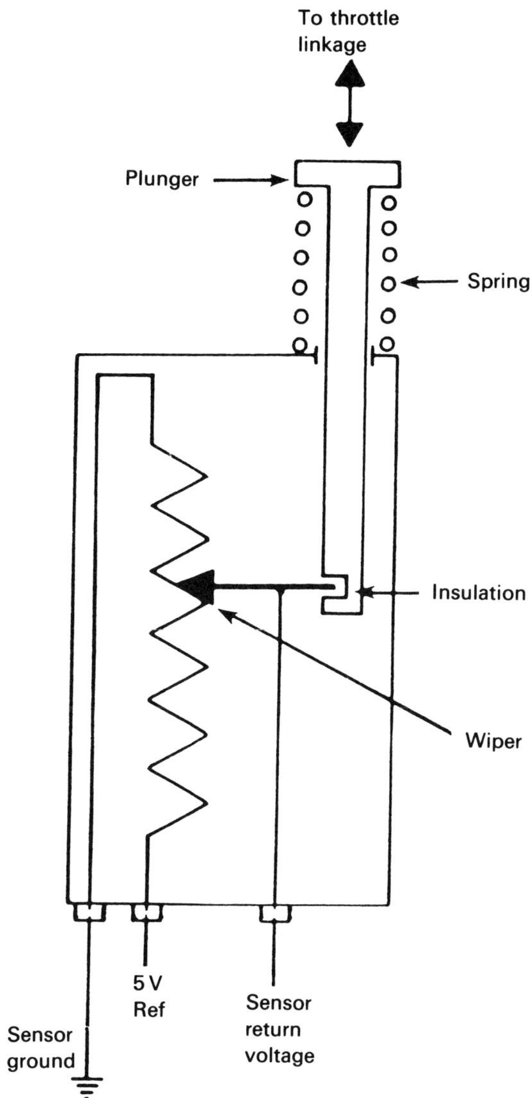

Figure 1.11 Potentiometric throttle position sensor.

wheel. So, by using the appropriate sensory system, an antilock braking system allows a driver to steer a vehicle during emergency braking.

Traction control system

Traction control is another accident avoidance system used to combat uncontrolled wheel spin and to keep vehicles from sliding sideways on slippery surfaces. In most vehicles with traction control, the system works only during low-speed acceleration and is then disconnected. In more expensive vehicles, the system operates at all speeds.

The same sensors that can detect wheel lock during braking with an ABS can be used to detect uncontrolled wheel spin. In operation, the traction control system senses when one driving wheel spins faster than the other and applies pulses of brake pressure to limit its slippage. This will then allow other wheels to receive and transmit engine power, thus improving steerability and traction. The operation of the system can be linked to engine control where, if the system is in operation for more than a given time (e.g. 2 s), the engine controller cuts fuel delivery to the engine cylinder on a selective basis to limit engine rev/min. This helps to minimize stresses on the brakes, transmission and engine prolonged traction control operation. A dashboard display can be provided to show when the system is in operation.

Tyre performance monitoring

A number of parameters affect the performance of road vehicle tyres. The operational adjustable parameters which can be monitored to ensure safe and economical driving are the tyre pressure and temperature. Other parameters, such as the tyre width, tyre deflection/oscillation and wheel speed in relation to other wheels on the vehicle, can also be measured in order to ascertain tyre performance status. The parameters to monitor depend on the types of sensor, system and measurement strategy used.

The most common sensory systems used for monitoring tyre performance are the pressure and temperature sensory systems. There are various systems available which can be used for the monitoring tasks during dynamic and static conditions [23, 24]. One of the methods of monitoring tyre performance is by the use of piezoresistive pressure sensors mounted on the wheel. The pressure sensor monitors the air pressure in the tyre. A sensor for monitoring the heat generated by the tyre can also be mounted on the wheel. As the wheel rotates relative to the axle, a means must be provided to power up the sensor and transmit the sensor signal for processing and display. This can be done by a signal pick-up device such as a circular antenna mounted on the wheel. As shown in Figure 1.12, to interface with the antenna mounted on the wheel, a fixed antenna is mounted on the brake caliper with a corresponding signal decoder. This type of system operates in two phases. In the first phase, the power supply is sent to the sensor through the two antennas. This is stored in a capacitor and used in the second phase to power the sensors as they measure temperature and pressure of the tyre; there is a decoder for analysis and display. This system can be used for continuous and reliable verification of the actual tyre pressure and temperature to an accuracy of a few per cent. With efficient software, the information received can be updated every second and used for dynamic and safe control of the vehicle.

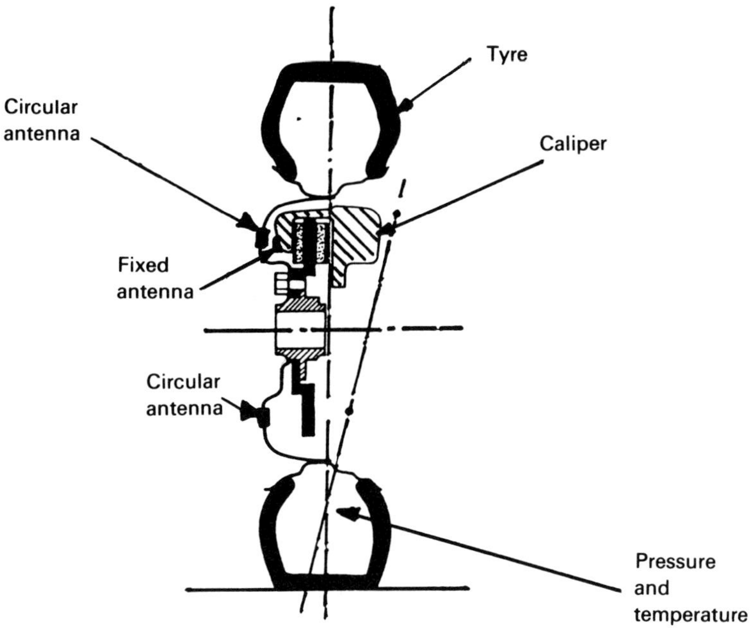

Figure 1.12 Tyre performance monitoring system.

1.5 BASIC AUTOMOTIVE OPTICAL SENSORS

As the advances in other types of sensory system for automotive application continue, so also does the growth in the use of optical sensors and systems for road vehicles. To complete the discussion on the current systems, some applications of optical sensors and systems for sensing linear and angular displacements will now be described.

Three major optical sensors and systems will be described; these will include the technique which provides the signal that may be processed by other integrated systems such as a computer processing system.

1.5.1 Linear position sensor

Linear displacement sensing has a wide range of applications in the automotive field, such as suspension systems, automated seat adjustments, and gear lever position monitoring. There have been several developments of optical sensors which can sense linear positions, one of which is shown in Figure 1.13.

The sensor consists of a pair of detectors, one located at the focal point of a lens, and the other in the same plane but shifted vertically. A light source is placed at the other end of the system and is free to float with respect to the pair of detectors and lens. The detector at the focal point has been chosen

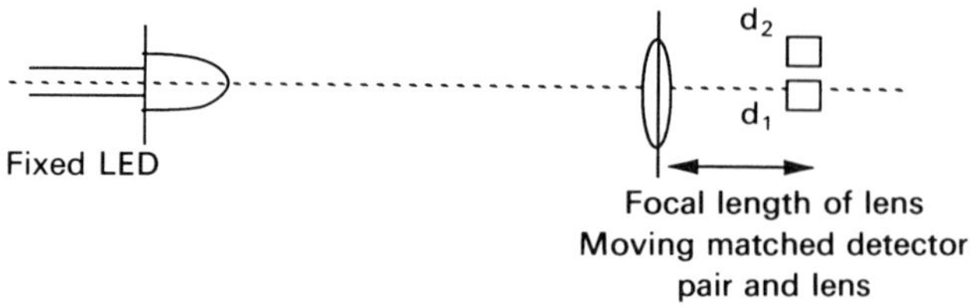

Figure 1.13 Optical linear position sensor (courtesy SIRA Ltd): d_1 = on-axis overfilled detector; d_2 = off-axis underfilled detector.

such that it is overfilled by the light that is focused onto it by the lens. The other detector is placed such that no light falls on it when the source is furthest away; as the light source moves closer, the amount of out-of-focus light falling on this detector increases.

The signal from the detector placed at the focal point is independent of the light source distance from the lens. This signal can be used as a normalizing reference. The output from the other detector is related to the inverse square of the separation between the light source and the lens, and hence it is used as the basis of the output signal.

The working principle described has the disadvantage that a light source and the detectors have to be fitted to different ends of the moving part. This can be overcome by using a retroflective device, but such a device must be kept clean and free of dirt.

1.5.2 Steering wheel rotation sensor

The type of optical sensor used for monitoring the rotation of the steering wheel is normally determined by the resolution of the steering wheel turn (e.g. 2° down to 0.3°), the space in which the sensory system will be fitted, the type of signal output to be provided, etc.

The optical sensor to be described was developed by SIRA [25] and is integrated with the steering column as shown in Figure 1.14. It consists of two types of optical sensor to take account of different styles of steering columns. The optical sensing method is the same in both types, the difference being the method of delivering light onto the sensor. One type uses a transmissive method and the other uses a reflective method. As shown in Figure 1.15, the transmissive method uses a perspex wheel fitted around the steering column. A series of darkened notches are etched on the wheel such that the wheel acts as a chopper. This provides output signals that represent right and left turns of the steering wheel. This is similar to the optical encoder to be described in section 1.4.3.

The optical head has a sandwich of flat light guides, each with an independent detector. This provides an angular resolution of number of

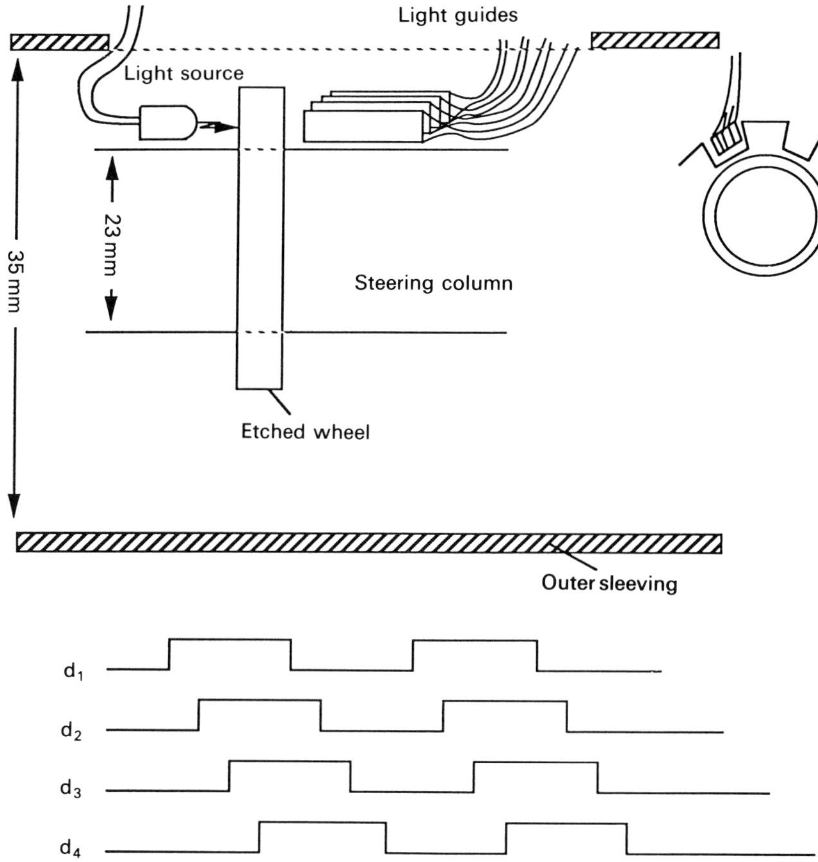

Figure 1.14 Steering wheel position sensor (courtesy SIRA Ltd).

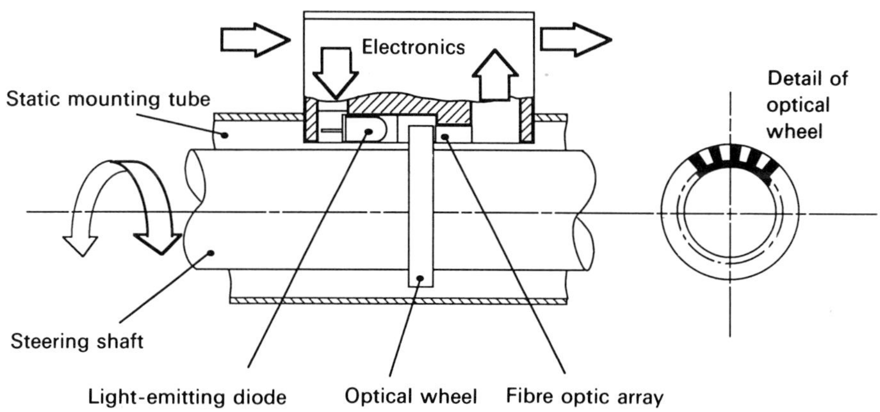

Figure 1.15 General assembly of steering wheel angular optical sensor (courtesy SIRA Ltd).

notches $*N$ where N is the number of the light guides in the sandwich that fit into a single notch gap. The effective resolution is limited by the width of the light guides used.

A reflective method can be used, where the perspex wheel is replaced by a reflective tape, with black lines stuck coaxially onto the steering column. The light guide sandwich consists of alternate illuminating and detecting elements which are directed normally to the surface of the steering column.

The sensory signal can be in a pulse train or analogue format. With the digital pulse output, the pulse train is decoded to provide a pair of output signals that represent left and right rotations of the steering column.

1.5.3 Optical shaft encoders

To sense the angular displacement and velocity of shafts such as steering columns, propeller shafts and drive shafts (road speed sensing), adapted optical shaft encoders may be used. An optical shaft encoder is an arrangement where a circular glass disk is encoded by etching rows of broken concentric arcs of transparent and opaque regions in the form of a binary pattern. As shown in Figure 1.16, a light source is assigned to each row with a corresponding detector on the opposite side of the disk. The arcs and sensors are arranged so that, as the light is projected through the disk, the pattern of the activated light sensors is a unique encoding of the position of the shaft to within a given angular resolution. The accuracy of the encoder disk is a function of how many light sensors (binary digits) are used to make up the binary code; the number of such sensors can range up to 12.

The pattern of the activated sensors is in computer-readable binary code. The binary code generated at each of the radial positions on the disk (Figure 1.16b) is generated by a unique combination of light sensors that are 'on' or 'off'. However, problems of accuracy can occur if the angular shaft position is represented in the standard set of sequential binary numbers, as given by the encoder disk (Figure 1.16b). Such problems can occur during the change of state between the sensors when the system becomes skewed from the radial line. Another problem can occur during the change of state between the sensors when the systems become unstable. To overcome this problem the natural binary code is modified so that, at any transition point, a change in only one binary digit is required. The resulting code is known as the Gray code (Figure 1.16c); absolute encoders usually employ this type of Gray code. Gray code disk increase reliability but require the use of additional decodes circuitry to complete the sensory systems.

1.6 FUTURE DEVELOPMENTS

This section discusses some of the progress being made in the use of sensory systems for vehicle automation. As sensory system application continues to

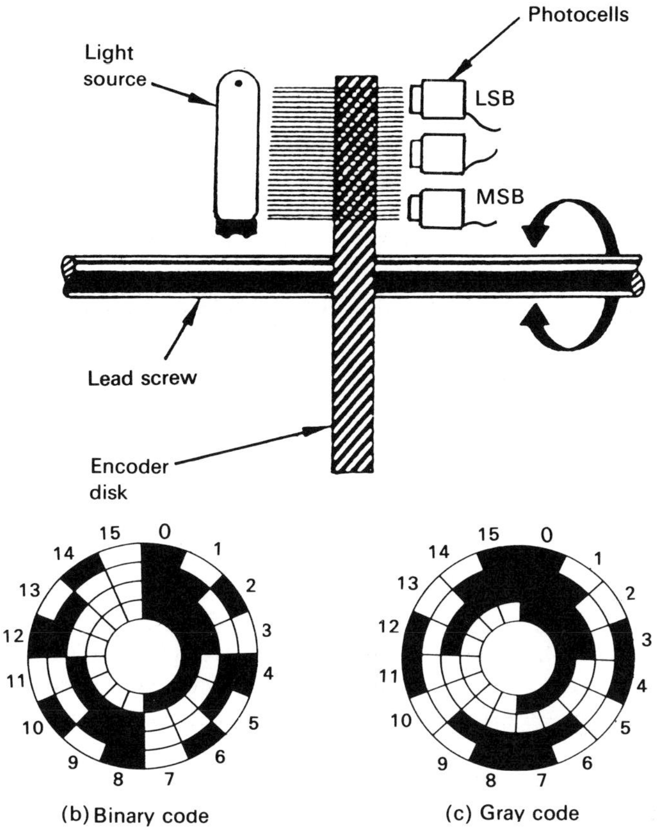

Figure 1.16 Optical shaft encoder.

expand, the fast rate of such expansion makes it difficult to precisely predict what the vehicle of the future will be without falling into the trap of quite rightly predicting the future vehicle as a smart road vehicle with some form of intelligence able to self-diagnose, manage and take complex decisions on its performance and vehicle navigation. It is easier to categorize the use of sensory systems for road vehicles into their near-term technology and more adventurous and futuristic technology expected to mature into production at the end of the decade. Smart sensors and systems will play important roles in the near-term and future road vehicle.

The near-term sensory technology will include intelligent cruise control systems, automatic route guidance systems and enhanced vision systems. Intelligent cruise control systems will include the use of radar sensors linked to electronic devices which will maintain a set time interval of separation from the vehicle in front with active control of the engine and transmission systems. The use of sensory systems for safe, active and passive safety

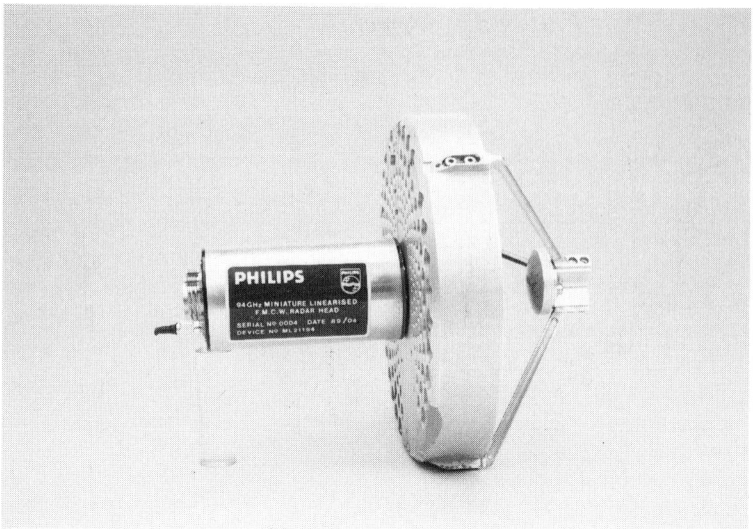

(a)

(b)

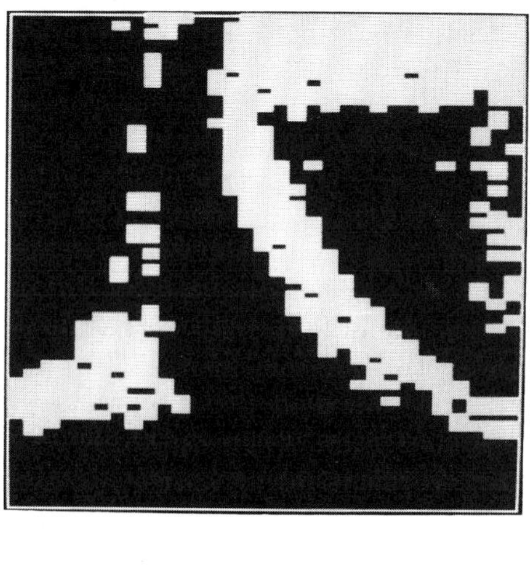

(c)

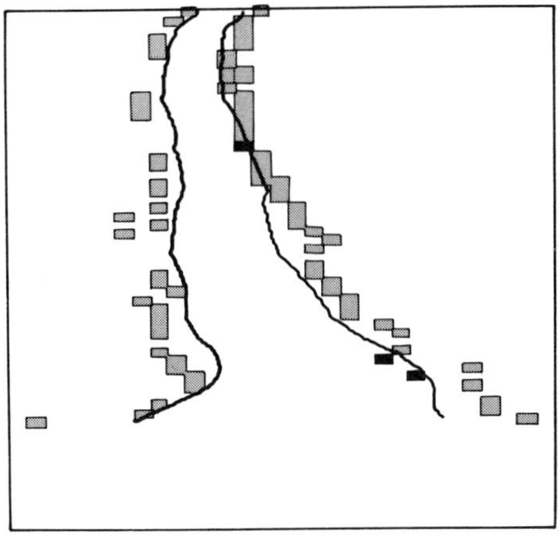

(d)

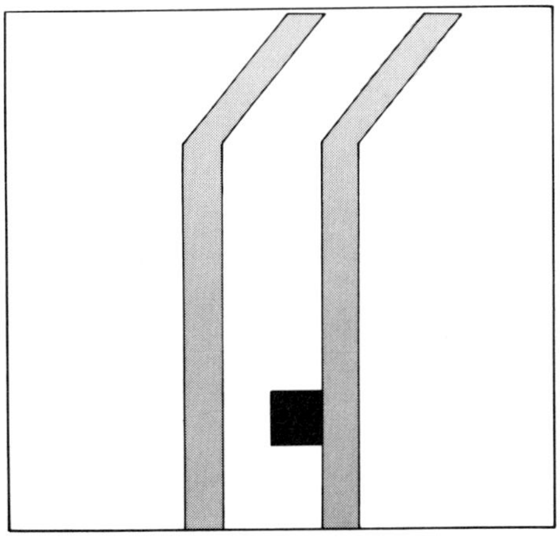

(e)

Figure 1.17 A vehicle obstacle avoidance radar system: (a) a radar antenna mounted on a vehicle; (b) country road being monitored by radar; (c) drawing of a raw radar map produced by the sensory system; (d) drawing of a processed radar map produced by the sensory system; (e) drawing of an enhanced map with blip to show target within roadway. (Courtesy Philips Research Laboratory, Redhill, UK.)

technology will obviously increase; some such safety systems will include the use of optical and visual sensory systems.

Systems which can be categorized as a long-term technology will involve the use of near infra-red illumination to increase the driver's depth of vision in conditions of low contrast, and at night. Some of the sensory technology will involve the use of thermal imaging sensors that utilize far infra-red cameras which can be located in the vicinity of the windscreen to provide long-distance pictures of the road and its contents under poor visibility in rain, fog and night conditions. Safety hazards such as the presence of pedestrians at a far distance on the road will show up as white spots on the driver information display system.

The development of integrated collision avoidance systems with various sensory systems utilizing optical infra-red, radar and ultrasonic devices to detect where the road lies in poor visibility will continue. These systems will be capable of locating obstacles such as vehicles, people and objects that are likely to cross the driver's path. The system will have the intelligence to assess what threat is posed by the obstructing object and what help should be provided to the driver to select an appropriate course of action. Such a system may have the power of controlling the vehicle autonomously in the event of danger and accident. There are various basic systems being developed across the world at present. A typical example is the car collision avoidance system currently being developed at the Philips Research Laboratories. The sensory system uses a radar technique which draws the driver's attention to obstacles which are potentially hazardous. Shown in Figure 1.17a is the radar antenna which is mounted on the vehicle. This continuously transmits a signal which is acquired by an on-board computer system. In the event of any obstacle, the computer calculates the distance of the obstacle from the time taken for the radar signal to return from it. The obstacle's size is also calculated from the change in the signal frequency. Figure 1.17b shows a typical country road, and a drawing of a radar map of it is shown in Figure 1.17c. The map is processed in colour. Figure 1.17d shows a similar drawing in which the outlines of the road edges are depicted by black lines, with targets such as the road hedges (Figure 1.17a) illustrated by the shaded boxes (Figure 1.17d). The solid boxes at the right-hand road edges of Figure 1.17d are targets within the roadway. A concept for a driver information display format is shown in Figure 1.17e. The radar system may link with antilock braking or traction control to automatically trigger braking when a crash is imminent, thus minimizing or avoiding impact.

There will be a growth of sensory systems for co-operative driving which will allow for inter-vehicle communication. Such a system will provide the devices which will interact with cellular phone networks, satcom, satellite ground stations and satellite to track road vehicles. The integrated communication will form part of route planning and information exchange systems.

The vast number of sensory systems which will be made available for road vehicles will demand that expert systems are developed for multi-sensor fusion technology. The use of a neural network system which will monitor, through the multi-sensory systems, driver behaviour/reaction, vehicle performance, and control and navigation systems will be developed. This will require advanced data processing techniques to rationalize the sensory information and possibly to take necessary system decisions. There will be a move towards smart sensory systems which will form part of a specialized distributed processing and task distribution system networked into a standard bus.

Other future systems could include those that use fuzzy logic and neural networks to monitor steering patterns, sounding an alarm when events such as no steering or unusually large movements indicate a driver may have dozed off or be intoxicated. There will be further developments in intoxication and sleep sensory systems; these can be linked to various interlock systems in the vehicle.

1.7 CONCLUSION

Undoubtedly, the automation of road vehicles has offered one of the greatest opportunities in the use of automotive sensors. This chapter has only presented introductory information on some of the sensory systems available. The description of some of the sensors has been cursory. These are the sensors which will be covered in more detail in subsequent chapters. Some of the sensory systems not covered in subsequent chapters have been discussed in more detail in this chapter. The future developments in automotive sensory systems have been discussed; there is no doubt that these sensors will be instrumental in revolutionizing and enhancing the performance and safety standards of road vehicles.

REFERENCES

1. Paulsen, J.J. (1989) The state of automotive electronics in the year 2000: a perspective of the North America marketplace. *I. Mech. E.*, C39/KNI, 1–63.
2. Rivard, J.G. (1986) *Automotive Electronics in Year 2000*. SAE Technical Paper No. 861027.
3. Wolber, W.G. (1978) *A Worldwide Overview of Automotive Engine Control Sensors Technology*. SAE Technical Paper No. 780207.
4. Fleming, W.J. (1982) *Engine Sensors: State of Art*. SAE Technical Paper No. 820904.
5. Westbrooks, M.H. (1988) Automotive transducers: an overview. *IEE Proceedings*, **135**, pt D5, 339–47.
6. Bertuol, B. (1991) Sensors as key component for automotive systems. *Sensors and Actuators*.
7. Blauhaut, R.B., Horton, M.J. and Wilkinson, A.C.N. (1983) A knock detection system using spark plug ionization current, 4th International Conference on Automotive Electronics (late paper not included in the proceedings), IEE, London.
8. Nwagboso, C.O. and Pendlebury, M. (1992) Fibre optics for combustion process monitoring of CNG engine. *Internal Communication VSRC*, **3**.
9. Rhodes, D.M. and Weson, M.J. (1983) The petrol engine: a case of closed loop control. *IEE Conference*, **229**, 62–6.
10. May, M.G. (1984) Flame Arrival Sensing Fast Response Double Closed Loop Engine Management. SAE Technical Paper No. 840441.
11. Nwagboso, C.O. (1992) Integrated sensing technique for intelligent gear changing control systems, in *Advanced Vehicle Control*, Proceedings of the Society of Automotive Engineers of Japan, JSAE, pp. 164–169.
12. Wallentowitz, H. (1991) Powertrain and chassis control systems integration. *Automotive Engineering*, **99** (1), 15–18.

13. Mizuguchi, M. *et al.* (1984) *Chassis Electronic Control Systems for the Mitsubishi Galant*. SAE Technical Paper No. 840341.
14. Nwagboso, C.O. (1991) *Vehicle Dynamics*, Lecture Notes, BEng Automobile Engineering, Bolton Institute, UK.
15. Karnopp, D. (1983) Active damping in road vehicle suspension systems. *Vehicle System Dynamics*, **12**, 291–311.
16. Dominy, J. and Bulman, D.N. (1985) An active suspension for a Formula 1 Grand Prix Car. *ASME Transactions, Journal of Dynamic Systems, Measurement and Control*, **107**, 73–8.
17. Poyser, J. (1987) Development of a computer controlled suspension system. *International Journal of Vehicle Design*, **8**, 74–86.
18. Cho, D. and Hedrick, J.K. (1985) Pneumatic actuators for vehicle active suspension applications. *ASME Transactions, Journal of Dynamic Systems, Measurement and Control*, **107**, 67–72.
19. Yokoya, Y. (1974) *Toyota Electronics Modulated Suspension (TEMS) System for the 1983 Soarer*. SAE Technical Paper No. 840341.
20. Milliken, W.F. (1988) *Active Suspension*. SAE Technical Paper No. 880779.
21. Goto, T. *et al.* (1990) Toyota active control suspension system, in *22nd International Symposium on Automotive Technology (ISATA)*, Italy, pp. 857–64.
22. Prosser, S.J. and Moore, J.H. (1992) Advances in automotive sensors, in *Sensors and their Application V*, Institute of Physics, 1992.
23. Anon (1990) MTM – the intelligent monitor. *SAE Australian Journal*, May/June, 20–1.
24. Gay, P. and Bugnot, D. (1989) Real Time Tyre Pressure Electronic Monitoring System with Dashboard Possibility. SAE 221, Autotechnologies.
25. Goodyer, E.N. (1989) An overview of a range of novel automotive sensors. *I. Mech. E. Proc. Conf.* **C391**.

CHAPTER 2

Sensors in automobile applications

2.1 INTRODUCTION

The availability of more and more powerful and less expensive microelectronic devices for the vehicle, as in many other fields of application, represents the strongest innovative thrust able to satisfy and harmonize the demands of the individual with those of the community in terms of performance, safety, reliability, protection of the environment and resources.

Not only are technological developments in microelectronics and computer science fundamental elements in achieving these targets, but so also is the availability of sensors suited to control and perform system diagnostics. The development of new reliable and low-cost sensors planned for use in the automobile field has been made necessary. The progress which took place both in technologies and materials has allowed the acquisition of new sensing by means of new transducing principles or others which are already well known and used in other fields.

A thorough analysis of these components is quite complex, just as it is risky to actually classify the range of materials or technologies for future automotive sensors. However, it seems possible to foresee, with a certain degree of confidence, that there will be a greater appreciation of materials and technologies, allowing total or at least partial integration of sensors with conditioning electronics.

2.2 ENGINE CONTROL SENSORS

The necessity to check exhaust gas composition to limit urban traffic pollution and to reduce consumption forced the adoption of an engine electronic control system.

At the very beginning, the electronic control system included separate systems checking supply, timing, ignition and recirculation of exhaust gases. The current control system has evolved towards an integrated system in which each single subsystem is a different part of the same unit. The measurements made by sensors, representing the inputs of this type of system, are:

- airflow quantity;
- drive shaft speed;
- drive shaft angular position;
- oxygen concentration in exhaust gases;
- coolant temperature;
- throttle valve position.

The main sensors performing these measurements will be described, particularly those that are being studied in research and development laboratories.

2.2.1 Airflow quantity

Airflow measurement can be made both indirectly by using a pressure sensor placed in the intake manifold integrated with a temperature sensor and directly by means of a mass airflow sensor. In the first case the manifold pressure and temperature measurements are used to calculate air density (δ) which, associated with engine rev/min (w), the airflow volume at each engine cycle (D) and engine volumetric efficiency (N_v), allows the calculation of the quantity of air injected into the cylinder according to the following equation:

$$Q = (wD/2)\delta N_v \tag{2.1}$$

Pressure sensors currently used are based on measurement of a membrane deformation using, for instance, piezoresistant elements. This measurement principle has been implemented in the past with electromechanical devices

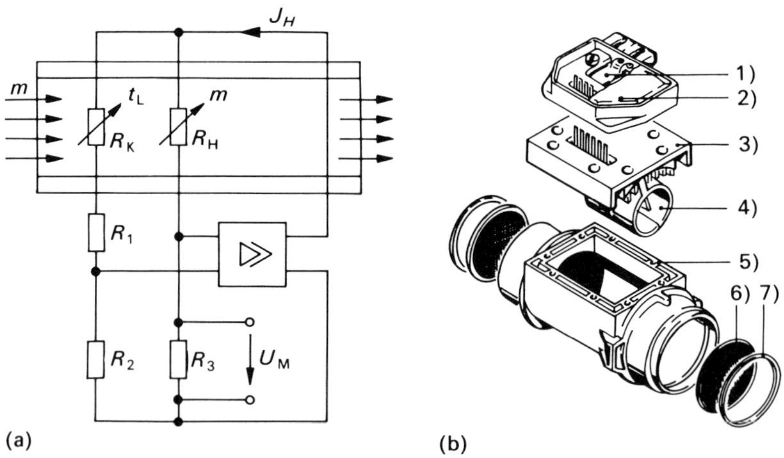

Figure 2.1 Hot-wire air mass flow sensor: (a) circuit: R_K = measurement resistor; R_H = hot wire (film); U_M = output; (b) sensor: 1, hybrid circuit; 2, 3, case; 4, Venturi tube with hot wire (film); 5, duct; 6, grid; 7, O-ring.

[1], and has evolved with the application of thin and thick technologies [2] and micromachining of silicon [3].

However, the electronic system producers are currently oriented towards a direct airflow measurement allowing higher measurement precision and a reduction in response times. The measurement principle on which these sensors are based is a variant of the classical airflow sensor known as a hot-wire anemometer used in meteorology to measure the wind speed. In the sensor (Figure 2.1b), the hot wire is replaced by a hot film on a ceramic substrate. The film must be manufactured with a metal whose coefficient of resistance for thermal variation is high, typically platinum. The airflow is regulated on its inlet side by an element which generates a laminar flow on the sensitive film. The resistive element can be mounted or realized on the same substrate as the conditioning electronics.

The hot film is heated electrically at a slightly higher temperature than intaken air. Temperature is detected with a sensor or with a second film whose resistance is measured. Hot-film and temperature sensors are inserted into a Wheatstone bridge circuit. The bridge supply voltage is provided by an amplifier. When air laps the hot-film element, this tends to cool. The amount of heat removed from moving air depends directly on the mass capacity. The variation in the film temperature affects the circuit electrical resistance.

The bridge circuit (Figure 2.1a) is unbalanced and generates a voltage at the amplifier terminals. The amplifier output is connected to the bridge supply voltage so as to balance it again. Thus the bridge supply voltage is a signal which depends directly on the airflow mass. Current devices [4, 5] usually have a frequency output which, even for the normal microcontroller with an 8-bit A/D converter, allows detection of the signal with the necessary precision throughout its variation range, which can reach 90:1 in the case of turbocharged engines.

Other existing principles for airflow mass measurement have been used to build experimental prototypes or laboratory instruments. Among these are the vane sensors which are based on the displacement measurement of a vane directly exposed to airflow and ultrasound sensors [6] which make use of the flight time measurement inside the flow.

2.2.2 Drive shaft speed and angular position

The position of the drive shaft and its speed are essential information for engine control because they constitute its timing base.

These two quantities are measured by detecting the passage, for a fixed point relative to the engine, of reference elements on ferromagnetic disks rotating with the shaft. Detection is by means of suitable sensors which supply an impulse signal whose frequency is proportional to the speed.

Currently the most commonly used sensors are the inductive sensors with a permanent magnet. Magnetoresistance [7] or Hall-effect [8] sensors are also used, always with a permanent magnet. The latter allows reference detection in almost static conditions and moreover the signal amplitude does not depend on drive shaft rotation speed.

An innovative technique [9] is the placement of magnetic references directly on the material of the rotary component. The main advantage of this method is the non-use of toothed wheels. In this case detection must be made with a Hall-effect sensor without a permanent magnet.

2.2.3 Oxygen concentration in exhaust gases

In order to satisfy the ever-increasing strict standards on polluting emissions, it has been necessary to adopt feedback engine control systems.

There are two main strategies of feedback engine control currently adopted on vehicles. One is well known as the three-way catalytic converter system. It was introduced by Volvo [10] and it is the most widespread in Europe, America and Asia.

In this system the air/fuel (A/F) ratio in the exhaust gases is detected by an oxygen sensor and controlled in such a way as to be always around the stoichiometric ratio of 14.7 (Figure 2.2). In these conditions the three-way catalytic converter is at its maximum efficiency and removes the polluting compounds (HC, CO, NO_x) by making them react chemically with the oxygen; therefore, the oxygen sensor response must be very fast and precise, particularly in stoichiometric ratio concentration conditions.

The other engine system control strategy, called the lean combustion system, was put into practical use by some Japanese manufacturers [11, 12].

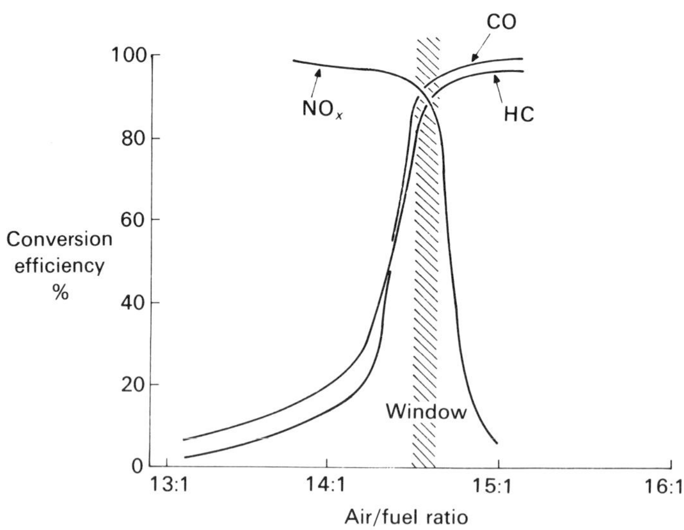

Figure 2.2 Catalytic converter efficiency versus A/F ratio.

Its purpose is to improve fuel economy and to keep the emission level below the admissible limit. It is based on the reduction of polluting components at high A/F ratios. It is necessary to keep the A/F ratio within a limited range in order to maintain the emissions below the limits and to avoid misfiring. For this purpose an oxygen sensor is used. From the operating features viewpoint, the oxygen sensors can be divided into three groups:

1. semiconductor oxide or chemioresistive sensors;
2. concentration cell sensors;
3. electrochemical pump sensors.

From the applications standpoint, they can be grouped in two sets only:

1. stoichiometric sensors;
2. lean mixture sensors.

Stoichiometric sensors are used in catalytic systems and typically give a nearly on–off response when the exhaust gas concentration ratio goes from a value lower than 14.7 to a higher one. Sensors used in the lean combustion systems have a proportional response in the range between 15 and 23 A/F ratio. Chemioresistive and concentration cell sensors are of stoichiometric type while electrochemical ones are for lean mixtures.

Chemioresistive sensors (*Figure 2.3*)

The electrical resistance of several semiconductor oxides varies with the oxygen partial pressure. This phenomenon is due to the oxidation or reduction of oxide by O_2. Oxides behave like p- or n-type semiconductors according to their chemical characteristics and can be doped with materials that function as donors or acceptors or can function as reaction catalysts between oxide and oxygen so as to lower the operating temperature.

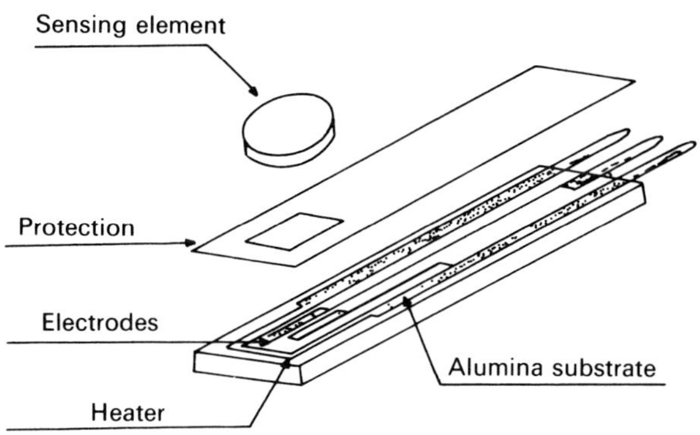

Figure 2.3 Chemioresistive thick film gas exhaust oxygen sensor.

In the case of n-type semiconductors, in a total absence of oxygen, electrons are free to move, allowing the material to have quite a low electrical resistance. These characteristics can be obtained only at high temperature (about 700 °C) so as to allow the electron transition from valency to conduction band.

When even small oxygen concentrations are present in the gas mixture to which the material is exposed, there is an immediate and considerable increase in resistivity. This phenomenon is due to charge carrier immobilization on the material surface by the oxygen molecules. By monitoring the granulometry of the sensitive element [13] it is possible to vary sensor sensitivity. When reducing gases and oxygen are present in a gas mixture they react with each other. Only the excess oxygen reacts with the semiconductor oxide, determining its resistivity. On the contrary, if the reducing gases are in excess, the semiconductor resistance will be that of the material in the absence of oxygen. Therefore, in the change from a reducing gases excess condition (rich) to an oxygen excess condition (poor) the semiconductor oxide resistivity varies remarkably.

These materials make possible the realization of devices with very simple, small-size, low-cost structures using thick or thin film technology on ceramic substrates.

At the present state of the art, only titania sensors (TiO_2), as shown in Figure 2.4, have been completely developed and are on the market even though there has not been a great spread of them until now. Besides having good chemioresistive characteristics, this material has shown itself to be more resistant to lead pollution [14] . Initially these kinds of sensors were made up of two electrodes and a TiO_2 deposition [15]. Then sensors with a heating element were developed [16].

The performance, in terms of cost and easy implementation, that this type of sensor allows us to achieve keeps alive the interest in this measurement principle. Recently, alternative materials, such as Ga_2O_3 [17], $SrTiO_3$ and $BaTiO_3$ [18, 19], have been developed in thin film form.

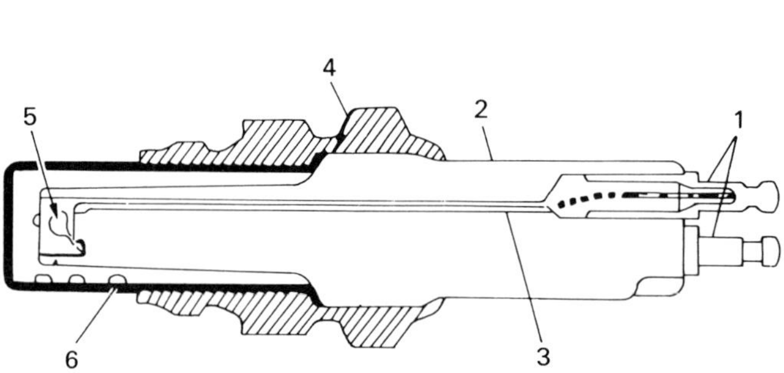

Figure 2.4 Titania gas exhaust oxygen sensor: 1, contacts; 2, ceramic insulator; 3, conductors (wires); 4, case (steel); 5, sensing element (TiO_2); 6, protection.

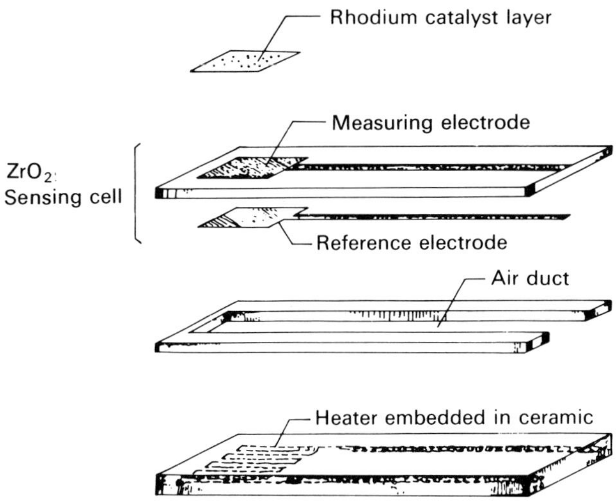

Figure 2.5 Zirconia thick film lambda sensor.

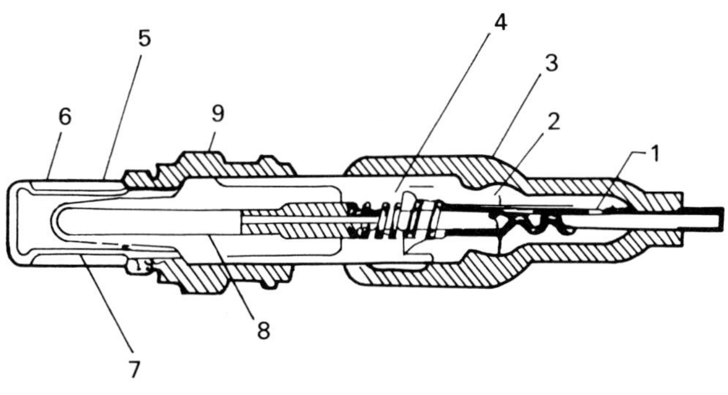

Figure 2.6 Zirconia lambda sensor: 1, wire; 2, insulator; 3, protection; 4, electric contact (spring); 5, ceramic sensing element (ZrO_2); 6, protection; 7, platinum film and protective coating; 8, internal conductor; 9, case (steel).

Concentration cell sensors (Figures 2.5 and 2.6)

Concentration cell sensors are based on the electrochemical properties of zirconia (ZrO_2) [20]. This kind of sensor was the first to be presented on the market and is the one most widely used. The sensor is also known as a lambda probe; this name comes from the ratio:

$$\lambda = \frac{\mathrm{A/F\,(real)}}{\mathrm{A/F\,(stoichiometric)}} \tag{2.2}$$

that is commonly used in the automotive field.

If a zirconia oxide element has two surfaces exposed to gases with different oxygen concentrations, a voltage between them is generated. The one exposed to a greater oxygen concentration is at negative potential while the other is at positive potential. This behaviour is increased at high temperatures. The physics of this phenomenon are still not well understood in all details [21, 22].

The structure of this kind of sensor was initially simple, being made up of two electrodes on a zirconia substrate [23, 24]. Subsequently a heater [25] was added to this structure. More recent developments have led to the manufacture of heated zirconia sensors with thick film technology on a ceramic substrate [26].

Electrochemical pump sensors

When a voltage is applied to a zirconia oxide element, it pumps oxygen through the solid electrolyte from the cathode to the anode. Sensors based on this phenomenon are made up of a zirconia ceramic element with two electrodes and an element which limits the diffusion of oxygen from the outside to the cathode. The working principle allows oxygen concentration measurement even in high A/F conditions and this sensor is therefore suited for use in lean combustion control systems.

Sensors used in the automotive field [27, 28, 29] are manufactured with thick film technology and are made up of a zirconia element with electrodes on both sides and a porous element placed between the cathode and the gases. The anode is in contact with the outside air. Applying a voltage to the electrodes causes an electric current to flow, which corresponds to the oxygen transfer from the gases to the exterior. The current reaches the saturation value for a given voltage when the quantity of pumped oxygen is limited by the diffusion processes through the porous septum. In these conditions the electric current is a function of the oxygen concentration in the gases.

2.2.4 Knock sensors

Unleaded petrols tend more easily to give rise to irregular combustion at high compression ratios and the high loads mainly found in turbocharged engines.

To avoid such phenomena it is no longer enough to foresee security margins in the advance adjustment because knocking for unleaded petrols becomes critical, depending on engine and fuel conditions. Therefore, a feedback control of the knock, optimizing engine performances, becomes necessary. Sensors typically used are wide-band accelerometers of the piezoelectric type placed on the crankcase [30]. Sensors have been developed which use other transducing principles such as ionization and pressure measurement in the combustion chamber. The measurement can be made using a suitable spark plug. This method has given good results [31] allowing the identification of knocking on each cylinder with a signal/noise ratio which is higher than that of accelerometric sensors, as they are free from mechanical interference.

The knock phenomenon can also be detected with more sophisticated methods such as direct pressure measurement in the combustion chamber. It allows the calculation of other quantities which are used in very advanced engine control systems. Recently, low-cost sensors of this type have been developed [32]. Knock detection and calculation of other parameters related to combustion can also be obtained by means of piezoelectric rings placed under spark plug [33] or under the engine head studs.

However, the trend, mainly due to the increase in calculation capabilities of recent microelectronic products, is to use one accelerometric sensor suitably placed on the crankcase or on the engine head. Its signal can be processed with even very sophisticated algorithms.

2.2.5 Diesel engine sensors

The development and introduction of electronic control systems for diesel engines has been much slower than for petrol engines. The first application of electronic controlled injection pumps appeared on the market only in 1984, first for the very heavy engines (in line injection pumps) and then for car engines (rotative pumps with distributor).

The mechanical revolution regulator which determines the pump shaft position has been replaced with a linear position sensor, generally of the linear variable displacement transformer (LVDT) kind [34]. The variables which constitute the inputs of the pump shaft position control system are:

- accelerator position;
- air, coolant and fuel temperature;
- turbocharging and atmospheric pressures;
- drive shaft speed and angular position.

The injection phasing adjustment is done by comparing the real injection start value (SOI) with the optimal reference value. The SOI is detected by a suitable sensor including a piezoceramic element placed in the injector high-pressure chamber or near its fixing to the engine head (washer).

Other systems for pollutant emission reduction have been introduced for this kind of engine, especially for particulates. Traps in ceramic material have already been adopted and are still being developed. The system using them needs a pressure sensor measuring their clogging for regeneration control. A second method to reduce the nitrogen oxide emissions is known as EGR (exhaust gas recirculation). This system can be controlled electronically; the use of an airflow mass sensor with feedback and diagnostic function is planned.

In the future, systems with injectors driven by an electrical signal acting on a solenoid will be adopted. This determines the length of the injection and also the quantity of injected fuel.

Table 2.1 Market for chassis electronic control systems: percentage of new vehicle production

Systems	*1990/1991*			*2000*			*Range costs** ($)
	USA	*Japan*	*Europe*	*USA*	*Japan*	*Europe*	
Vehicle control							
Transmission	3	25	8	25	50	30	600–2500
Antiskid	10	10	12	95	95	90	250–800
Steering	5	15	5	55	52	50	400–2000
Suspension	2	2	2	7	18	18	800–3000

* Range costs indicate the cost difference due to the added electronic control system (sensors, actuators and electronics).

In these systems the manifold pressure is independent of engine conditions. Smoke sensors at the exhaust will be adopted to perform feedback control. Several prototypes and measurement methods [35–39] have now been developed. However, a reliable automotive kind of sensor has not yet been finalized.

2.3 VEHICLE CONTROL SENSORS

The electronic systems market for vehicle control is rapidly developing and forecasts say that by the end of the 1990s the market share will be about 30% of the automotive electronic systems. This trend is shown in Table 2.1.

In the area of engine control, the development of new systems is accelerated by the need to make the new products comply with the law. But in the

Table 2.2 Sensors required by chassis control systems

ABS	*4WS*	*4WD*	*AS*
Wheel speed*	Steering wheel angle	Brake on-off*	Actuator stroke*
Longitudinal acceleration	Rear steering angle*	Wheel speed	Actuator stroke*
Lateral acceleration	Yaw rate	Throttle angle	Yaw rate
	Lateral acceleration	Steering wheel angle	Vehicle speed
	Vehicle speed		Longitudinal acceleration Hub vertical acceleration

*Minimum-set of sensors enabling fault-tolerant-mode operation.

vehicle systems the main reason for the search for innovative solutions is to be found in the competition amongst various manufacturers.

Antiskid (ABS), four-wheel steering (4WS), four-wheel drive (4WD) and active suspensions (AS) represent the main systems that can tailor the vehicle dynamic behaviour to customer expectations. These systems tend to improve not only performance, but also, and above all, vehicle comfort and safety.

Benefits which come from each single system can be increased by different functions being integrated into a single system.

Further research on sensors, actuators and microcontroller design is required to improve overall performance and reduce costs.

Table 2.2 lists the sensors required by the various control systems; the additional sensors are not necessary to the system strategies but are useful for diagnostic capacities and control performance improvement.

A summary of the main vehicle sensors is presented.

2.3.1 Displacement and angle sensors

Transduction techniques for these kinds of sensor are of two types:

- with mechanical part in contact;
- without contact.

The potentiometric sensors are the most used ones in vehicle systems. One of the most interesting techniques is the use of conductive plastic elements with a multiple contact cursor [40].

The high immunity to parameters such as temperature and electromagnetic interferences (EMI) due to passive and ratiometric functioning makes these sensors particularly suited to automotive use.

Much progress has been made in the resistance of sliding contact wear. Some manufacturers guarantee over 100 million life cycles for their products.

Despite these improvements, the emerging trend in the automotive field is to use new types of sensors which do not require mechanical parts in contact. There exist different transducing techniques based on different physical principles [41]:

- magnetic (Hall effect, magnetoresistive, LVDT);
- optical (incremental, absolute encoders);
- capacitive.

All listed transducers need specific signal-conditioning electronics and so they are more sensitive to temperature and electromagnetic interference compared to potentiometric devices.

Thin and thick film deposition techniques for magnetoresistive materials seem to be very promising for the development of linear position and angular sensors which can be of the absolute or incremental type.

Automotive applications examples of these devices are sensors that use thin film deposition of ferromagnetic materials (MRE), which have interesting characteristics of stability, linearity and resistance in rough environmental conditions [42].

The severe environmental conditions of vehicle systems make the optical or capacitive solution less suitable for applications in such an area, even though there are some studies on automotive capacitive position sensors [43].

The use of no-contact devices remains in any case dependent on the achievement of costs competitive in the automotive market.

2.3.2 Accelerometers

Acceleration represents one of the physical quantities which is most useful to strategies of different vehicle control systems. The most widespread constructive technology makes use of silicon cantilever structures [44] or cantilever on alumina substrate [45]. Lever arm deformation is detected by capacitive or piezoresistive techniques. The range of interest for vehicle applications is $\pm 2\,G$ with a bandwidth of 50 Hz.

Some devices of particular interest make use of feedback measurement methods [46], acting on the seismic mass with electrostatic forces which are in opposition to the movement forced on it by the acceleration. Therefore high resolutions, low time response and low cross-sensitivity are obtained, which represent some of the main requirements for vehicle application.

2.3.3 Gyrometers

In the strategies of 4WS and AS control systems, a role of great interest is played by the yaw and roll rate measurements of the vehicle. The dynamic range of these parameters is very wide (0.1–100 deg/s), and a high sensor accuracy is required.

The development of gyrometers for on-board applications has had an increasing growth in the last few years. As well as traditional solutions based on mechanical devices (Coriolis effect) or optical ones (Sagnac effect) [47], alternative solutions have been developed in order to satisfy the low-cost requirements for large-scale production.

Research carried out by Renault has led to the realization of an acoustic gyrometer prototype [48] utilizing the measurement of the Coriolis effect on a gas contained in a cylindrical cavity.

Another interesting transduction principle is based on the displacement of the electrical field direction on piezoelectric elements, due to the Coriolis effect.

The structure is based on three piezoelectric ceramic elements placed on the sides of a rod of triangular equilateral section [49]. One element works as a resonance frequency oscillator while the other two detect the distribution of the electrical field in the bar's section. The electrical field vector direction is correlated with rotational speed of the structure.

Absence of moving parts, the possibility of miniaturization and high sensitivity make these solutions very attractive for future automotive applications.

2.4 SAFETY SYSTEMS SENSORS

Safety systems can be divided into active and passive ones. The first serve to prevent accidents and the second to reduce the effects.

Only some passive systems have reached the reliability and cost level necessary for the automotive market. Recently some car manufacturers have introduced the airbag [50, 51] whose drive is activated by the sudden deceleration after the crash. The sensor activating the system is made up of one or more accelerometers with high-reliability features to minimize the probabilities of malfunctions.

Active safety systems are more complex and have not yet reached a high enough reliability level to make them suitable for the automotive market. The development of these sensors and systems is one of the topics of the Prometheus European project [52]. In this project the state of the research on the following functions is presented.

- road conditions identification;
- obstacle recognition;
- vehicle localization on the road network.

2.4.1 Ice on the road detection sensors

The reflection coefficient of the road surface is detected by means of an infrared light beam, a solid-state laser directed towards the road and a photodetector which measures the reflected radiation. The wavelength is chosen within an absorbing band of the ice which is slightly different from that of water or steam, allowing ice presence detection on a wet road.

2.4.2 Road roughness sensor

Different techniques for road roughness measurement have been studied, based on the reflection of ultrasounds, light or microwaves. The most promising methodology uses signals coming from measurement of the distance between vehicle and road surface by means of an optical position sensor [53].

However, the costs of the present devices are still too high for the automotive market.

2.4.3 Anticollision sensors (Figure 2.7)

Anticollision systems are generally based on distance measurement to define the safety zone around the vehicle and the criticalities of surrounding objects.

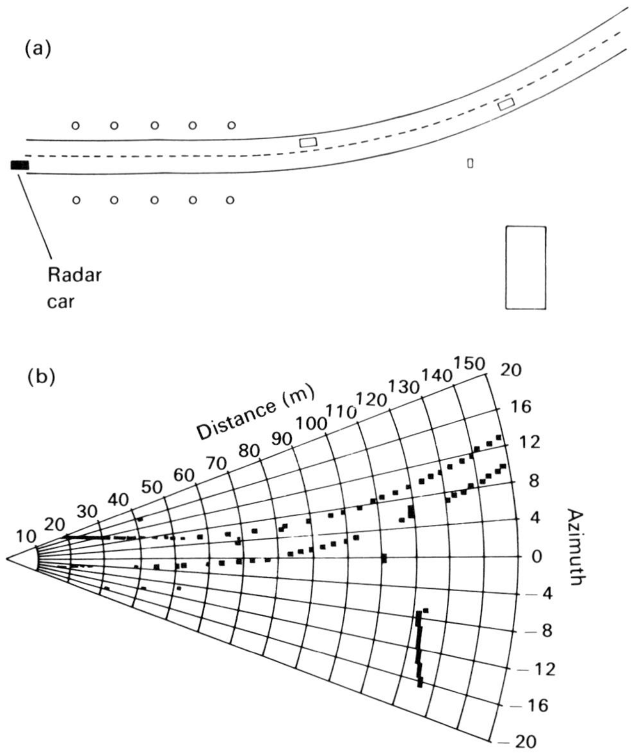

Figure 2.7 Typical radar map.

Speeds and relative accelerations can be derived from this kind of information.

The main measurement techniques are optical (lidar), microwave (radar) and ultrasonic.

Systems based on scanned laser beams allow a high angular and distance definition and seem to be suitable for cruise control applications. They cannot be used in adverse weather such as snow, fog and rain. To overcome this limitation, microwave radars based on two modulation techniques are being studied: frequency modulation and impulse modulation. For both, the choice of the carrier frequency is of great importance because if determines the antenna dimensions and the beam divergence. To obtain limited dimensions and low divergencies, the system must work at frequencies higher than 20 GHz. It is foreseen that evolution of the MMIC (microwave monolithic integrated circuit) will allow future devices to operate at frequencies higher than 35 GHz with large-scale production.

The beam scan [54] can be realized in three different ways:

1. mechanical, with direct movement of the antenna or reflecting element;
2. with frequency scan (only for impulse radar);
3. phased array.

Ultrasonic sensors find applications in parking systems which give an adequate support to the driver in conditions of limited visual angle. These systems generally use two or three ultrasonic sensors, mounted near the bumper, to give the best coverage of the controlled area [55]. The signal processing allows the calculation of the minimum distance between the vehicle and the obstacle, warning the driver when it reaches a fixed safety threshold. Systems of this kind are available on the market for both motor vehicles and industrial vehicles.

2.5 NAVIGATION SYSTEMS SENSORS

In these systems the information to give the driver is the vehicle position on the road network, the shortest way to reach a certain destination, data regarding available parking space, weather conditions and traffic jams or queues.

The means to collect information are based on infrastructural systems such as satellites, radio communication systems (RDS), cellular telephones and autonomous techniques (dead reckoning) and make use of detailed digital maps to solve problems of ambiguity, recalibration and routing [56].

Sensors used to give useful information to be integrated with the data contained in the digital maps are:

- odometers;
- magnetic compass (fluxgate);
- gyrometers.

2.6 SURROUNDING COMFORT SENSORS

Electronic control systems for air-conditioning allowing higher levels of efficiency and comfort have been developed. The more complex systems make use of temperature, solar irradiation, humidity and airflow sensors.

Another important function is the level control of exhaust gas pollutants (HC, CO, NO_x) inside the vehicle due to the traffic. In this case one or more gas sensors must be integrated within the conditioning systems to detect the presence of such gases and to allow the electronic unit to act accordingly to a specific strategy [57].

These sensors must be able to detect concentrations of harmful gases lower than 10 p.p.m. There are commercial devices with thick film deposition of SnO_2 doped with noble metals and containing a heating element which keeps the sensitive element at the functioning temperature (400 °C).

2.7 CONCLUSIONS

The availability of low-cost and reliable sensors is probably the most critical factor in electronic control systems realization. For this reason, integration can be a solution which allows the sensors to be shared among different systems, with a reduction in the number of required sensors, and consequently cost reduction. In these devices the specific conditioning and self-diagnosis electronics will be integrated into the sensors (smart sensors) [58, 59]. In future, sensors will be used in systems with strategies known as 'embedded simulation' [60]. The availability of a real-time simulation of the system will make possible the comparison between the real behaviour of the system detected with sensors and the ideal one derived from the 'embedded simulation'.

REFERENCES

1. Wolber, W.G. (1980) *Automotive Engine Control Sensors '80*. SAE Technical Paper No. 800121.
2. Dell'Acqua, R. *et al.* (1982) *Characteristics and performances of Thick Film Pressure Sensors for Automotive Applications.* SAE Technical Paper No. 820319.
3. Kress, H.J. *et al.* (1991) Silicon pressure sensor with integrated CMOS signal conditioning and compensation of temperature coefficient. *Sensors and Actuators*, **A25–7**, 21–6.
4. Takahashi, K. *et al.* (1990) *Hot Wire Air Flow Meter for Engine Control Systems.* SAE Technical Paper No. 900258.
5. Sauer, R. (1988) *Hot Film Air Mass Meter – A Low Cost Approach to Intake Air Measurement.* Technical Paper No. 880560.
6. Magori, V. and Jena, A.V. (1991) Automotive ultrasound air mass flow meter, in *Proceedings of Eurosensors V*, Rome 1991.
7. Yoshimi, Y. *et al.* (1987) *MRE Rotation Sensor: High Accuracy, High Sensitivity Magnetic Sensor for Automotive Use.* SAE Technical Paper No. 870470.
8. Wheeler, W. (1978) *Precision Position Sensors in Automotive Applications.* SAE Technical Paper No. 780209.
9. Ansaldi, E., Malvicino, C. and Palazzetti, M. (1991) A method of detecting the position of a mechanical part made of ferromagnetic material. European patent application 91830146.6.
10. Engh, G.T. (1977) *Development of the Volvo Lambda-sond System.* SAE Technical Paper No. 770295.
11. Matsushita, S. (1985) *Development of the Toyota Lean Burn System.* SAE Technical Paper No. 850044.
12. Yamaguchi, T. and Nogi, T. (1987) *Engine Control System for Lean Burn Combustion.* SAE Technical Paper No. 870291.
13. Xu, C. *et al.* (1991) Grain size effect on gas sensitivity of porous SnO_2-based elements. *Sensors and Actuators*, **B3**, 147–55.
14. Takani, A. *et al.* (1985) *Progress in Lead Tolerant Titania Exhaust Gas Oxygen Sensors.* SAE Technical Paper No. 850381.
15. Howarth, D.S. and Micheli, A.L. (1984) *A Simple Titania Thick Film Exhaust Gas Oxygen Sensor.* SAE Technical Paper No. 840140.
16. Pfeifer, J.L. *et al.* (1984) *Heated Thick Film Titania Exhaust Gas Oxygen Sensor.* SAE Technical Paper No. 840142.
17. Fleischer, M. and Meixner, H. (1991) Gallium oxide thin films a new material for high temperature oxygen sensors. *Sensors and Actuators*, **B4**, 437–41.
18. Lampe, V. *et al.* (1991) Comparison of transient responses of exhaust-gas sensors based on thin films of selected metal oxides, in *Proceedings of Eurosensors V*, Rome 1991.

19. Hanrieder, W. *et al.* (1991) Causes of fast degradation of thin film lambda probes in engine vehicle exhaust, in *Proceedings of Eurosensors V*, Rome 1991.
20. Takeuchi, T. *et al.* (1983) *Oxygen Sensor for A/F Control.* SAE Technical Paper No. 830929.
21. Fleming, W. (1977) Physical principles governing non-ideal behaviour of the zirconia oxygen sensors. *Journal of the Electrochemical Society*, **124**, 1443–46.
22. Anderson, J.E. and Graves, Y.B. (1981) Steady state characteristics of oxygen concentration cell sensors subjected to non equilibrium gas mixtures. *Journal of the Electrochemical Society*, **128**, 294–300.
23. Dueker, H., Friese, K.H. and Haecker, W.D. (1975) *Ceramic Aspects of Bosch Lambda Sensor.* SAE Technical Paper No. 750223.
24. Haman, E., Manager, H. and Steirke, L. (1977) *Lambda Sensor with Y_2O_3–Stabilized Ceramic for Application in Automotive Emission Control Systems.* SAE Technical Paper No. 770401.
25. Wiedenmann, M. *et al.* (1984) *Heated Zirconia Oxygen Sensor for Stoichiometric and Lean Air-Fuel Ratio.* SAE Technical Paper No. 840382.
26. Higuchi, N. *et al.* (1985) *Heated Zirconia Exhaust Gas Oxygen Sensor Having a Sheet-Shaped Sensing Element.* SAE Technical Paper No. 850382.
27. Soejima, S. and Maze, S. (1985) *Multilayered Zirconia Oxygen Sensor for Lean Burn Engine Application.* SAE Technical Paper No. 850378.
28. Suzuki, S. *et al.* (1985) *Thick Film Zirconia Air-fuel Ratio Sensor With a Heater for Lean Mixture Control System* . SAE Technical Paper No. 850379.
29. Kamo, T. *et al.* (1985) *Lean Mixture Sensor.* SAE Technical Paper No. 850380.
30. Dues, S.M. (1990) *Combustion Knock Sensing: Sensor Selection and Application Issues.* SAE Technical Paper No. 900488.
31. Detecting systems for knocking in IC engine. European patent No. EP 394–234–a.
32. Mock, R. and Meixner, H. (1991) A miniaturized high temperature pressure sensor for the combustion chamber of a spark ignition engine. *Sensors and Actuators* **A25–27**, 103–106.
33. Morris, J.E. and Chi, L. (1987) *Improved Intra-cylinder Combustion Pressure Sensor.* SAE Technical Paper No. 870374.
34. Zabler, E. and Heinz, F. (1991) Mechatronic sensors in integrated vehicle architecture, in *Proceedings of Eurosensors V*, Rome 1991.
35. Collings, N. and Baker, N. (1986) *Real Time Smoke Sensor for Diesel Engine.* SAE Technical Paper No. 860157.
36. Schweiner, G.W. (1986) *Ion Probe in the Exhaust Manifold of Diesel Engine.* SAE Technical Paper No. 860012.
37. The technical trend of diesel electronics in Japan, in *International Symposium of Future Automotive Microelectronics*, ATA 1988, paper No. 8.
38. Smoke sensor. Japanese patent n. JP–A–60–123757.
39. Measuring soot content diesel engine exhaust-gas directly measuring oxygen concentration before and after filtering. Patent No. DE–3935–149–A.
40. Riley, R.E. (1989) *High Performance Resistive Position Sensors.* SAE Technical Paper No. 890302.
41. Wells, R.F. (1990) *Automotive Steering Sensors.* SAE Technical Paper No. 900493.
42. Yoshimo, Y. *et al.* (1988) *MRE-Type Non Contact Position Sensor.* SAE Technical Paper No. 880408.
43. Kobayashi, H. *et al.* (1990) A study of capacitance-type position sensor for automotive use. *Sensors and Actuators*, **A24**.
44. Dell'Acqua, R. (1989) *A Bright Future For Thick Film Sensor.* SAE Technical Paper No. 890480.
45. Cotignoli, G. (1989) *A Thick Film Accelerometer For Electronic Suspension Control.* SAE Technical Paper No. 890481.
46. *Semiconductor Capacitance-Type Accelerometer with PWM Electrostatic Servo Technique.* SAE Technical Paper No. 910274.

47. Oho, S. *et al.* (1990) *Optical Fiber Gyroscopes For Automobiles.* SAE Technical Paper No. 900490.
48. Leblond, H. *et al.* (1990) A New Gyrometer For Automotive Applications. 23 Fisita Congress n. 905081.
49. Nakamura, T. (1990) Vibration gyroscope employs piezoelectric vibrator. *JEE*, September.
50. Macdonald, G.A. (1990) A review for low cost accelerometer for vehicle dynamics. *Sensors and Actuators*, **A21–23**, 303–7.
51. Adams, T.C. (1990) *The Development of an Advanced Air-bag Crash Sensing System.* SAE Technical Paper No. 905140.
52. Glathe, H.P. *et al.* (1990) *The Prometheus Programme – Objectives, Concepts and Technology for Future Road Traffic.* SAE Technical Paper No. 905180.
53. Chaka, R.J. (1978) *The Design and Development of a Highway Speed Road Profilometer.* SAE Technical Paper No. 780064.
54. Daniels, D.J. (1988) *Radar for hazard warning.* ISATA Paper 88058.
55. Noll, M. *et al.* (1991) Ultrasonic sensor for reverse driving applications, in *Proceedings of Eurosensors V*, Rome 1991.
56. Sistemi Elettronici e Satellitari per il Posiziona mento e la Guida di Veicoli Terrestri Civili, Notiziario Tecnico AMMA, pp. 3–10, No. 2/86.
57. Sensing system for controlling air circulation valves in engine vehicles. USA patent No. 4930407.
58. Bertuol, B. (1991) Sensors as a key component for automotive systems. *Sensors and actuators*, **A25–7**, 95–102.
59. Kleinschimdt, P. and Schmidt, F. (1991) How many sensors does a car need? in *Proceedings of Eurosensors V*, Rome 1991.
60. Westbrook, M. (1991) Automotive sensors. *Sensor Review*, **3** (11), 3.

PART TWO

Powerplant and Transmission

CHAPTER 3

Development of non-contacting capacitive torque sensor and system for vehicle application

3.1 INTRODUCTION

The torque/speed relation of an engine is an important parameter for fuel-consumption economy. Optimum fuel economy would be obtained by keeping the engine in the lowest possible gear during acceleration by changing the gear ratio to give increased vehicle speed instead of operating the throttle to increase the engine speed. The throttle should only be operated to increase the power at the lowest possible engine speed. Obviously, the engine speed is not a suitable parameter for adjusting the transmission in such a strategy, so the torque should be used to adjust the transmission ratio for operation within the mechanical limitations [1]. The engine crankshaft, or another axle where the torque is to be measured, is rotating, which strongly favours the non-contact measurement of this torque. An electrical contact with the rotating axle would require the use of slip-rings with a sensor system. Another boundary condition that restricts the range of the possibilities originates from structural limitations. The required mechanical properties of the axle usually prohibit the milling of slots for placing the sensor electronics or a local stricture of the axle for realizing a larger twist angle in the sensor. The milling of slots would be required in non-contact strain-gauge-based torque measurement systems. The electrical power for operation of such a sensor can be supplied to the axle by inductive means and the telemetry of sensor data is also quite feasible. However, practical constraints do not usually permit the weakening of the axle. Moreover, the relatively high engine speed would require a careful mass balancing in the axle to avoid vibrations caused by inertia. Nevertheless, such systems have been implemented in ship axles, where the mass and diameter of the drive shaft and the low number of revolutions per minute permits the mounting of a strain-gauge-based sensor and readout electronics on the axle. A local stricture of the axle can be used to enlarge the torque-induced twist angle. This would facilitate the measurement of the torque using two angular displacement sensors; one on either side of the stricture. As the torque-induced twist angle is inversely

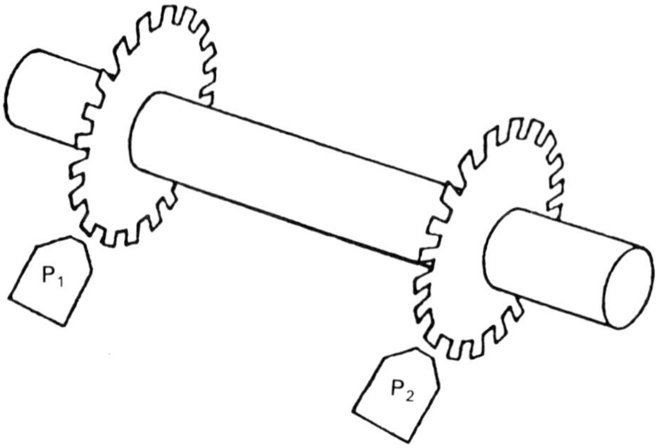

Figure 3.1 Measurement of the torque in an axle using magnetic pick-ups.

proportional to the fourth power of the axle diameter, a significant gain in angular displacement can be obtained; however, structural problems usually prohibit the implementation of these techniques.

The relation between the twist angle φ, over an axle length L in an axle of uniform diameter D and a modulus of rigidity G, at an applied torque T, can be described by:

$$\frac{\varphi}{T} = \frac{32L}{\pi G D^2}\left[\frac{\text{rad}}{\text{Nm}}\right] \tag{3.1}$$

which results in a 10 mm thick steel axle ($G_{\text{steel}} = 8 \times 10^{10}\,\text{N/m}^2$) and a sensing distance $L = 100$ mm in $\varphi/T = 10^{-3}$ rad/Nm. The torque can therefore be determined using two angular displacement sensors spaced a distance L apart to measure this twist angle. For direct torque sensing, with an inaccuracy smaller than 1 Nm and without stricture of the mechanical structure, displacement sensors are required with a circumferential inaccuracy smaller than 6 μm. The simplest differential non-contact angular displacement sensing technique is based on the mutual displacements of two flanges that are clamped on the axle spaced a certain distance apart. Optical, magnetic and capacitive displacement sensing techniques can be applied.

A magnetic torque sensor is shown schematically in Figure 3.1. Assume the teeth of the clamped wheels to be magnetic and the pick-ups P_1 and P_2 to be magnetic sensors. A torque causes the responses of the Hall plates to be out of phase. These signals can be converted into a duty-cycle modulation, which can be detected in the readout circuits.

An optical torque sensor is shown schematically in Figure 3.2. Two disks, each with slits, are mounted on an axle a distance L apart. The twist angle controls the overlapping between the slits and, thus, pulsewidth modulates the transmission from a LED light source to a photodetector. Disks are

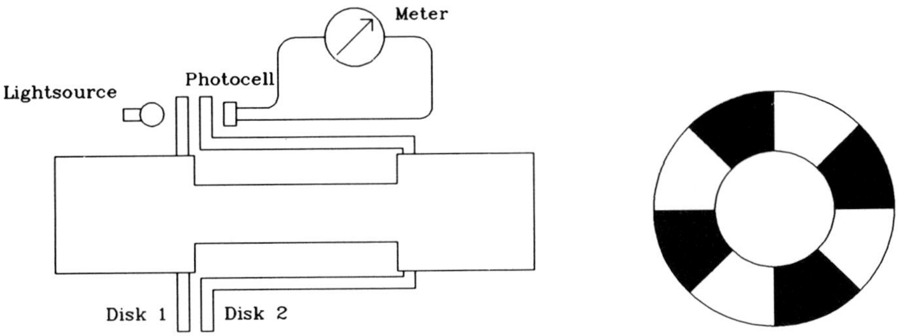

Figure 3.2 Optical torque sensing using two disks with a torque-modulated slit width.

implemented for pursuing a mechanical amplification of the torque-induced angular displacement in this sensor in order to enable torque sensing using conventional displacement sensors. The use of disks is restricted by the available space, which is limited by mechanical boundary conditions, such as those imposed by bearings.

The twist angle in an axle also affects the magnetic properties. A non-contact magnetic torque sensor is possible, based on the magnetostrictive effect, which is basically a strain-induced magnetic field line deflection. The torque-induced compressive and tensile stresses distort the magnetic field lines as shown in Figure 3.3. Applying a magnetic field in the axial direction using external coils makes it possible to measure the perpendicular field component. A second pair of coils can be used for non-contact sensing of the perpendicular field and thus enable the non-contact sensing of the torque. A disadvantage of this method is the power dissipation and the dependence of the sensitivity of the type of steel used. It is possible to get around the latter disadvantage by using the ferroelectric properties of amorphous ribbons instead of magnetostriction of the axle itself. Groups of amorphous iron-based ribbons can be bonded to the axle with the longitudinal direction of one half of the ribbons at 45° with respect to the axial direction and the other half perpendicular to that as shown in Figure 3.4 [2]. When applying a torque to the axle, half of the ribbons will be subjected to tensile stress and the other half to compressive stress. The magnetostriction changes the permeability of the ribbons and causes a change in the mutual inductances between the drive coil, that drives both groups of ribbons, and the two sense coils, that detect the torque-induced magnetic anisotropy in the two perpendicular groups of ribbons. This method, therefore, enables a non-contact torque sensing [3, 4].

A disadvantage of the differential angular displacement method for measuring the twist angle is that the difference between the output signals of two sensors is generally used to determine the twist angle in the readout

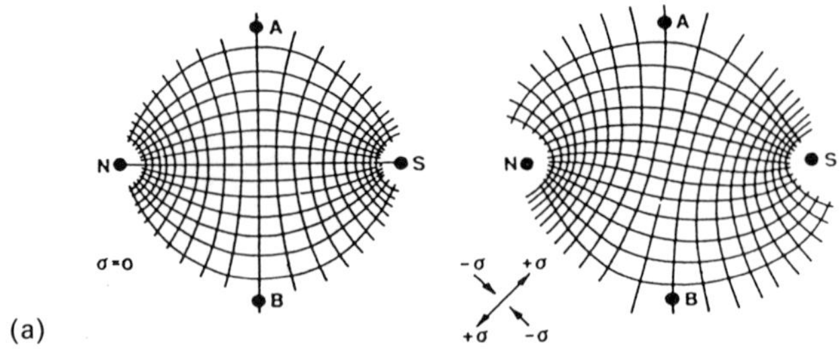

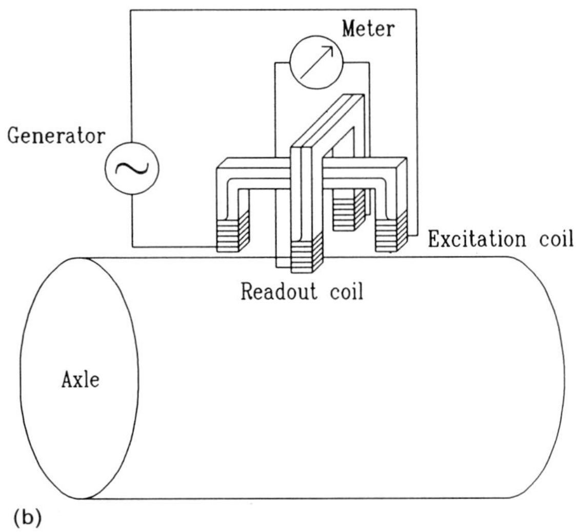

Figure 3.3 Operating principle of the magnetostrictive torque sensor (a) and basic sensor structure (b).

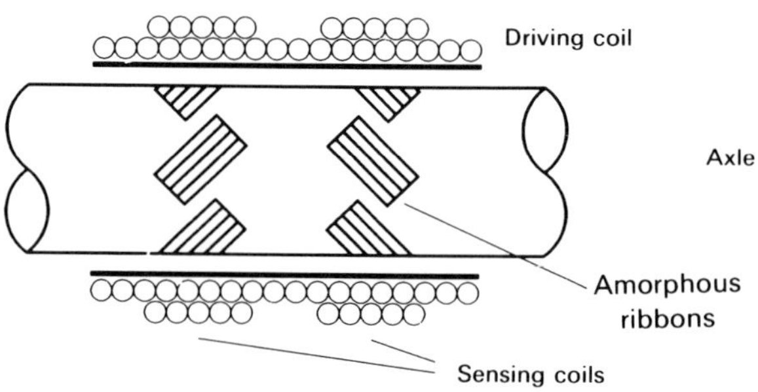

Figure 3.4 Magnetic torque sensor using amorphous ribbons.

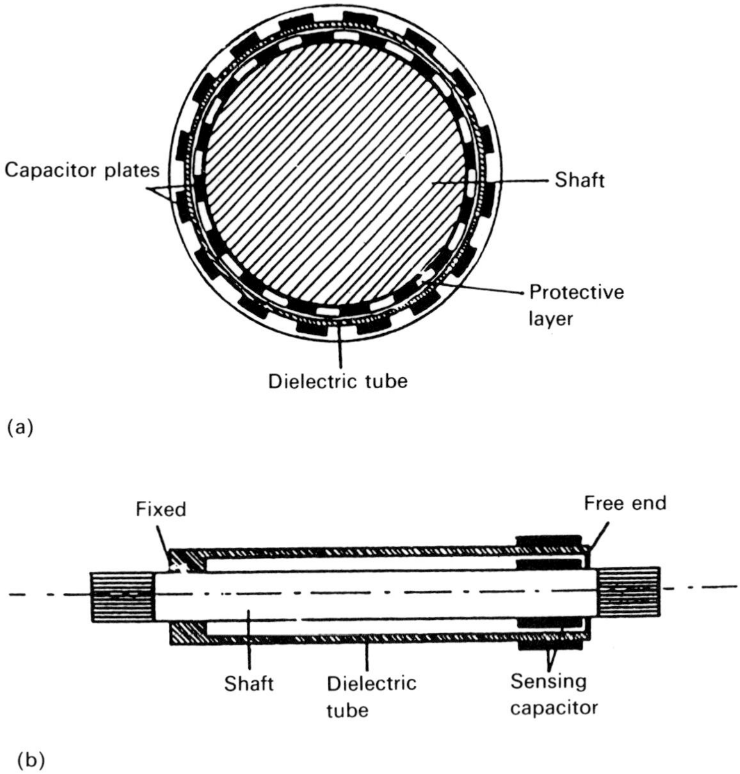

Figure 3.5 Capacitive torque sensor described by Turner (5): (a) cross-section; (b) longitudinal section.

circuitry rather than the response of one differential sensor. A very interesting non-contact capacitive torque sensor that overcomes this drawback has recently been reported in the literature and is described as a non-contact torque sensor based on a differential capacitive displacement sensor. This sensor is shown in Figure 3.5 [5] and consists of two sets of serrated teeth. One set is applied to the outside to a cylindrical tube of dielectric material. The tube is clamped around the shaft at a distance L from the teeth on the axle in such a way that the teeth on the axle and those on the tube are overlapping. A twist angle results in a movement of one set of teeth with respect to the other, which enables the measurement of the torque by measuring the change in capacitance. The readout is based on the change of an LC product using a grid dip principle. Although this method results in a non-contact, low-cost and reliable sensor, it requires a rather complicated mechanical structure composed of two coaxial cylinders with a rubber bearing at the free end. Apart from this sensor, research is directed towards inductive torque sensors based on amorphous ribbons.

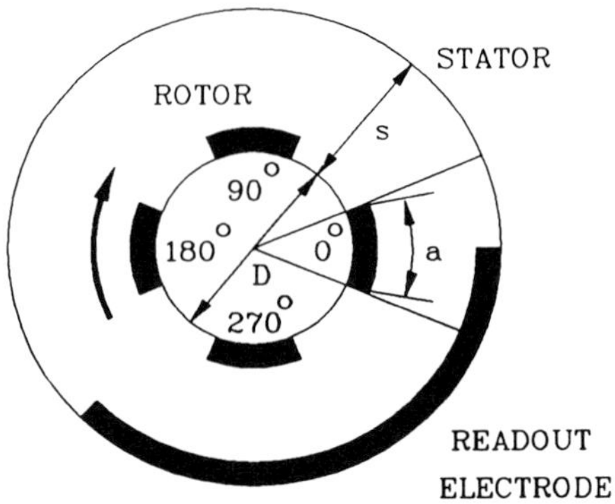

Figure 3.6 Basic capacitive angular displacement sensor.

The response of the capacitive torque sensor discussed here is solely determined by the overlap between two electrode patterns that are directly connected to their respective substrates, the axle and the casing, and is, therefore, based on a very simple mechanical structure, which is tolerant to the spatial limitations that are imposed by, for example, bearings.

3.2 PRINCIPLE OF OPERATION

The capacitive torque sensor is basically a differential angular displacement sensor and is composed of two capacitive displacement sensors mounted on the axle and spaced a certain distance apart in order to enable the measurement of the twist angle. Each of the capacitive angular displacement sensors is composed of two parts, a non-moving part, the stator, and a rotating part, the rotor. The rotor consists of four driving electrodes which are connected to four sinusoidal voltages with equal amplitudes and relative phase differences of 90°, as shown in Figure 3.6. The voltages are capacitively coupled from the stator to the rotor. The capacitive coupling of the four driving voltages back to the stator results in a sinusoidal voltage on the readout electrode on the stator which can be measured. The phase of this voltage depends on the rotation angle ϕ. In this way all the connections are made to the stator and the rotor can rotate freely, so that a contactless measurement is possible.

The electrical equivalent of the sensor is shown in Figure 3.7, in which C_1, C_2, C_3 and C_4 are the four capacitances of the rotor with the 0°, 90°, 180° and 270° driving electrode respectively, C_r is the capacitance of the pick-up electrode with the rotor, and C_a the parasitic capacitance of the rotor and the pick-up electrode to ground. When assuming $C_a \gg C_r, C_1, C_2, C_3, C_4$,

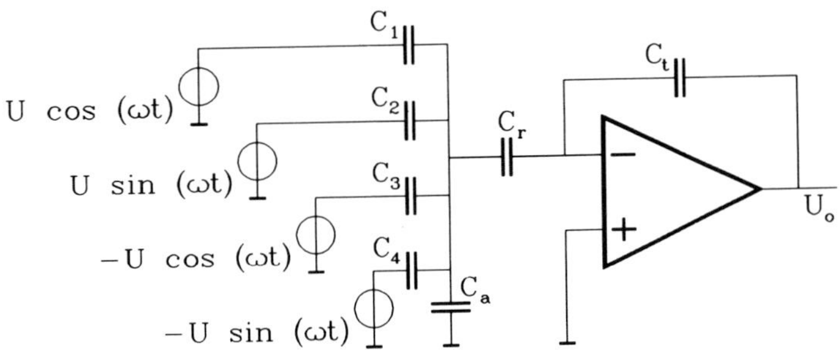

Figure 3.7 Equivalent electrical circuit of the sensor capacitors with the parasitics and the readout charge amplifier.

an output voltage U_0 is generated that can be described by:

$$U_0 = -\frac{C_r}{C_t C_a}[(C_2 - C_4)U \sin \omega t + (C_1 - C_3)U \cos \omega t] \tag{3.2}$$

The phase ψ of this output voltage satisfies:

$$\begin{aligned}\psi &= \arg\{C_1 V_1 + C_2 V_2 + C_3 V_3 + C_4 V_4\}\\ &= \arg\left\{C_1 + C_2 \exp\left(j\frac{\pi}{2}\right) + C_3 \exp(j\pi) + C_4 \exp\left(j\frac{3\pi}{2}\right)\right\}\\ &= \arg\{C_1 - C_3 + j(C_2 - C_4)\}\\ &= \arctan\left(\frac{C_2 - C_4}{C_1 - C_3}\right)\end{aligned} \tag{3.3}$$

Using the parallel-plate approximation with $C = \varepsilon_0 \varepsilon_r A/d$, in which d is the thickness of the dielectric material, $\varepsilon_0 \varepsilon_r$ is the permittivity of the dielectric material and A is the overlapping area, and assuming that only the overlapping area A changes, Equation 3.3 reduces to:

$$\psi = \arctan\left(\frac{A_2 - A_4}{A_1 - A_3}\right) \tag{3.4}$$

in which A_1, A_2, A_3 and A_4 are the overlapping areas of the rotor electrode with the 0°, 90°, 180° and 270° driving electrode respectively.

3.3 THE SENSOR SYSTEM

As mentioned before, the capacitive torque sensor is basically a differential angular displacement sensor and is composed of two capacitive angular displacement sensors mounted on the axle and spaced a certain distance apart. The operation of each of the capacitive displacement sensors is based on the combined capacitive coupling of four sinewaves to the readout electrode. The rotor is composed of an array of electrodes organized as a bar-space type of grating and the stator consists of a single bar-shaped readout electrode. The rotor electrodes are connected to sinewaves with the same amplitude; however, adjacent electrodes have a phase difference of 90°. In this way the repetitive phase pattern $-0°-90°-180°-270°-0°-\cdots$ is generated along the rotor with a periodicity over four electrodes of the array. The phase of the sinewave on the stator electrode is determined by the superimposed coupling between the individual stator electrodes and the readout bar, which depends on the mutual spacings, and is, therefore, a direct measure of the position of the stator with respect to the rotor [6].

The fringing fields can be disregarded in case of a rotor-to-stator spacing much smaller than the width of the electrode bars and at relatively low angular velocities. For a 10 mm diameter axle with a spacing of about 1 mm this implies that only one set of four driving electrodes is allowed along the circumference. An almost linear relation between the output phase and the angular position can be obtained when using a rotor pattern with a spacing equal to the bar width, a, and a stator electrode width, b, equal to

$$b = 3a\frac{D + 2s}{D} \tag{3.5}$$

as shown in cross-section in Figure 3.6. The electrical phase at the readout electrode changes over 360° when the rotor completes one full mechanical revolution. This sensor, therefore, enables the dimensionless transduction from mechanical rotation over a certain angle into an electrical phase shift. The sensor is intended for application in the 0–100 Nm torque range. The associated maximum angle of rotation between the two angular displacement sensors spaced a distance $L = 100$ mm apart can easily be derived from Equation 3.1 and is equal to $\varphi = 5.7°$. A minimum resolvable torque equal to 0.2 Nm can be detected when using a phase meter with a 0.01° resolution. This performance is more than adequate for practical torque measurement in the intended application area; however, the eventual objective of the torque sensor presented here is the integration of the sensor readout circuitry with a simple AD converter in a single chip. This 'smart sensor' will enable on-chip phase readout with 0.1° resolution without having to resort to external professional equipment, by using a phase-to-pulsewidth conversion to drive an integrated counter for a gated AD conversion based on counting. This miniaturization is pursued at the expense of a reduced resolution of the phase measurement and an unmodified sensor configuration would yield a torque resolution of only 2 Nm. To compensate for this detrimental effect a different sensor structure should be designed with an intrinsically better resolution. Such an objective can be met by changing the number of electrodes and the electrode dimensions, which results in a sensor where the effect of fringing fields can no longer be disregarded.

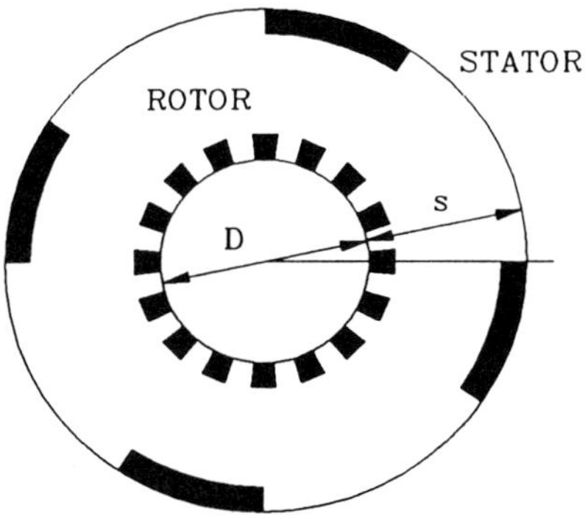

Figure 3.8 Capacitive angular displacement sensor with an enhanced resolution.

The resulting sensor structure would be composed of 16 rotor electrodes and four stator electrodes as shown schematically in cross-section in Figure 3.8. In this way a phase pattern of four times 360° is generated along the circumference of the rotor. Basically, this method implies the transformation of the mechanical rotation over an angle φ into an electrical phase change equal to 4φ. The four readout bars are distributed equidistant over the stator. The detected phases are, therefore, synchronous and the four readout strips can be connected in parallel in order to increase the nominal transducer capacitance and, thus, also the signal amplitude of the sinewave at the input of the readout circuitry is enhanced. This concept can be extended to $4n$ rotor bars and n stator bars. Photolithographic constraints limit the bar width, a, and the bar spacing to about $a_{min} = 0.4$ mm. This limits n to

$$n_{max} = \frac{\pi(D+s)}{2na_{min}} = \frac{\pi \times 11}{8 \times 0.4} = 10 \tag{3.6}$$

This results in a torque resolution in the practical sensor exceeding 0.2 Nm, which is well within the range of the target values. Unfortunately, extreme accuracy requirements should be imposed on the tolerances and eccentricity of the rotor-to-stator spacing, s, for obtaining synchronized phases at all readout bars. A cancellation of the readout signals will occur instead of reinforcement if such requirements are not met and the signal level enhancement, pursued by the synchronization, is largely undone by the tolerances in a practical structure.

For this reason a sensor is constructed with several electrical periods over the rotor circumference, but with only one stator readout strip irrespective of the number of rotor electrodes. The bar width is $a = 0.4$ mm and the rotor-to-stator spacing $s = 1$ mm. These dimensions indicate that the parallel-plate approximation is utterly inadequate for describing the sensor characteristics. The angular displacement sensor with improved resolution

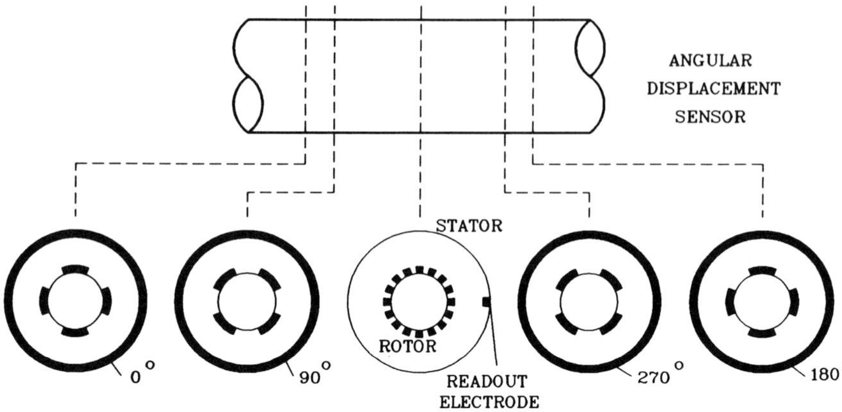

Figure 3.9 Practical capacitive angular displacement sensor for different cross-sections along the axial direction.

is basically of the incremental type. Therefore, an ambiguity can occur in the readout of extremely high torques. A small torque is indicated if the angular displacement exceeds four rotor strips. The practical torque sensor is designed to exhibit a sufficiently large safety margin.

So the angular displacement sensor is composed of a coaxial stator and rotor electrode as depicted at several cross-sections along the axial direction in Figure 3.9. As mentioned in section 3.2, the relation between the phase of the output voltage and the angular information is given by Equation 3.3. As the fringing fields can no longer be disregarded, it is not possible to derive a simple analytical expression for the relation between $\arg[U_0]$ and the angle of rotation. Equation 3.3 has been solved numerically and the result is shown for one period of the repetitive stator pattern in Figure 3.10. The torque sensor is composed of two sets of these angular displacement sensors spaced a well-known distance apart and $\{\arg[U_{01}]-\arg[U_{02}]\}$ is used as a measure of the torque-induced twist angle φ. As arg $[U_0]$ is not a linear function of the angle of rotation, a modulation of $\{\arg[U_{01}]-\arg[U_{02}]\}$ occurs when measuring the torque on a rotating axle. This property prevents the direct torque measurement and the average value of $\{\arg[U_{01}]-\arg[U_{02}]\}$ is used instead. The measurement of the average of the difference in phase can easily be implemented in the readout circuits, using a phase-to-pulsewidth modulation and a counter for the pulsewidth-to-digital conversion, by counting over the time that is needed to travel over at least one set of four bars. The measurement time depends on the angular velocity and gives a minimum measurement delay equal to:

$$\Delta = \frac{\text{time per revolution}}{n} \tag{3.7}$$

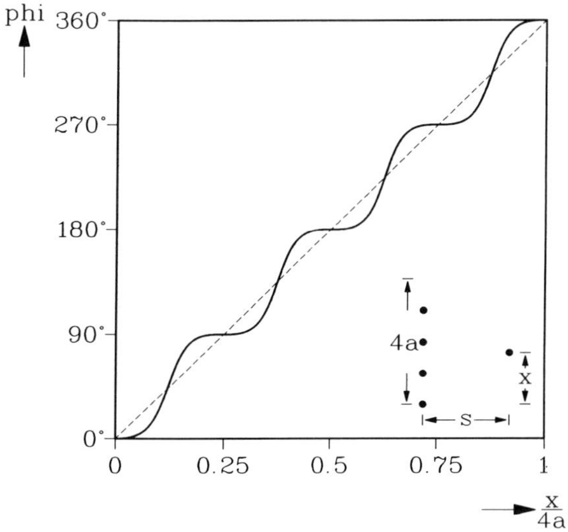

Figure 3.10 Calculated response of an angular displacement sensor with a rotor-to-stator spacing, *s*, much larger than the electrode spacing, *a*. The phase angle on the readout electrode is shown versus the position X.

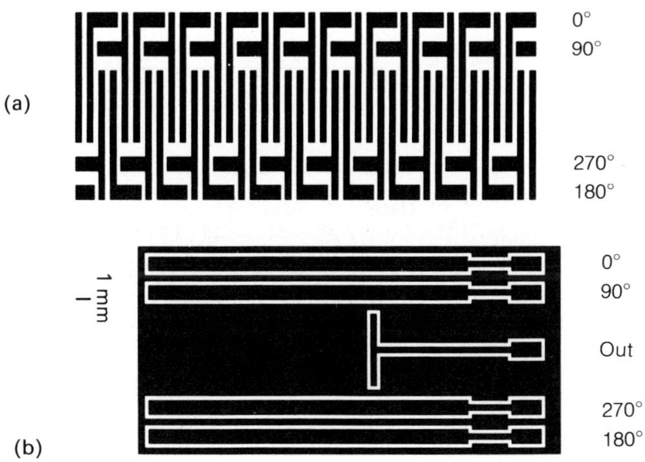

Figure 3.11 Rotor pattern (a) and stator pattern (b) used for the angular displacement sensors that are implemented in the practical torque sensor.

An improved signal-to-noise ratio can be obtained by counting over n/m revolutions; however, the increased measurement time will give an extra delay, which might affect the stability of a closed-loop system that is based on this sensor.

A sensor using the design shown in Figure 3.9 has been built. The electrodes of the practical sensor are realized on flexible, printed circuit board material and are attached to their respective substrates using a special cement that is customarily used for creep-resistant bonding of strain gauges. A dielectric is required in between the rotor electrode and the axle with a thickness of the same order of magnitude as the rotor–stator spacing. This measure prevents the loss capacitance, C_a, between the interdigitated rotor electrode and the conductive axle from severely reducing the amplitude of the AC voltage before return coupling to the detector. The stator and rotor electrode patterns of the capacitive angular displacement sensor are shown in Figure 3.11. The rotor pattern basically consists of a structure with large-area pads for the capacitive coupling of the AC voltages to the rotor. The driving and readout electrodes are separated using a ground electrode in between, to prevent direct coupling to the stator and to minimize stray field coupling between stator and rotor.

3.4 SENSOR PERFORMANCE

The sensor performance for static torques has been measured using the sensor with a 10 mm diameter steel axle that is clamped on one side, as depicted

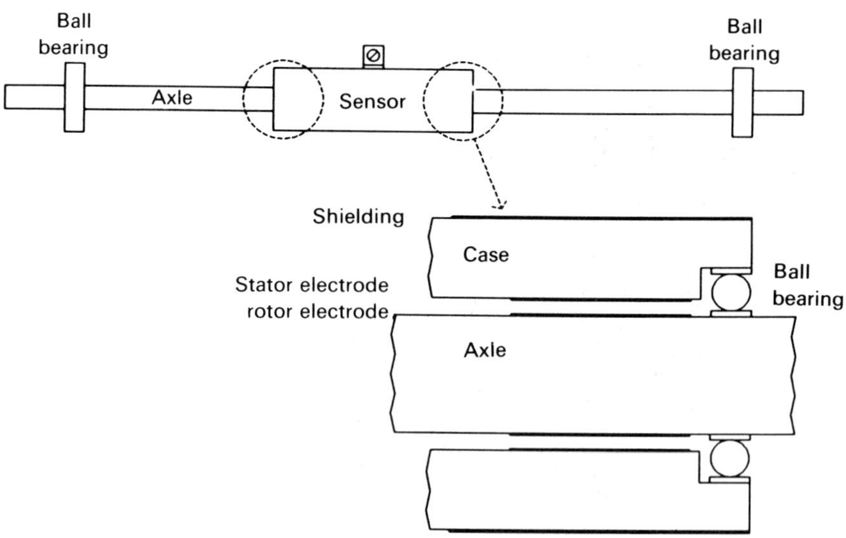

Figure 3.12 Measurement set-up for the capacitive torque sensor.

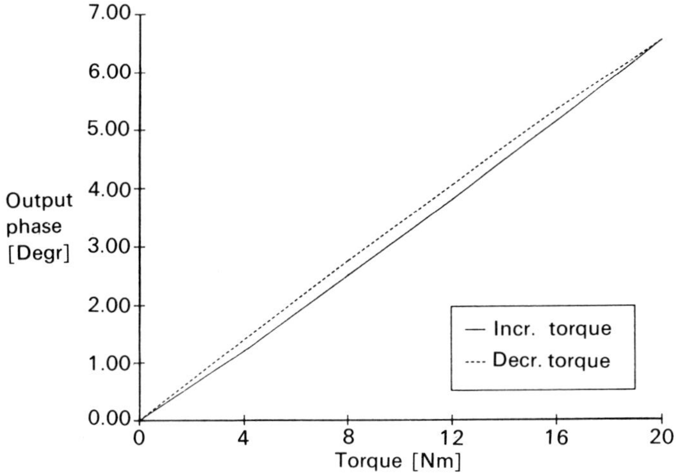

Figure 3.13 Results of measurements performed on the capacitive torque sensor at increasing and decreasing torque.

in Figure 3.12. On the other side a lever is connected to the axle. A charge amplifier has been used for the readout of the output phase. A torque can be applied to the axle using standard weights connected to the end of the lever. The response of the sensor to a torque increasing from 0 Nm to 20 Nm and subsequently decreasing again to 0 Nm is shown in Figure 3.13. Preliminary measurements have been performed on a rotating axle at low angular velocities. The results are in agreement with the static response curves; however, the accuracy of these measurements was not yet sufficient to give quantitative results. The measurements indicate a hysteresis. Loading an axle from the unloaded situation and subsequently reducing the load again, at a certain value of the applied torque, does not immediately give the initial phase angle. The error decreases after a few minutes. This behaviour is characteristic for creep. The bonding technique used to cement the rotor electrode to the axle is critical for this effect and a substantial improvement can be expected when using a thick film printing technique for the realization of the rotor electrodes. As no saturation occurs, this sensor is suitable for a wider range of torques. In the present measurement setup it is not yet possible to apply such torques.

3.5 CONCLUSIONS

Non-contact capacitive torque sensing is possible, based solely on the capacitive coupling between a stator and a rotor electrode pattern. A very robust and simple torque sensor can be constructed in this way. A sensitive angular displacement-to-phase angle conversion has been implemented for the readout with a sensitivity exceeding 0.3°/Nm. Present prototypes suffer from hysteresis, which can be reduced when using a different technique for the bonding of the rotor electrode. As the average phase difference between two capacitive displacement transducers is measured, the torque indication is not affected by the angular velocity of the axle. A fixed angular mismatch between the two displacement sensors gives rise to an offset, which can easily be compensated by an initial torque measurement prior to the mechanical loading of the axle. The mechanical power supplied by the axle can also be determined when using the rate of change in the phase of one of the displacement sensors as the velocity signal. Future research will be focused on a reduction of the creep behaviour and a widening of the operating range by testing different bonding techniques as well as by improving the electrode geometry. The research will also aim at the realization of a custom-built silicon chip that will contain the electric circuitry required for driving and readout of the sensor.

REFERENCES

1. Westbrook, M.H. (1985) Sensors for automotive application. *Journal of Physics E: Scientific Instrumentation*, **18**, 751–8.
2. Sasada, I., Sakai, E., Uramoto, S. and Harada, K. (1985) Noncontact torque sensor employing synchronized switching process, *in Proceedings of the 5th Sensor Symposium*, Japan, pp. 115–20.
3. Hase, H. and Wakamiya, M. (1989) Torque sensor, *in Proceedings of the 8th Sensor Symposium*, Japan, pp. 279–82.
4. Nishibe, Y., Nonomura, Y., Abe, M., Tsukada, K., Takeuchi, M. and Igarashi, I. (1989) Real time measurement of instantaneous torque with high accuracy using magnetostrictive sensor, *in Proceedings of the 8th Sensor Symposium*, Japan, pp. 111–14.
5. Turner, J.D. (1989) The development of a thick-film non-contact shaft torque sensor for automotive applications. *Journal of Physics E: Scientific Instrumentation*, **22**, 82–8.
6. Klaassen, K.B. and van Peppen, J.C.L. (1982) Linear capacitive displacement transduction using phase readout. *Sensors and Actuators*, **3**, 209–20.

CHAPTER 4

Sensors and systems for crankshaft position measurement

4.1 INTRODUCTION

Crankshaft angular position measurements are fundamental to all modern automotive gasoline-fuelled engines. These measurements are required to control fuel-injection timing as well as ignition timing. However, many other functions can be performed from such measurements through the use of advanced signal processing. These additional functions are essentially diagnostic in nature although there is potential for substitution of primary fuel and ignition control functions. This chapter introduces various sensors and systems for the measurement of crankshaft position and velocity, and explains a number of applications of these measurements, describing the basic instrumentation, the signal-processing hardware and the algorithms for each application. Although not all of the functions discussed in the chapter are found commercially today, they are included here in anticipation of their potential use. The applications discussed in the chapter include: (1) ignition and injection timing; (2) measurement of average and instantaneous engine angular velocity; (3) estimation of engine average and instantaneous indicated torque; and (4) detection of engine misfires.

The measurement of relative and absolute engine position is an important component of the control strategy in an electronically controlled engine, being indispensable in the computation of ignition and injection timing. Similarly, the measurement of engine speed provides an important performance variable. These applications will be discussed in sections 4.3 and 4.4, respectively. In addition to the above measurements, estimates of engine indicated or brake torque are also useful for advanced control and diagnostic purposes. Advanced control applications include adaptive fuelling in which fuel delivery can be adjusted to correct for deficiencies in the fuelling system. Another important application of the indicated torque estimate is diagnosis of misfire. Whenever there is a complete lack of combustion in any cylinder during a given engine cycle due to fuelling or ignition failure, there is no torque production associated with that portion of the engine cycle associated with the power stroke of the misfiring cylinder. The associated misfire can be detected using the indicated torque estimate obtained from crankshaft

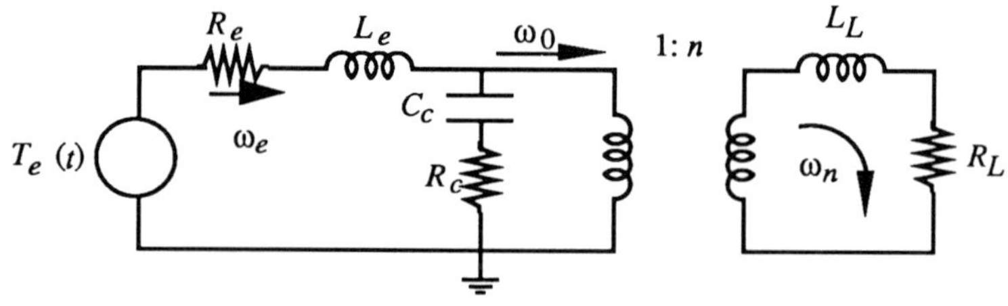

Figure 4.1 Equivalent circuit for IC engine.

angular position measurements. One of the objectives of this chapter is to illustrate how estimates of engine indicated or brake torque may be obtained from a measurement of crankshaft position/velocity.

In order to illustrate the relationship between engine torque and the crankshaft angular velocity signal, it will be useful to introduce a dynamic model for the engine/drivetrain, which relates the instantaneous engine angular velocity to the torque applied to the crankshaft. This relationship is perhaps best understood from an approximate equivalent circuit for the engine/manual transmission and from an associated mathematical model described below and by Ribbens [1] and Connolly and Rizzoni [8]. In the equivalent circuit shown in Figure 4.1, voltages is analogous to torque, current is analogous to angular speed, and circuit impedance elements are analogous to parameters of the mechanical system:

$$\text{inductance } L \rightarrow \text{moment of inertia } J$$

$$\text{capacitance}^{-1}\, C^{-1} \rightarrow \text{effective torsional spring rate}$$

$$\text{resistance } R \rightarrow \text{friction}$$

Further, the generation voltage $T_e(t)$ represents the torque applied to the crankshaft, where

$$T_e[\theta(t)] = T_i[\theta(t)] - T_r[\theta(t)] - T_{fp}[\theta(t)] \tag{4.1}$$

$$T_i[\theta(t)] = \text{indicated torque}$$

$$T_r[\theta(t)] = \text{torque associated with inertia of reciprocating components}$$

$$T_{fp}[\theta(t)] = \text{torque loss associated with piston/ring friction and pumping}$$

$$[\theta(t)] = \text{instantaneous crankshaft angular position}$$

The instantaneous crankshaft angular velocity is denoted $\omega_e(t)$, and the angular speed of the transmission output shaft is denoted $\omega_n(t)$ and $\omega_o(t)$ is the instantaneous angular speed of the transmission input shaft. For the purposes of the present discussion, the differential or the instantaneous speed at the drive axles will not be modelled, although such modelling is straightforward. The transmission is represented by an ideal (DC coupled) transformer in which the turns ratio $l:n$ corresponds to the drivetrain gear ratio. Here, once again, for simplicity, we neglect transmission losses, although such modelling is possible. Assuming an engine coupled to a manual transmission, the parameters of this equivalent circuit include:

L_e = moment of inertia of engine rotating components (including crankshaft, flywheel and clutch pressure plate)

R_e = frictional loss parameter for engine rotation

L_L = moment of inertia at transmission output load

R_L = power absorption parameter of load on the transmission

C_c = compliance of clutch plate torsional springs

R_c = damping parameter for clutch plate

The model of Figure 4.1 neglects many effects which influence the relationship between T_i and crankshaft angular motion (e.g. road-induced torsional oscillations in $\omega_o(t)$); however, it captures the essential relationship between engine-produced torque and crankshaft angular position and velocity. This relationship will be exploited in various applications, to be discussed later in this chapter. It can be seen with reference to Figure 4.1 that the instantaneous crankshaft angular velocity is dynamically related to the instantaneous torque applied to the crankshaft by each cylinder. This torque consists of the sum of an average component, plus a pulsating component superposed thereon, associated with each power stroke. As a result of this torque the crankshaft angular velocity, $\omega(t)$, given by:

$$\omega(t) = \Omega + \delta\omega(t) \tag{4.2}$$

where Ω is the average speed (proportional to engine speed) and $\delta\omega(t)$ is the speed variation due to the torque pulsations. The output voltage of any crankshaft position/velocity sensor at constant (average) engine speed is of the form

$$v(t) = V\sin(\theta(t)) \tag{4.3}$$

where $\theta(t)$ is the instantaneous crankshaft position relative to reference line fixed to the engine block. The instantaneous sensor frequency, $\omega_s(t)$, is proportional to the crankshaft angular speed:

$$\omega_s(t) = M\omega(t) = M\Omega + M\delta\omega(t) \tag{4.4}$$

where the integer M depends on the nature of the position-sensing method (e.g. M is the number of lugs on the ferromagnetic disk for a magnetic sensor, or the number of optical lines for an optical sensor). The actual detail of the position-sensing methods will be discussed in the next section. At this stage it will suffice to remark that the form of the crankshaft angular velocity signal is that of a frequency-modulated carrier where

$$M\Omega = \text{carrier frequency}$$

$$M\delta\omega(t) = \text{modulating signal}$$

The signal model presented here will be a useful reference in the discussion of various applications in later sections of this chapter.

4.2 CRANKSHAFT POSITION AND VELOCITY SENSORS

There are two primary sensor categories which find common applications in crankshaft position measurements; these are magnetic and optical. Each of these categories can have a variety of configurations and implementations. For example, magnetic sensors can be realized in a variety of ways, each having its own relative advantages. In the remainder of this section the principles of operation of inductive, magnetoresistive, Hall effect and optical sensors will be discussed.

4.2.1 Inductive sensor

One very common magnetic sensor for crankshaft position measurement is the inductive sensor illustrated in Figure 4.2. In this sensor configuration, a permanent magnet having a coil wound around it is placed in close proximity to a ferromagnetic disk. The ferromagnetic disk is attached to the crankshaft and rotates with it. The disk has M teeth, as illustrated, which cause the magnetic flux linkage λ in the coil to vary with the angular position of the

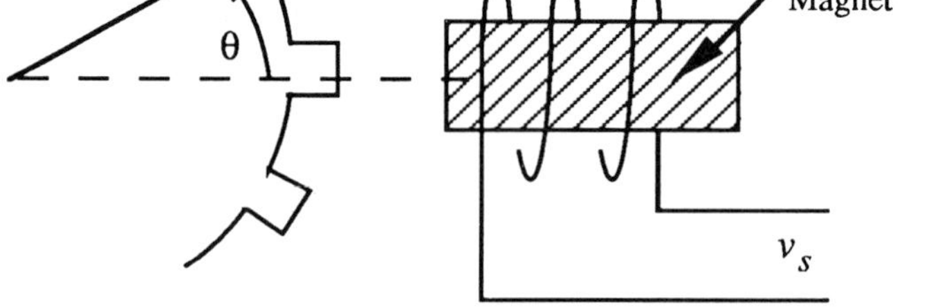

Figure 4.2 Inductive angular position sensor.

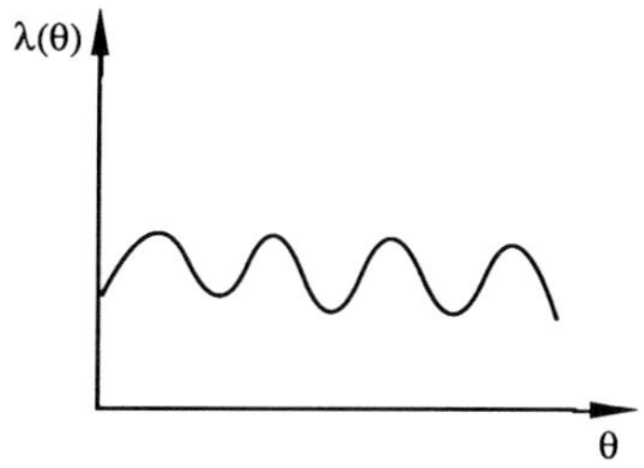

Figure 4.3 Flux linkage as a function of angular displacement.

disk, θ. The coil voltage is given by

$$\begin{aligned} v &= \frac{\mathrm{d}\lambda}{\mathrm{d}t} \\ &= \frac{\mathrm{d}\lambda}{\mathrm{d}\theta}\frac{\mathrm{d}\theta}{\mathrm{d}t} \\ &= \omega\frac{\mathrm{d}\lambda}{\mathrm{d}\theta} \end{aligned} \tag{4.5}$$

There is one cycle of sensor voltage produced for each disk tooth crossing the coil axis. Crankshaft angular position measurements can be made from such a configuration since the function $\lambda(\theta)$ is fixed by the physical configuration. This function has M essentially identical cycles as shown in Figure 4.3.

The active magnetic field sensor described above has certain deficiencies for crankshaft position measurement. Perhaps the most serious of these is the speed dependence of the output. The amplitude of the sensor output is proportional to crankshaft angular speed. In particular, at zero speed (or near zero speed), the sensor generates no output. Consequently, this sensor cannot be used for static crankshaft position measurements such as are often used for end-of-assembly line calibration purposes. In additon, the speed dependence generates a once per revolution low-frequency 'drift' signal due to any run-out in the ferromagnetic disk assembly. This run-out causes systematic timing errors whenever zero crossing points are used to define crankshaft position. Furthermore, this active magnetic field sensor has a relatively complicated output impedance. The impedance is responsible for a speed-dependent phase shift of the sensor output voltage. This phase shift is also responsible for speed-dependent timing errors whenever zero crossing points are used for absolute crankshaft position measurements. Compensation for this phase shift through relatively straightforward means is feasible.

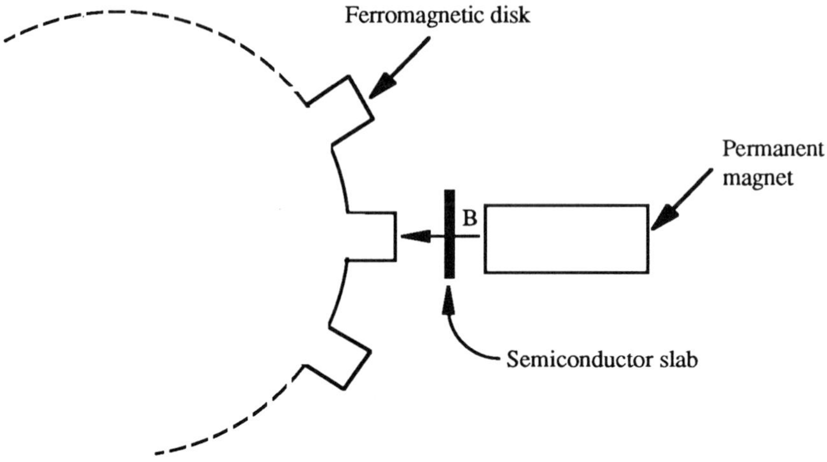

Figure 4.4 Magnetoresistive sensor configuration.

4.2.2 Magnetoresistive sensor

An alternative magnetic sensor which is free of the speed-dependent difficulties of the active magnetic field system sensor described above is the magnetoresistive sensor. This type of sensor incorporates a thin slab of semiconductor material which can move through a magnetic field. Operation of the sensor is based upon a property of semiconductors called magneto-resistivity, in which the material resistivity (ρ) is a function of applied magnetic field:

$$\rho(B) = \rho_o(1 + \gamma B^2) \tag{4.6}$$

where B is the magnetic flux density in the material. A configuration for this sensor which is potentially applicable to crankshaft position sensing is depicted in Figure 4.4. This sensor element is typically connected into a bridge circuit such that an output voltage proportional to B is obtained, as shown in Figure 4.5. In the bridge circuit of Figure 4.5, the resistance of three legs of the bridge is fixed, and is given by

$$R_o = R(B)|_{B=0} \tag{4.7}$$

As the ferromagnetic disk rotates, the magnetic flux density fluctuates cyclically and the bridge voltage v_0 fluctuates with crankshaft angle θ correspondingly as shown in Figure 4.5. There is one cycle for each tooth on the ferromagnetic disk. The magnetoresistive position sensor has the relative advantage of an output independent of crankshaft speed. In fact, the voltage shown above can be obtained for static crankshaft position.

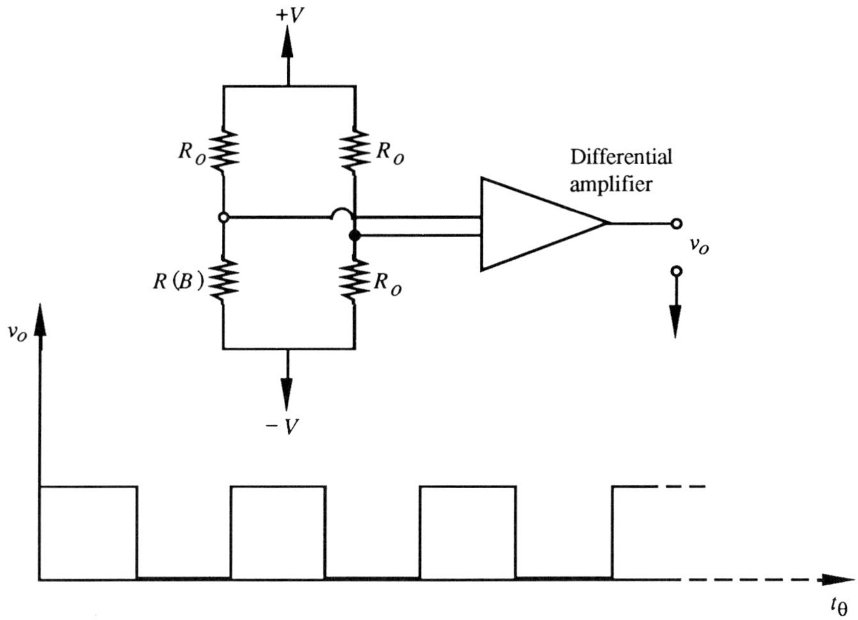

Figure 4.5 Bridge configuration and output waveform of magnetoresistive sensor.

4.2.3 Hall-effect sensor

One of the primary advantages of the magnetoresistive sensor relative to the active magnetic induction sensor is its ability to measure crankshaft angular velocity at all engine speeds. Another sensor offering the same advantage is the Hall-effect sensor. The Hall-effect sensor is fabricated from a thin slab of semiconductor material and is located for crankshaft position measurements in the gap between a permanent magnet and a ferromagnetic disk, in the same configuration which was depicted in Figure 4.4. The operation of this sensor is based upon a phenomenon known as the 'Hall effect'. Whenever an electric current is passed through the slab of semiconductor material by means of an external circuit, and in the presence of magnetic field, the electrons are deflected perpendicularly to the direction of current flow by the magnetic field.

In Figure 4.6, the current, I, is represented by electrons, e, which have negative charge, flowing from left to right. The magnetic field, B, is perpendicular to the page, pointing into the page. Whenever an electron moves through a magnetic field, a force, called the Lorentz force, is exerted on the electron, which is proportional to the electron velocity and to the strength of the magnetic flux. The direction of this force is perpendicular to the direction of the magnetic flux lines and to the direction in which the electron is moving.

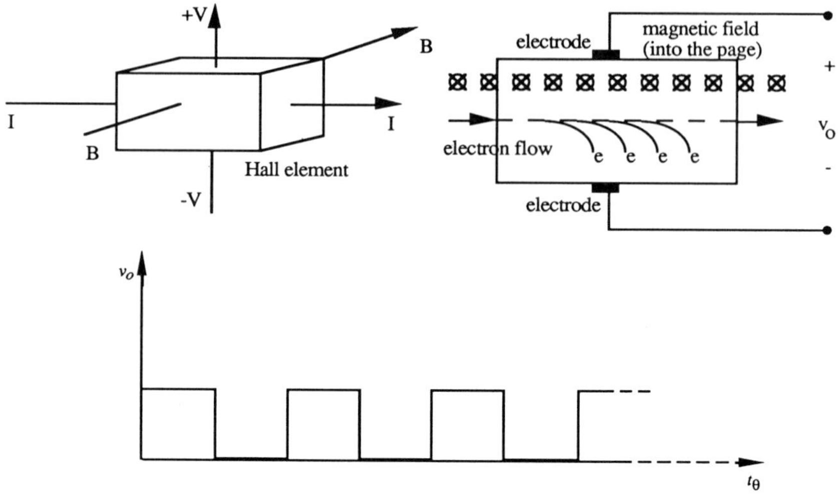

Figure 4.6 Hall-effect sensor configuration and waveform.

In Figure 4.6, the Lorentz force direction is such that the electrons are deflected towards the lower sense electrode. Thus, this electrode is more negative than the upper electrode and an electric field, E, is created which is proportional to B. This electric field generates a potential, v_o, between the external contacts, which is also proportional to B. The voltage waveform which is produced by the Hall-effect sensor is shown in Figure 4.6. Since v_o is proportional to the magnetic flux density, it reaches a maximum when any of the tabs is symmetrically located between the magnet pole pieces. If the disk is driven by the crankshaft, then the disk will have a number of tabs which provide reference points for crankshaft position measurement, as illustrated in Figure 4.6.

4.2.4 Optical position sensor

Optical tachometers and position encoders are finding increasing applications in automotive systems. The basic optical tachometer is shown in Figure 4.7, depicting a slotted disk rigidly mounted at one end of a rotating shaft; the number of slots, or lines M, which are uniformly spaced around the perimeter of the disk, varies depending on the desired degree of resolution. Typical values of M range from approximately 100 to 1000. In recent years, due to the evolution of digital signal processing hardware and software, it has become more advantageous to select $M = 2^p$, p integer, so as to permit the application of fast-Fourier transform-based processing algorithms. Traditional optical tachometer/position encoder systems, used in engine laboratory testing and research applications, are based on 360° or 720° slotted wheels

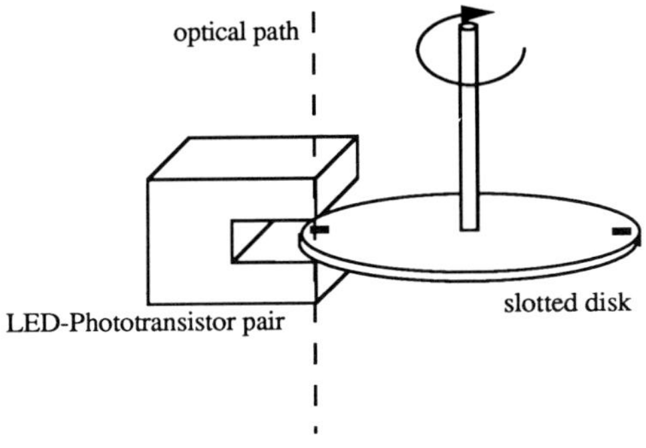

Figure 4.7 Optical position/velocity sensor.

(i.e. $M = 360$ or 720). Modern chemical etching techniques permit the fabrication of much higher resolution optical tachometers (up to 36 000 lines/revolution). The basic principle of operation of such tachometers is the same regardless of the line density (spatial resolution); the only added requirement of very high resolution tachometers is an additional optical grating, which is normally placed between the slotted disk and the optical sensor whenever the physical dimension of the optical sensing element is greater than the spacing between adjacent lines on the disk. Such gratings are usually employed when the number of lines, M, is of the order of 500, or greater. Commercially available systems often employ a prepackaged and prealigned light source–sensor pair, consisting of a light-emitting diode (LED)–phototransistor pair. The LED acts as a light source, and the phototransistor as a sensor. In such arrangements the phototransistor is usually positioned behind a slit of width comparable with the line spacing on the disk. Thus, the phototransistor can resolve changes in light intensity caused by the passage of adjacent lines. The optical tachometer described above generates an electrical output which switches at a frequency f_{tach} given by

$$f_{\text{tach}} = \frac{M\Omega}{60} = \frac{M\omega}{2\pi} \tag{4.8}$$

where Ω is the speed of the rotating shaft in rev/min, ω the same speed in rad/s, and M the number of lines.

The sensing element, which is a phototransistor, responds to light intensity, that is to the power density of an optical beam (W/m^2), over some optical bandwidth dependent on the semiconductor material employed, typically covering parts of the visible and near infrared spectrum. These devices are therefore compatible with both LED and incandescent light sources. A phototransistor has a structure similar to that of a bipolar transistor, with

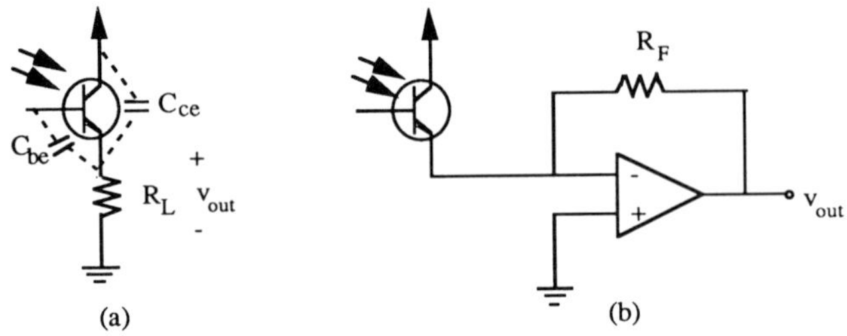

Figure 4.8 Phototransistor circuits.

no external electrical connections to the base region. Minority carriers are created in the base region by absorption of photons from the incident light beam in the base region. The corresponding collector current is proportional to this minority carrier concentration – much like in a conventional transistor – and is therefore proportional to the optical intensity.

Figure 4.8a depicts a typical circuit arrangement for the phototransistor. Note that the light intensity in effect generates base current which is amplified by the transistor. In addition to the choice of an appropriate light source, which depends on the application and on the shielding to infrared and visible radiation which can be provided for the sensor, an important consideration is the sensor bandwidth. The circuit bandwidth of the phototransistor is limited by the parasitic collector-emitter capacitance, C_{ce}, and base-emitter capacitance, C_{be}. These give rise to two time constants:

$$\tau_{be} = r_{be} C_{be} \tag{4.9a}$$

$$\tau_{ce} = R_L C_{ce} \tag{4.9b}$$

where r_{be} is the base-emitter junction resistance and R_L the load resistance. The base-emitter junction time constant, τ_{ce}, can be greatly reduced by means of the op-amp circuit of Figure 4.8b which reduces the effective load resistance seen by the phototransistor to zero (ideally). Thus, the lower limit on the sensor bandwidth is imposed by τ_{be}. The bandwidth required of the sensor can be determined by considering the maximum speed of rotation of the shaft and the number of lines, to determine the maximum sensor frequency, f_{max}, according to:

$$f_{max} = \frac{M\omega}{2\pi} \tag{4.10}$$

For example, for an engine with maximum speed of 7000 rev/min and a tachometer with 256 lines, the required sensor bandwidth would be close to 30 kHz, well within the bandwidth of a typical phototransistor.

Other optical sensors which find application in crankshaft position sensing are the optical shaft encoders shown in Figure 4.9. These encoders exploit an

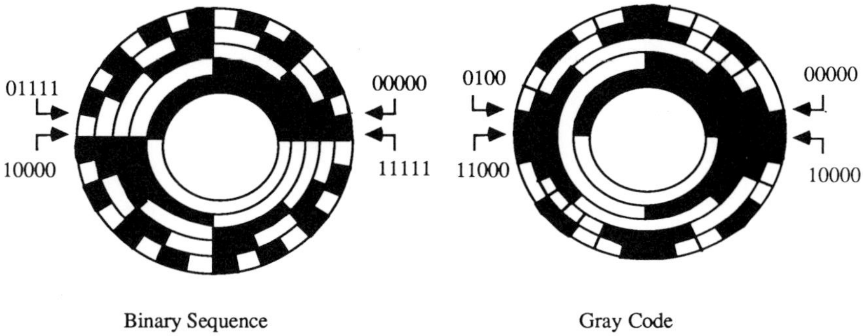

Figure 4.9 Angular position encoders.

array of LEDs and phototransistors to output an absolute shaft position digital signal. Figure 4.9 depicts a 5-bit encoder which would permit a position resolution of

$$\Delta\theta = \frac{1}{2^5} \times 360° = 11.25°$$

An 8-bit encoder would permit a resolution of approximately 1.4°. While the slotted disk tachometer is naturally suited to the measurement of shaft velocity, the encoder of Figure 4.9 finds application in absolute position sensing, such as would be required in ignition or fuel injection timing.

A number of applications of the sensors discussed above, and the related signal-processing approaches, will be discussed in the following sections.

4.3 APPLICATIONS OF CRANKSHAFT POSITION MEASUREMENT

4.3.1 Ignition and fuel-injection timing

The measurement of crankshaft absolute position is critical in ignition and fuel-injection application. In principle, any of the above sensors can be used, although those capable of static (zero speed) measurement are preferable to simplify end-of-assembly engine calibration procedures. Let us denote the output of an appropriate position-measuring sensor by $v[\theta(t)]$. Crankshaft position can be measured at a fixed point in each cycle by, for example, detecting zero crossings of sensor voltage $v[\theta(t)]$; for example, the zero crossings can be detected by means of an analog comparator, such as is shown in Figure 4.10. The comparator switches from logic low to logic high for each zero crossing of the sensor voltage. Each transition serves as an index for a specific crankshaft angular position. Absolute crankshaft position

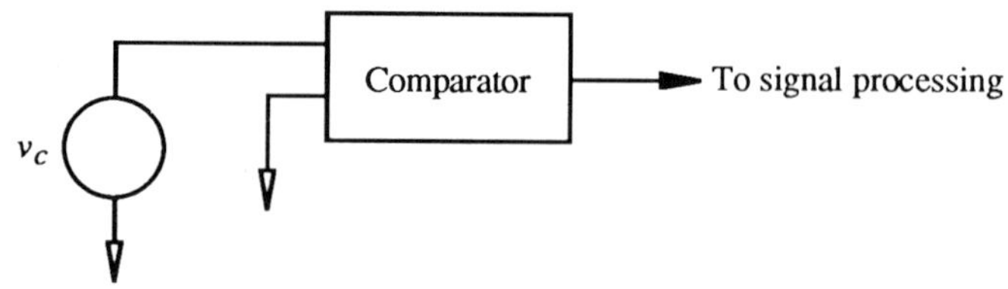

Figure 4.10 Zero crossing detector.

relative to a reference (e.g. top dead centre of number 1 cycle) determination can be measured in a variety of ways, employing either magnetic or optical sensors. For example, a separate ferromagnetic disk having a single lug, in conjunction with a magnetoresistive or Hall effect sensor, would generate a pulse at a fixed point in each engine cycle. Then absolute engine position is given at each of the lugs on the crankshaft disk.

An alternative and less costly means of obtaining absolute angular position involves modification to an existing crankshaft ferromagnetic disk with multiple lugs. A typical scheme involves removing one or more lugs at specific positions on the ferromagnetic disk. The sensor output will have 'gaps' or missing cycles for the missing teeth. It is relatively straightforward for the engine control computer to identify such gaps and the following sequence of normally spaced sensor cycles. Typically, there will be one such gap for each cylinder event during any engine cycle (two crankshaft revolutions).

In the following paragraphs a crankshaft position-measuring system is discussed, in which a finite state machine performs the functions necessary to determine crankshaft position for ignition timing purposes. This system has been experimentally evaluated and found to perform successfully in generating ignition timing signals for a production engine.

System description

The structure of the instrument for the fault-tolerant detection of crankshaft angular position is shown in block diagram form in Figure 4.11. The system is composed of three major blocks: sensor/interface, signal conditioning, and state table.

Sensor/interface

This sensor is employed in the system of Figure 4.11 of the magnetoresistive type. In addition to the advantages generally afforded by magnetic sensors, e.g. immunity to environmental factors and ruggedness, the magnetoresistive sensor also provides an output which is independent of speed, unlike transducers of the inductive type. The sensor is magnetically coupled to the starter

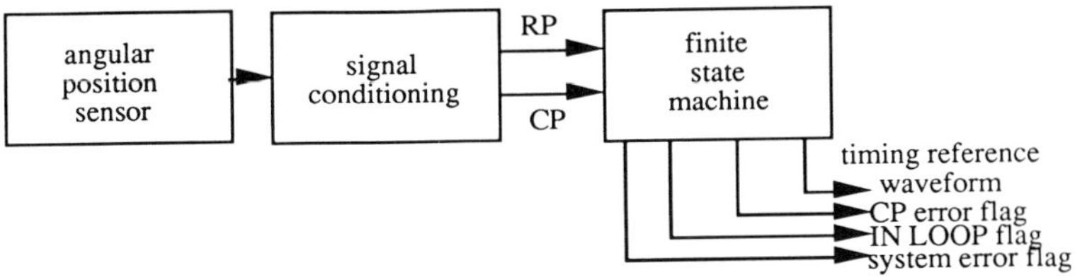

Figure 4.11 Crankshaft position measurement system.

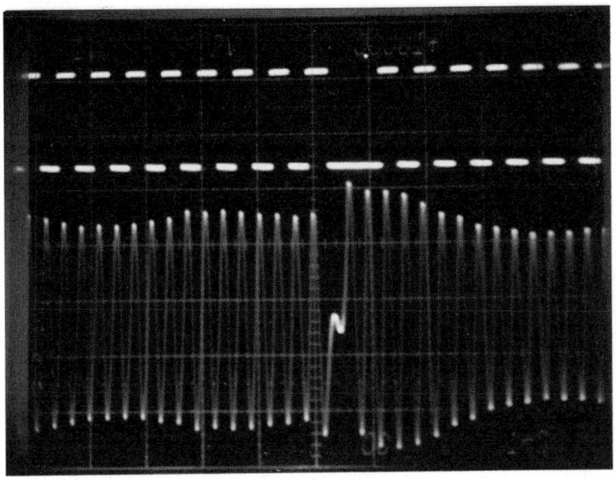

Figure 4.12 Sensor and logic waveforms.

ring gear on a modified flywheel. The output waveform is quasi-sinusoidal and provides a pair of zero crossings for each tooth on the ring gear. A small number of teeth are modified to provide a reference signal at known angles of rotation of the crankshaft. Figure 4.12 shows the raw output of the sensor, and the corresponing logic waveform including one of the reference signals.

Signal conditioning

The signal-conditioning circuit converts the raw sensor output into TTL-level pulse waveforms. These consist of a clock waveform at a frequency of M cycles per revolution, where M is the number of ring gear teeth, and a reference pulse waveform consisting of N unevenly spaced pulses per revolution. The reason for the uneven spacing is that a unique sequence of clock pulses can be generated in this fashion for each engine revolution. The system was developed for use with a six-cylinder engine, with $M = 144$ and

$N = 6$. Note that the first waveform (clock waveform, CP) provides a relatively high resolution measurement of *relative* angular position, while the second (reference waveform, RP) yields six references of *absolute* engine position.

State table

The third block performs the computations and error-checking functions of the system. It consists of logic and memory circuits which process the information provided by the signal-conditioning circuit. The relative and absolute crankshaft angular position signals are processed to obtain accurate engine position information, which can, for example, be employed in implementing a spark ignition strategy. The inputs to the finite state machine are the clock waveform (CP) and the reference pulses (RP); the outputs can be tailored according to need to provide the required engine position information as well as a number of error-checking and self-diagnostic signals. A RESET input is connected to the ignition circuit to ensure correct initialization of the system on start-up.

The finite state machine can access the unique sequence generated by the CP and RP waveforms in a state table, and uses the clock signal (CP) to cycle through the states as the crankshaft rotates. Thus, any desired output related to crankshaft angular position can be generated by suitable programming of the finite state machine. For example, a typical output might consist of a reference pulse to mark a fixed baseline value of spark advance angle for each cylinder, to be used in computing the actual ignition timing based on operating conditions.

A state table stored in the finite state machine represents each increment of engine revolution by one fixed state, and generates a one-to-one correspondence between states and degrees of engine rotation relative to a fixed reference, e.g. TDC position for cylinder number 1. The duration of each state is determined by the number of zero crossings in the sensor waveform, i.e. by the number of teeth in the ring gear. In the experimental system, each state was represented by one tooth (two zero crossings), i.e. 2.5° (exactly) of engine rotation. Different applications may require varying levels of resolution. To be emphasized is that the mapping between engine angular position and states in the table is unique, and repeats exactly for each revolution of the crankshaft. For ignition timing applications, it is essential that after initialization (i.e. ignition) the system start up as quickly as possible. The present technique ensures that, on the average, the system is fully operational after at most one third of one crankshaft revolution.

Figure 4.13 is a flowchart indicating the sequence of events following the ignition.

1. The system is reset. State stable 'address' is set at zero position. State table awaits first RP. 'IN-LOOP' output reads logic 0.
2. First RP arrives. State table starts cycling according to state clock, CP.
3. Second RP is detected. Based on the (unique) number of clock cycles

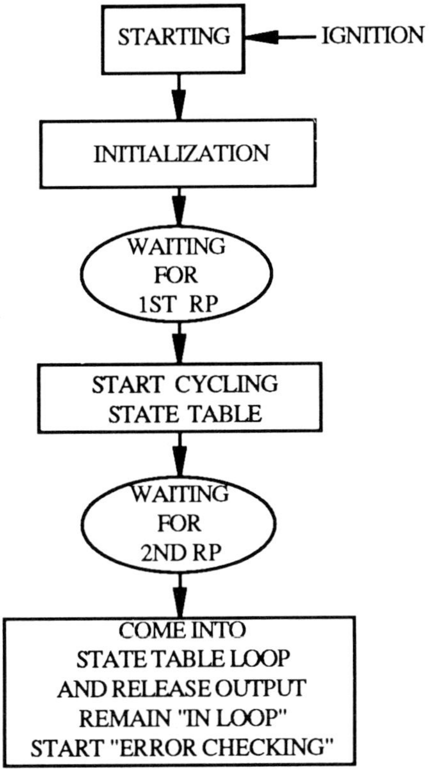

Figure 4.13 Flowchart of crankshaft position measurement system.

between first and second RP, the state table can compute the exact angular position of the crankshaft.

4. The state table enters the loop corresponding to normal operation. From this point on, barring any sensor errors, the state table cycles through its states without the need to receive any absolute position information. The only input required is CP.
   At this point, the 'IN-LOOP' flag registers a logic high.
5. The RP signal is continuously monitored to detect any discrepancies between the state table and sensor inputs. This feature enables the implementation of self-diagnostics and error-correcting strategies.

Error–correcting capabilities

The status of the system is at any time defined by one of three conditions: (1) no error; (2) CP error; and (3) RP error. When no error takes place in the inputs provided by the sensor, the output of the state table provides at all times the correct engine position information, thanks to the repeatability of the unique pulse sequence generated by the sensor. If the sensor signal were to fade momentarily and one or more clock pulses were missed (CP error), then an instantaneous discrepancy would be detected between the sensor output and the engine position state table. This discrepancy would be detected between the sensor output and the engine position state, at the occurrence of the first RP following the missing CP. The state table would then automatically correct for the missed CP by advancing to the state corresponding to the engine position.

Experimental results

The concepts described above were implemented in practice by duplicating the information provided by an existing commercial crankshaft position

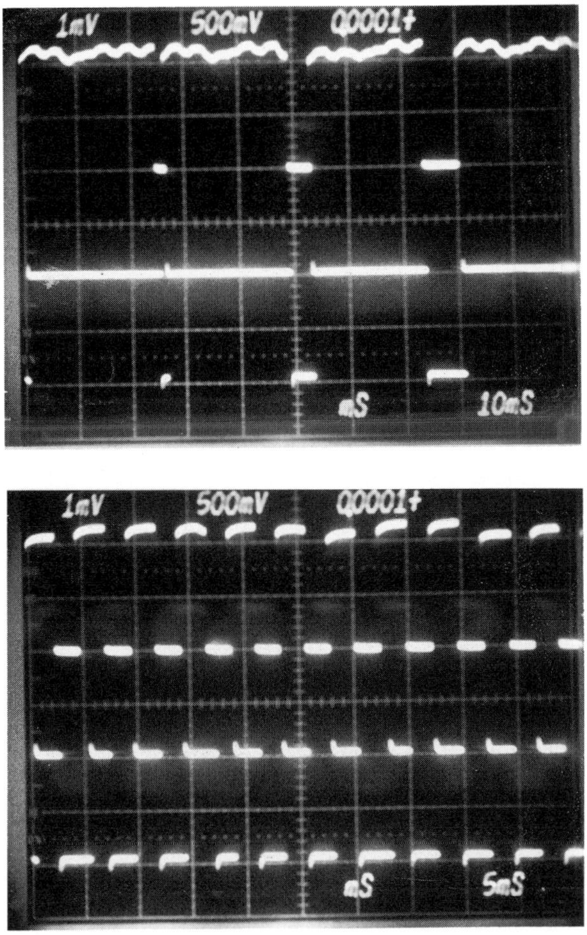

Figure 4.14 Experimental crankshaft position system waveforms.

sensor system on a production vehicle, equipped with a 3.8-litre V6 fuel injected engine and a distributorless ignition system. The crankshaft position information obtained by means of the state-table-based system was used as an input to the ignition timing control module, in place of the existing system. The vehicle successfully ran with the experimental ignition system. Figure 4.14 displays oscilloscope photographs of timing waveforms simultaneously obtained from the commercial and the experimental systems. The choice of waveforms was dictated exclusively by the availability of a commercial system which could be compared to the experimental system.

Among the potential sources of error limiting the accuracy of such a position measurement system, we considered primarily those due to the initial positioning of the ring gear and to the variation in the relation between zero crossing and crankshaft position due to tooth-by-tooth variations in ring gear geometry. This last error may be due, for example, to imperfections in the ring gear manufacturing process or to tooth wear and damage. The tooth aperiodicity error, ε, may be viewed as a pseudo-random process and analysed as such. It is clearly bounded:

$$|\varepsilon| \leqslant \frac{\delta\theta}{2} \tag{4.11}$$

where $\delta\theta$ is the mean distance in degrees between zero crossings. In the case $M = 144$, $\varepsilon_{max} \leqslant 0.625^\circ$ although, in practice, the actual error is much smaller. With current manufacturing technologies, it is feasible to obtain an accuracy in the measurement of engine position of the order of a quarter of a degree.

It is important that the signal-conditioning circuit be designed for error-free operation, to prevent the occurrence of an excessive CP count. If a missed detection occurred in the case of a RP, the engine position output generated by the state table would not be affected since that is determined exclusively by the CP signal once the system is 'IN LOOP'. However, this could be used as a self-diagnostic feature since it might identify a failure in the sensor subsystem. A false alarm in RP could not at this point be detected in a failsafe manner. In the experimental system, we designed the signal-conditioning/sensor interface circuit in order to minimize the possibility of a false alarm in the RP waveform. Note, however, that an occasional error in this signal would cause only one, or at most two, misfires before the finite state machine entered the IN LOOP state again and started cycling correctly.

4.4 APPLICATION OF CRANKSHAFT VELOCITY MEASUREMENT

4.4.1 Measurement of average crankshaft velocity

The sensors and measurement systems described in section 4.2 can provide a very accurate and reliable measurement of shaft average angular velocity,

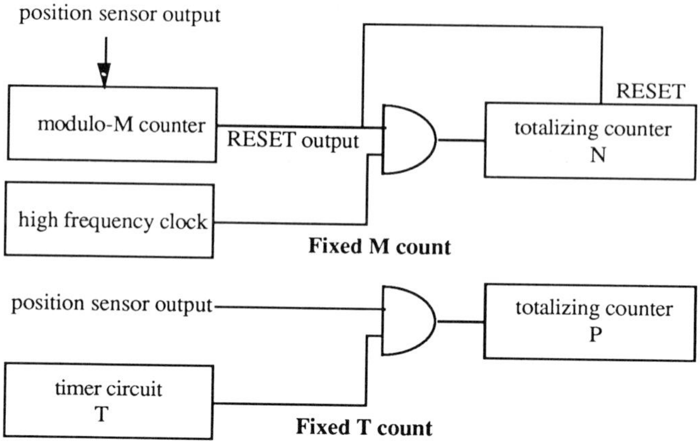

Figure 4.15 Two tachometer signal-processing schemes.

when employed in one of the modes illustrated in Figure 4.15. In the first mode ('fixed M count'), a high-frequency clock (oscillator), running at a frequency f_{CLOCK} is 'ANDed' with the RESET output of a modulo-M counter, where $M \leqslant 1$, and then used to increment a totalizing counter. Since the modulo-M counter resets itself and the totalizing counter at the completion of each shaft revolution, the count, K, recorded by the totalizing counter at the end of each revolution is related to the shaft's average angular velocity according to the following relationship:

$$Kf_{CLOCK} = \frac{\Omega}{2\pi} \tag{4.12}$$

so that the average angular velocity of the shaft is given by

$$\Omega = K2\pi f_{CLOCK} \text{ rad/s}$$

or

$$\Omega = 60Kf_{CLOCK} \text{ rev/min}$$

In the second mode ('fixed T count') a timing generator is used to gate pulses from the position sensor through an AND gate for a fixed period of time. Let the number of pulses totalized over T be equal to P. Then the count P is related to the sensor signal frequency, f_S, by the following expression:

$$Pf_S = T \tag{4.13}$$

and since the period T is known, the average angular velocity may be computed from

$$\Omega = \frac{2\pi f_S}{M} = 2\pi \frac{T}{PM} \tag{4.14}$$

It is very important to remark that if the measurement of angular velocity is performed on the basis of an integer number of revolutions of the disk (i.e. according to the first mode), errors due to uneven spacing of the lines (or of the lugs), or to misalignment of the disk, will not affect the measurement. This fact should be considered in the design of a practical tachometer. If, on the other hand, the intended application requires that the angular velocity be measured with finer resolution than once per revolution, then one must take into account the sources of error which affect such a measurement. These can be broadly classified into two groups:

1. line pattern imperfections (optical sensors) or irregularity in lug spacing (magnetic sensors);
2. errors due to misalignment.

The former can be essentially neglected in optical sensors thanks to the excellent quality achieved in etching techniques; however, misalignment errors must be considered whenever high accuracy is required. Misalignment of the disk can occur due to either imperfect concentricity of the disk or run-out. These two cases are illustrated in Figure 4.16.

The effect of concentricity error is that the centre of rotation of the disk is displaced with respect to the centre of the line pattern by an amount ΔR, causing the disk line or lug pattern to travel at different velocities relative to the sensor. It can be shown that the instantaneous frequency of the tachometer in the presence of concentricity error is given by

$$f_{\text{tach}} = \frac{\omega N}{2\pi}\left(1 + 2\frac{\Delta R}{R}\sin\theta\right) \tag{4.15}$$

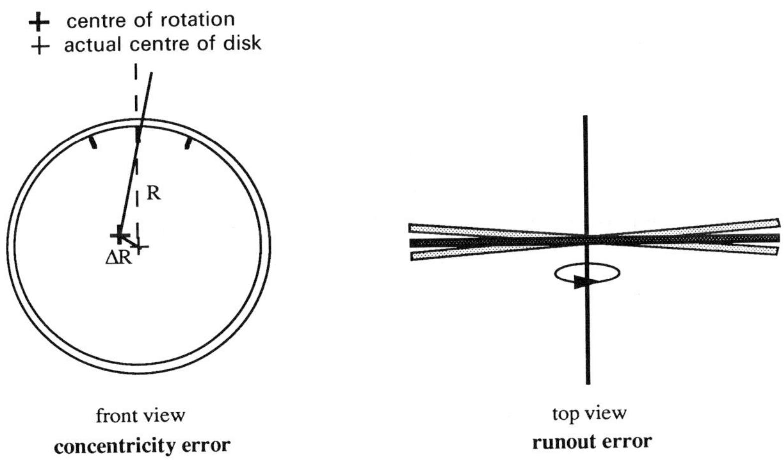

Figure 4.16 Mechanical errors in position-measuring systems.

where:

R = radius of line pattern
ΔR = concentricity error
θ = shaft angle
ω = instantaneous angular velocity of shaft

Thus, the instantaneous frequency of the sensor signal sees an additional *frequency modulation* due to the concentricity error. This error can be eliminated if a second sensor is placed 180° apart from the original sensor, and the outputs of the two sensors are summed.

The error due to run-out (wobble) causes an *amplitude modulation* of the sensor output waveform, which has a once-per-revolution period. This error is less troublesome than the concentricity error, however, because it does not affect the zero crossing of the sensor waveforms, which are typically used as a reference of the instantaneous position of the disk.

Frequency-to-voltage conversion

Any of the tachometers described in the preceding section can be used to generate a sensor signal whose frequency varies in proportion to the instantaneous angular velocity of the crankshaft. The signal model which will be assumed for the purpose of the present discussion was described in the introductory section, with reference to the dynamic model for the engine shown in Figure 4.5. There it was shown that the sensor signal is a frequency-modulated waveform, of the form

$$v_s(t) = A\cos(\theta(t)) \tag{4.16}$$

where

$$\begin{aligned} v_s(t) &= \text{output of tachometer} \\ \theta(t) &= \text{instantaneous phase of } v_s(t) \\ \dot{\theta}(t) &= \text{instantaneous frequency} \\ &= \omega(t) \\ &= \text{instantaneous angular velocity} \end{aligned}$$

The instantaneous angular velocity could therefore be recovered by means of a frequency-to-voltage (F–V) conversion. Three techniques are commonly employed to accomplish F–V conversion:

1. digital count of time interval;
2. charge pump F–V conversion;
3. phase-locked demodulation.

Each of these techniques will now be discussed.

Digital count of time interval

With reference to the waveform of Figure 4.17, the instantaneous frequency of the FM signal may be approximated by computing the instantaneous period

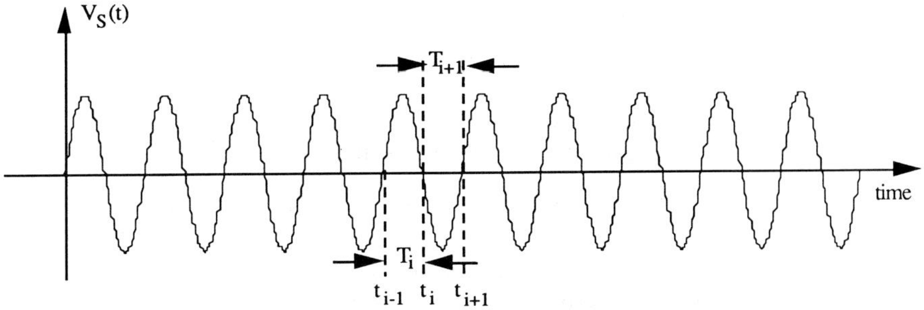

Figure 4.17 Waveform for F–V conversion.

between successive zero crossings. Let T_i represent the period between the ith and $(i+1)$th zero crossing, indicated by the times t_{i-1} and t_i; then a discrete approximation of the instantaneous frequency may be obtained by counting the number of clock cycles, n_i, of a high-frequency reference oscillator between successive zero crossings. Let T_{CLOCK} be the period of the clock oscillator (inverse of its frequency, f_{CLOCK}); then

$$T_i = n_l T_{\text{CLOCK}} = \frac{n_i}{f_{\text{CLOCK}}} \tag{4.17}$$

and the instantaneous frequency of the sensor waveform is

$$f_i = \frac{1}{2T_i} \tag{4.18}$$

This technique may be viewed as a higher speed version of the average speed measurement denoted as 'fixed M count' in the preceding section, where n_i = number of clock cylces in T_i. A digital-to-analog (D/A) converter may then be employed to obtain a stepwise analog approximation of the desired signal. The fundamental drawbacks of this method are: (1) the requirement for a very stable, very high frequency clock in order to obtain good resolution and accuracy in the speed measurement; (2) significant memory requirements in the practical implementation; and (3) the difficulty of computing an inverse $1/T_i$ efficiently. The advantage of such a scheme is that, if properly implemented, it has the potential to be a very accurate F–V conversion scheme. The digital count method is well suited to test and research applications where hardware cost is not a major issue.

Charge pump F–V conversion

The second method for F–V conversion utilizes a principle similar to that of the digital time interval measurement just discussed. If a pulse of fixed amplitude, A, and fixed duration, τ, is generated at each zero crossing of the sensor waveform, it can be readily shown that a voltage proportional to the instantaneous signal frequency may be obtained. This scheme is illustrated

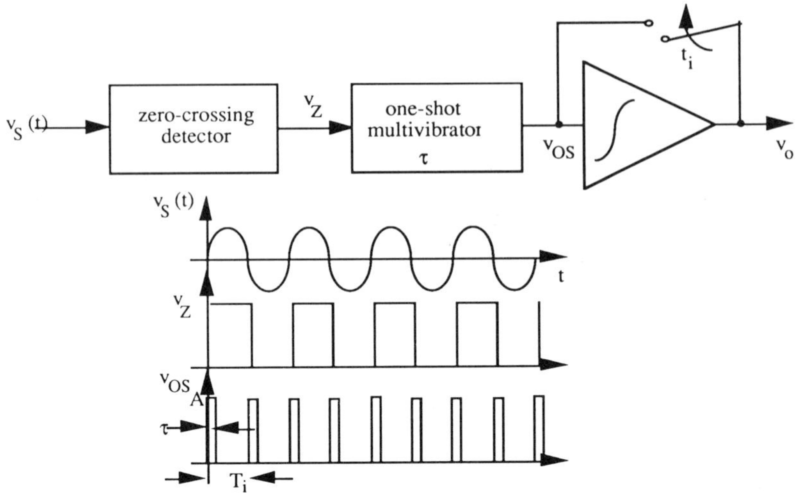

Figure 4.18 F–V conversion.

in Figure 4.18. Ideally F–V conversion could be obtained by computing the following integral:

$$\int_{t_{i-1}}^{t_i} V_{\text{os}}(t)\text{d}t = A\frac{\tau}{T_i} = 2A\tau f_i \tag{4.19}$$

yielding a voltage proportional to the frequency of $v_s(t)$ during the ith cycle. In practice, of course, it is quite difficult to reset the integrator of Figure 4.18 at each zero crossing, so practical F–V converters employ a low-pass filter in place of the ideal integrator. One drawback of this scheme is that the pulse train at the sensor carrier frequency, $M\Omega$, causes an unavoidable ripple which must be filtered. Such filtering can be a drawback, if fast response is desired.

Phase-locked demodulation

A third method for F–V conversion utilizes a phase-locked loop (PLL) as a frequency-to-voltage converter, or FM demodulator. The PLL, shown in Figure 4.19, can act as an FM demodulator once it is phase locked to an input signal waveform. When the PLL is in lock, any change in the input signal frequency generates an error voltage at the output of the phase detector, which can be either analog (a mixer), or digital, consisting of a pair of zero crossing detectors. Any difference between the instantaneous phase of $v_s(t)$ and of the signal generated by the voltage-controlled oscillator (VCO) causes an error signal at the output of the phase detector. The error signal is then approximately filtered and amplified in such a way as to change the frequency

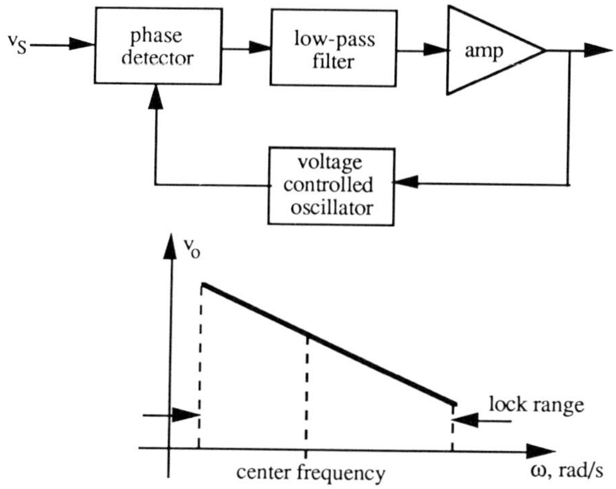

Figure 4.19 Phase-locked loop F–V conversion.

of the output signal of the VCO to match that of $v_s(t)$. Thus, the output voltage, $v_o(t)$, is the voltage which is required in order to maintain the VCO running at the same frequency as the input signal, and changes in the input signal frequency are matched by changes in $v_o(t)$. In this sense, the PLL acts as an F–V converter, with the input–output characteristic shown in Figure 4.19. Note that the PLL can offer only a finite lock range, and may not be applicable in situations where the range of crankshaft speeds exceeds a ratio of 6:1 or so. The transient operation of the PLL, and in particular the capture process by which the frequency of the VCO is brought closer to that of the input signal, is rather complex, since the PLL is a closed-loop feedback system. Its detailed description is beyond the scope of this chapter.

4.4.2 Estimate of indicated torque

Another interesting application of crankshaft position measurement is the estimation of individual cylinder indicated torque. The model for the engine dynamics presented in section 4.1 and summarized in the circuit model of Figure 4.1 describes the dynamic relationship between net crankshaft torque and crankshaft angular velocity in terms of an approximate dynamic model. The structure of the model, and the relationship between crankshaft torque and indicated torque, are shown in block diagram form in Figure 4.20, where the relationship between indicated pressure and indicated torque is given by the expression

$$T_i(\theta) = g(\theta) \cdot P_i(\theta) \tag{4.20}$$

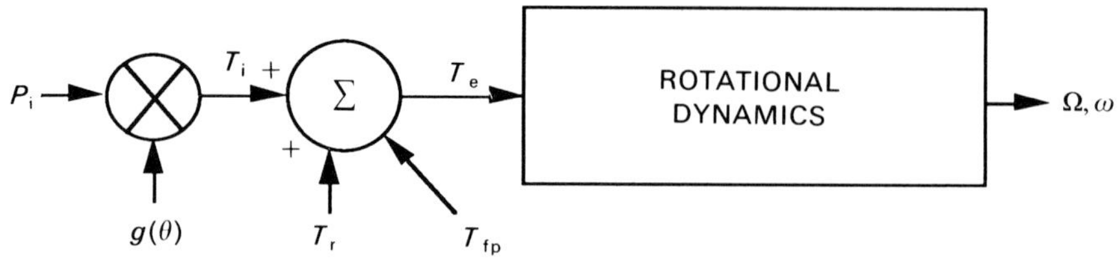

Figure 4.20 Model of engine dynamics.

with $g(\theta)$ being a known function of engine geometry. Let us establish some notation at this point which we will use throughout this section. Let each of the variables in Figure 4.20 be composed of average (or DC) and time varying (or AC) components; for example,

$$T_x(\theta) = T_x + \tau_x(\theta) \tag{4.21}$$

where lowercase Greek letters denote the AC components (x is any of the subscripts used here) and uppercase Roman letters symbolize the DC components. Thus,

$$T_e(\theta) = T_e + \tau_e(\theta) \tag{4.22}$$

$$T_e = T_i - |T_{fp}| \tag{4.23}$$

$$\tau_e(\theta) = \tau_i(\theta) + \tau_{fp}(\theta) + \tau_r(\theta) \tag{4.24}$$

According to the model of Figure 4.1, the angular velocity of the crankshaft, $\omega(\theta)$, is related to the AC component of crankshaft torque, $\tau_e(\theta)$. In particular, a precise relationship will be developed shortly between the acceleration of the crankshaft, $a(\theta)$, and both T_e and $\tau_e(\theta)$. We define here

$$a(\theta) = \frac{d\omega(\theta)}{dt} \tag{4.25a}$$

$$= \dot{\omega}(\theta) \tag{4.25b}$$

As explained in the introductory section, the net engine torque $T_e(\theta)$ causes the crankshaft to rotate at an angular speed $\omega(\theta)$, where

$$\omega(\theta) = \Omega + \delta\omega(\theta) \tag{4.26}$$

where:

$$\Omega = \text{average angular speed}$$

$$\delta\omega(\theta) = \text{speed variations about } \Omega$$

The underlying principle which permits the estimation of indicated torque from the measurement of crankshaft angular velocity is that, due to the nature of the pressure process, a relationship exists between P_i and $p_i(\theta)$, i.e.

between DC and AC components of cylinder pressure. Write

$$P_i = \boldsymbol{F}[p_i(\theta)] \tag{4.27}$$

This relation between average and time-varying components of cylinder pressure is a consequence of the combustion pressure development in an IC engine, where it can be empirically observed that an increase in average indicated pressure (P_i) during a power stroke is unavoidably associated with a corresponding increase in the amplitude of the pressure fluctuation ($p_i(\theta)$), due to the fact that the pressure signal is bounded at the lower end by the exhaust pressure. It then follows from the model of Figure 4.20 that we can express average indicated torque as a function of its fluctuations:

$$T_i = f(\tau_i) \tag{4.28}$$

The transformations $\boldsymbol{f}$ and $\boldsymbol{F}$ can be empirically identified and tabulated as a function of operating conditions for any given engine. Experimental results will be shown shortly to support these statements.

Now, let λ denote radian frequency, to avoid confusion with the symbol ω, denoting angular velocity, and define

$$\lambda_o = \text{frequency of rotation of crankshaft}$$

$$\lambda_f = \text{firing frequency}$$

It follows that, for a four-stroke cycle engine,

$$\lambda_f = \frac{N}{2}\lambda_o \tag{4.29}$$

where N is the number of cylinders. Let the notation $\cdot(j\lambda)$ represent the Fourier transform of the periodic function (θ), with representing any of the processes defined thus far. Then, the circuit model predicts that a relationship exists between the time-varying component of net engine torque, τ_e, and the acceleration of the crankshaft, a:

$$\tau_e(j\lambda) = H(j\lambda)\cdot a(j\lambda) \tag{4.30}$$

where $H(j\lambda)$ is the sinusoidal frequency response of the engine dynamics, as derived from the circuit model of Figure 4.1, or from any other suitable model which accounts for the relevant dynamics of the engine/drivetrain. Combining this result with the hypothesized relationship between AC and DC components of $T_i(\theta)$ we immediately have a very powerful statement:

a measurement of crankshaft velocity (and hence, acceleration) contains information regarding both average and time-varying crankshaft engine torque.

Formally,

$$\frac{a(j\lambda)}{H(j\lambda)} = \tau_e(j\lambda) = \tau_i(j\lambda) + \tau_r(j\lambda) + \tau_{fp}(j\lambda) \tag{4.31a}$$

or

$$\frac{a(j\lambda)}{H(j\lambda)} - \tau_r(j\lambda) - \tau_{fp}(j\lambda) = \tau_i(j\lambda) \tag{4.31b}$$

leading to:

$$T_i = f\left\{\frac{a(j\lambda)}{H(j\lambda)} - \tau_r(j\lambda) - \tau_{fp}(j\lambda)\right\} \tag{4.32}$$

That is, the average indicated torque produced during each stroke may be computed from a measurement of crankshaft acceleration. The next section will describe the experimental verification of this concept.

Experimental results

The experimental verification of the model was carried out in a dynamometer engine test cell. A four-cylinder in-line, spark-ignited, fuel-injected engine driven by an electric dynamometer was used in the experiments. The engine was coupled to the dynamometer by means of a manual transmission. The dynamometer was controlled by a solid-state dynamometer/throttle controller. Simultaneous measurements of indicated cylinder pressure, crankshaft angular velocity and brake torque were taken over the entire range of engine operating conditions. The measurements of pressure and velocity were made with reference to crankshaft angular position, by sampling the above-mentioned signals synchronously with the rotation of the engine, at intervals of 5°. DC motoring tests were employed to obtain an estimate of the average frictional and pumping loss. The results of the motoring tests are displayed in Figure 4.21 for the wide-open and closed-throttle cases, which correspond to the lowest and highest frictional loss, respectively. The two curves are fit by a regression line,

$$T = m\Omega + b \tag{4.33}$$

with

$$m = 0.232,\ b = 16.3 \quad \text{(closed throttle)}$$

$$m = 0.224,\ b = 11.9 \quad \text{(wide-open throttle)}$$

and a correlation coefficient of 0.994 for the wide-open throttle case and 0.983 for the closed-throttle case. Since the correlation coefficient is very high, a straight line approximation is used to model the total DC torque losses as

$$T_{loss} = T_{fp} + r_e\Omega \tag{4.34}$$

where T_{fp} is a function of load, as Figure 4.21 suggests. Note the separation between load- and speed-dependent effects. This feature of the engine model is very important, in that it allows for the modelling of speed-dependent effects as a constant, lumped friction parameter. Since the measurement of

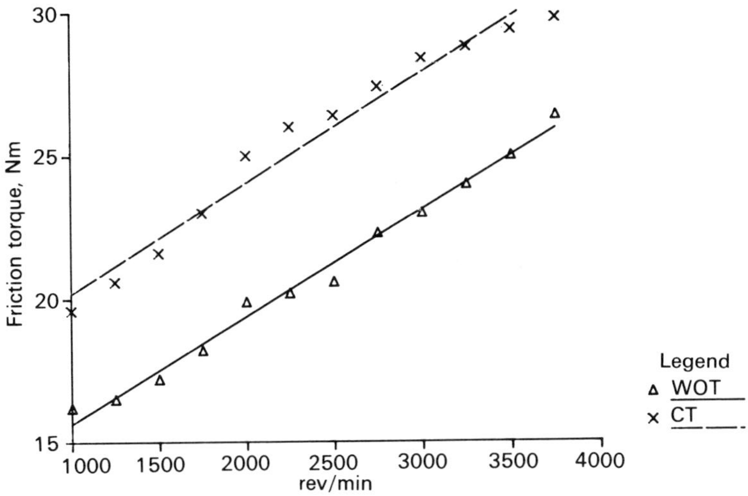

Figure 4.21 Engine frictional and pumping loss.

cylinder pressure provides all the information necessary to compute indicated torque (AC and DC), and since the reciprocating inertia torque component can be computed analytically from knowledge of engine geometry parameters, we may compute the approximate net engine torque using the equation below:

$$\tau'_{\mathrm{e}} = \tau_{\mathrm{i}}(\theta) + \tau_{\mathrm{r}}(\theta) \cong \tau_{\mathrm{e}}(\theta) \tag{4.35}$$

where the AC friction and pumping loss has been assumed to be negligible.

Recall how a relationship between T_{i}, the average indicated torque, and a transformation of $\tau_{\mathrm{i}} \boldsymbol{f}(\tau_{\mathrm{i}})$ had been postulated earlier. The particular form of $\boldsymbol{f}(\cdot)$ selected to test the torque sensor concept consists of the component of the discrete Fourier transform (DFT) at the engine firing frequency, λ_{f}. It is worth noting at this stage that due to the periodic motion of the engine, much of the essential information in the pressure torque and angular velocity processes is at the firing frequency, λ_{f}. Fourier analysis of indicated torque data reveals that more than 70% of the energy in the indicated torque waveform is concentrated at the DC and the fundamental components. Figure 4.22 displays $\boldsymbol{f}(\tau_{\mathrm{i}})$ vs T_{i} for a 25-point map in the T–Ω plane. The regression line for this curve is

$$\boldsymbol{f}(\tau_{\mathrm{i}}) = m_{\mathrm{i}}\, T_{\mathrm{i}} + b_{\mathrm{i}} \tag{4.36}$$

$$m_{\mathrm{i}} = 0.89$$

$$b_{\mathrm{i}} = 1.82$$

with a correlation coefficient of 0.998. This result states in essence that the

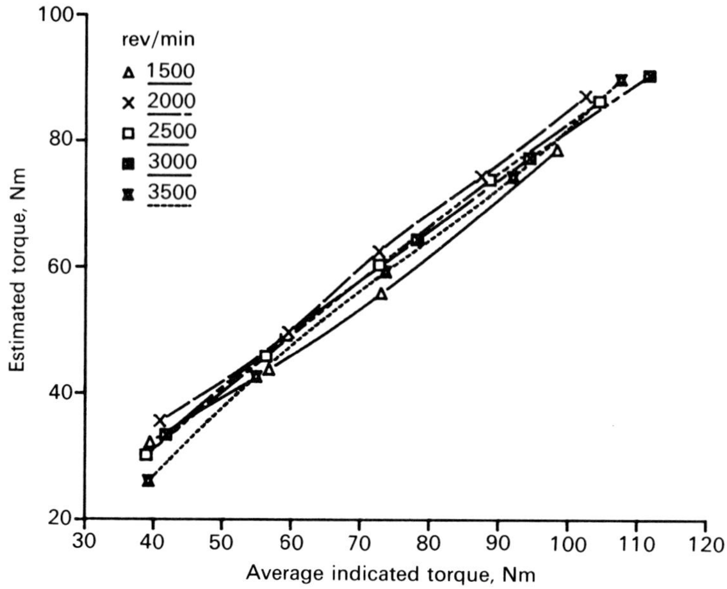

Figure 4.22 AC versus DC component of indicated torque.

first harmonic of $\tau_i(\theta)$ is linearly related to the mean indicated torque produced during each stroke.

Having chosen a form $f(\cdot)$ (which corresponds to computing the amplitude of the first harmonic of firing frequency) we can write:

$$T_i = f\left(\left|\frac{a(j\lambda_f)}{H(j\lambda_f)} - \tau_r(j\lambda_f) - \tau_{fp}(j\lambda_f)\right|\right) \tag{4.37}$$

Hence, the model predicts that measurement of crankshaft acceleration, deconvolved through the dynamics of the engine and corrected for reciprocating inertia torque, is a measure of average indicated torque. If the measured angular velocity signal is deconvolved through the dynamics described by the engine model, and corrected for the component of the reciprocating inertia torque at that frequency, an estimate of the AC indicated torque, $\hat{\tau}_i$, may be obtained. Figure 4.23 shows the $\hat{\tau}_i$ vs T_i curves obtained from the same experimental data of Figure 4.22. It should be noted that the data points of Figures 4.22 and 4.23 do not represent an average over several engine cycles, but correspond to the data acquired during one individual stroke (power stroke for cylinder number 1), confirming that if properly implemented this technique can be useful to estimate individual cylinder indicated torque. An application of this concept will be illustrated in the next section.

In summary, the measurement of crankshaft angular velocity can lead to

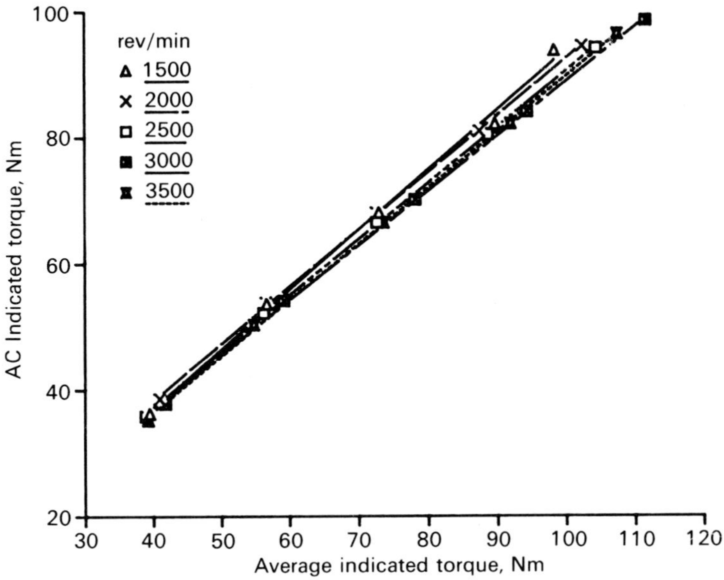

Figure 4.23 Estimated and measured indicated torque.

an estimate of average and individual cylinder indicated torque. The complexity of the method is moderate, if one considers that the operations which need to be performed in practice are: (1) the measurement of crankshaft angular velocity (with a resolution of 5–10°); (2) the computation of a single component of the DFT of the angular velocity waveform: (3) the multiplication of the spectral component of the angular velocity by one complex number, corresponding to the frequency response of the engine at the firing frequency (which could be prestored in a table); and (4) the correction (requiring a complex addition) for the effects of reciprocating inertia (which could be prestored in a table). The method described in this section has been successfully tested in various real-time implementations, and offers promise for applications in engine diagnostics and engine/transmission control.

4.4.3 Misfire detection

The long-term performance of automotive exhaust emission control systems is strongly influenced by the physical condition of the catalytic converter. Unfortunately, the catalytic converter is susceptible to irreversible damage from any number of factors. One of the important factors affecting catalyst health is engine misfire. It is well documented that a sufficient level of engine misfire can result in catalyst damage such that its performance is degraded to a point at which the vehicle no longer meets exhaust emission standards.

For the purpose of this discussion, a total misfire will be assumed to represent the complete lack of combustion (due to insufficient spark or incorrect mixture) in one cylinder, during one cycle.

The integrity of the exhaust emission system can best be maintained by monitoring its performance continuously on board the vehicle. It is with the intent of monitoring emission system performance that the California Air Resources Board has proposed regulations which will require vehicles to be equipped with on-board monitoring systems. These proposed regulations are known as OBDII and will probably be followed by similar US Environmental Protection Agency regulations.

Generally speaking, nearly all methods proposed for misfire detection have one of three conceptual bases: (1) individual cylinder torque estimation; (2) chemical analysis of exhaust gases; (3) optical measurements of in-cylinder combustion events. The second and third of these are probably impractical from a cost standpoint, owing to the necessity of having separate signals from each cylinder. The method discussed in the present section is based upon an estimate of instantaneous indicated torque derived from crankshaft angular speed measurement and only requires a measurement of crankshaft position/velocity.

Torque nonuniformity index

The estimate of individual cylinder indicated torque from a measurement of crankshaft position/velocity which was described in the preceding section forms the basis of the misfire detection method discussed herein. The individual cylinder torque estimates, representing the mean torque produced by each cylinder during each stroke, form an N-vector for each engine cycle, where N is the number of cylinders. Let the transpose of this N-dimensional torque vector be defined by

$$\mathbf{T}' = [T_1 T_2 T_3 \ldots T_N] \tag{4.38}$$

This vector may be computed for each engine cycle, and information regarding the torque nonuniformity in each engine cycle may be extracted from it according to the following method. The average torque produced during the cycle is defined as follows:

$$\langle \mathbf{T} \rangle = \frac{1}{N} \sum_{n=1}^{N} T_n \tag{4.39}$$

Then an N-dimensional vector representing torque nonuniformity is computed:

$$\mathbf{n} = \mathbf{T} - \langle \mathbf{T} \rangle \mu' \tag{4.40}$$

where $\mu' = [1, 1, 1, \ldots]$ is an N-dimensional vector.

If the torque produced by the engine during a given cycle were absolutely uniform, then $\mathbf{n}$ would be an N-dimensional null vector. In reality $\mathbf{n}$ is never a null vector owing to cyclic variability of combustion and to misfire, if present. A metric d is computed for representing torque nonuniformity and for detecting misfire, which we call the *torque nonuniformity metric*, or index. This metric is computed by evaluating the norm of a vector which indicates the deviation from ideal engine performance caused by any nonuniformity in the generation of torque. Any L_p norm is adequate for the purpose of detecting the presence of a misfire; two possible choices are:

$$d = \|\mathbf{n}\| 1 \quad L_1 \text{ norm} \tag{4.41}$$

or

$$d = \|\mathbf{n}\| 2 \quad L_2 \text{ norm} \tag{4.42}$$

Although either norm will suffice for misfire detection, the L_1 norm is normally preferred because it is more efficiently computed (it does not require squaring and square-root operations).

Detection of misfire is done on a statistical hypothesis testing basis. The method is best explained with reference to Figure 4.24, which is an experimentally obtained plot of the approximate probability density function for the random variable d under the three hypotheses:

H_0: normal engine operation
H_1: misfire in one cylinder (not necessarily the same for each cycle)
H_2: misfire in two cylinders (not necessarily the same for each cycle)

The histograms of Figure 4.24 were obtained by means of a real-time measurement of the nonuniformity index performed on board a vehicle equipped with a fuel-injected six-cylinder engine and an automatic transmission. The misfires were induced by interrupting the fuel to one or more fuel injectors.

The left distribution represents the statistics of the random fluctuations in random variable d for normal engine operation. This distribution results from normal cyclic variability in combustion. The distribution in the centre corresponds to the statistics in the event of one misfire in each engine cycle. The distribution is the same regardless of which cylinder misfires. The cylinder number of the single misfiring cylinder can be randomly changing from cycle to cycle. The distribution on the far right represents the extreme consequence of two simultaneous misfires, and represents an upper bound to the practical magnitude of the metric. This last case will not be discussed further.

The detection of misfire can be based on a variety of criteria according to statistical decision theory. One particularly simple test can be applied to the nonuniformity index d for each engine cycle by selecting a threshold d_t midway between the mean values $\bar{d}_{H0}$ of the probability distribution for normal operation, $p_{H0}(d)$, and $\bar{d}_{H1}$, of the probability distribution for misfiring operation, $p_{H1}(d)$:

$$d_t = \frac{\bar{d}_{H0} + \bar{d}_{H1}}{2} \tag{4.43}$$

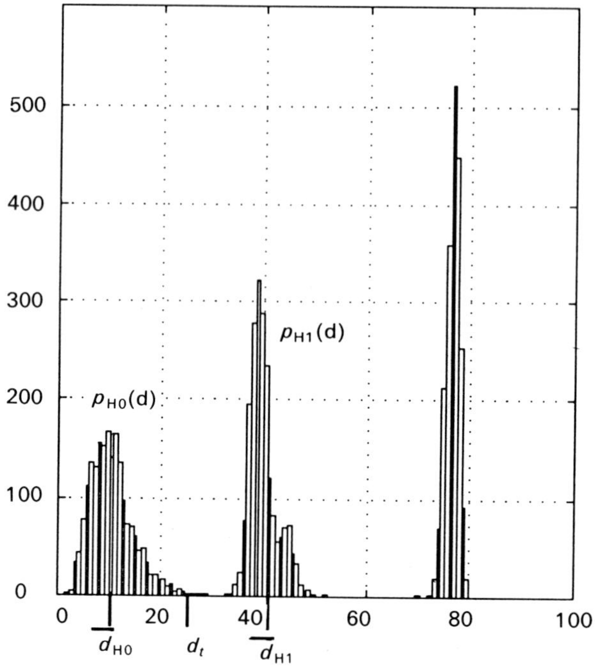

Figure 4.24 Histograms of nonuniformity index.

This particular choice of thresholds corresponds to assigning equal costs to false alarms and missed detections, and corresponds to the criterion

$$d > d_t \rightarrow \text{misfire}$$

$$d < d_t \rightarrow \text{normal operation}$$

Two types of errors are incurred in using such a method:

1. $d < d_t$ for a misfiring cycle (missed detection);
2. $d > d_t$ for a normal cycle (false alarm).

It should be apparent that the location of the threshold, d_t, allows for trading off one type of error against the other. The local probability of error (type 1 or type 2), P_e, may be computed as follows:

$$P_e = \int_{d_t}^{\infty} p_{H0}(d)\mathrm{d}d + \int_{0}^{d_t} p_{H1}(d)\mathrm{d}d \tag{4.44}$$

Experimental verification of this misfire detection concept has taken place both in the laboratory and on the road [3–8], demonstrating the feasibility of this misfire detection method based on a measurement of crankshaft angular position. Error rates measured during testing on rough and unpaved roads have been lower than 1%.

4.5 CONCLUSIONS

This chapter has introduced a variety of techniques for crankshaft angular position/velocity measurement and has discussed several applications of these measurements. The applications presented here are useful for engine control/operation as well as for measurement of engine performance, with potential for on- and off-board diagnostic applications. With the increasing presence of high-resolution crankshaft position measurement systems on board production vehicles, some of these methods may gain wider acceptance in the industry in the near future.

Most of these methods are fully discussed in refs 9–21.

REFERENCES

1. Ribbens, W.B. (1981) *A Non-contacting Torque Sensor for the IC Engine.* SAE Technical Paper No. 810155.
2. Rizzoni, G. (1986) A model for the dynamics of the internal combustion engine. *PhD dissertion,* Department of Electrical and Computer Engineering, the University of Michigan, Ann Arbor MI.
3. Ribbens, W.B. (1985) *Experimental Road Test of a Noncontacting Method of Measuring IC Engine Torque Nonuniformity.* SAE Technical Paper No. 840454.
4. Ribbens, W.B. and Gross, D. (1986) *Torque Nonuniformity Measurements in Gasoline Fueled Passenger Cars Equipped with Automatic Transmission–Theory and Experimental Results.* SAE Technical Paper No. 860414.
5. Rizzoni, G. (1987) A passenger vehicle on-board computer system for engine fault diagnosis, performance measurement and control, in *Proceedings of the 37th IEEE Vehicle Technology Conference,* Tampa, FL, Institute of Electrical and Electronic Engineers, pp. 450–7.
6. Ribbens, W.B. and Naaseh, M. (1987) *Incipient Failure Detection in Bus Engine Components.* SAE Technical Paper No. 872264.
7. Ribbens, W.B. and Rizzoni, G. (1990) *Onboard Diagnosis of Engine Misfires.* SAE Technical Paper No. 901768.
8. Connolly, F. and Rizzoni, G. (1991) Real time estimation of engine torque for the detection of engine misfires, in *2nd Symposium on Advanced Automotive Technologies,* ASME Winter Annual Meeting, Atlanta, GA, American Society of Mechanical Engineers.
9. Freestone, J.W. and Jenkins, E.G. (1985) *The Diagnosis of Cylinder Power Faults in Diesel Engines by Flywheel Speed Measurement.* IEE Paper No. C33/85.
10. Mihelc, W. and Citron, S. (1984) *An Online Engine Roughness Measurement Technique.* SAE Technical Paper No. 840136.
11. O'Higgins, J., Citron, S. and Chen, L.Y. (1989) *Cylinder by Cylinder Engine Pressure Torque Waveform Determination Utilizing Crankshaft Speed Fluctuations.* SAE Technical Paper No. 890486.
12. Rezeka, S.F. and Henein, N.A. (1987) *A Diagnostic Technique for the Identification of Misfiring Cylinder(s).* SAE Technical Paper No. 870546.

13. Ribbens, W.B. (1983) *A New Metric for Torque Nonuniformity*. SAE Technical Paper No. 830425.
14. Ribbens, W.B. (1985) *Experimental Road Test of a Noncontacting Method of Measuring IC Engine Torque Nonuniformity*. SAE Technical Paper No. 850454.
15. Ribbens, W.B. and Rizzoni, G. (1989) *Application of Precise Crankshaft Position Measurements for Engine Testing, Control and Diagnosis*. SAE Technical Paper No. 890885.
16. Rizzoni, G. (1989) Estimate of indicated torque from crankshaft speed fluctuations: a model for the dynamics of the IC engine, in *XXVII International Symposium on Automotive Technology and Automation*, Florence, June, 1987, ISATA.
17. Rizzoni, G. (1987) *Diagnosis of Individual Cylinder Misfires by Signature Analysis of Crankshaft Speed Fluctuations*. SAE Technical No. Paper 890884.
18. Rizzoni, G. (1989) Estimate of indicated torque from crankshaft speed fluctuations: a model for the dynamics of the IC engine. *IEEE Trans. Vehicular Technology*, **VT-38**, No. 3.
19. Rizzoni, G. (1989) *Diagnosis of Individual Cylinder Misfires by Signature Analysis of Crankshaft Speed Fluctuations*. SAE Technical Paper No. 890884.
20. Rizzoni, G. and Ribbens, W.B. (1989) Application of crankshaft position measurement for engine testing, diagnostics and control, in *Proceedings of the 39th IEEE Vehicle Technology Conference*, San Francisco, CA, Institute of Electrical and Electronic Engineers.
21. Rizzoni, G., Pipe, J.G., Riggins, R.N. and Van Oyen, M.P. (1988) Fault isolation and analysis for IC engine onboard diagnostics, in *Proceedings of the IEEE Vehicle Technology Conference*, Philadelphia, PA, Institute of Electrical and Electronic Engineers.

CHAPTER 5

Integrated sensors and control system for engines using gaseous fuels

5.1 INTRODUCTION

The majority of road vehicles are powered by internal combustion engines fuelled with petrol or diesel fuel. These 'traditional' automotive fuels have been abundantly available and have formed a basis for the worldwide expansion of road transport. Engine manufacturers have always concentrated on further development of the petrol and the diesel engine, which has led to impressive results both in fuel efficiency and in exhaust emission levels. It is expected that the petrol engine and the diesel engine will retain their position as the most important powerplants for road vehicles for several decades to come.

In a number of countries gaseous fuels have also been used in road vehicles, mainly liquefied petroleum gas (LPG) and compressed natural gas (CNG). In some cases the share of gaseous fuels in road transport is relatively important.

For example, in The Netherlands approximately 700 000 cars are running on LPG. They account for some 12% of the total car population; however, they cover almost 20% of the total annual mileage. Another example is Italy, with over 1 000 000 LPG-fuelled cars and 250 000 cars running on CNG. In several other countries, interest in LPG and CNG is growing, although at present the worldwide market share of gas-fuelled vehicles is still of the order of only 1–2%.

There are good reasons for further expansion of gaseous fuels in road transport.

First, the proven reserves of natural gas are already greater than those of petroleum. Moreover, natural gas reserves are more equally distributed over the world than petroleum reserves. This means that natural gas can take over at least part of the role of petrol and diesel.

Second, gaseous fuels have properties which enable them to achieve lower engine exhaust emission levels than are possible with petrol or diesel fuel.

The introduction of electronic controls and engine management systems has opened enormous possibilities for further improvements in performance and emission reduction of petrol and, to a lesser extent, diesel engines. Such conrol systems are becoming common on all modern engines in road vehicles. They are essential on engines which are operating in conjunction with a catalytic converter. Engines running on gaseous fuels will also demand this type of control in order to fully exploit the favourable combustion and emission characteristics of the fuel.

This chapter discusses the development of engine management systems for spark-ignited engines running on gaseous fuels, hereafter designated as gas engines.

An engine management system controls the behaviour of the engine under all conditions, thus including steady-state running as well as operation under transient conditions. It makes use of a number of signals from sensors. These sensors measure the value and changes in value of the different parameters which determine the combustion process in the engine. The sensor signals are the input for an electronic control module (ECM). The ECM processes the input signals and subsequently produces an output signal to one or more actuators, which control a number of functions, the main ones being air/fuel ratio and ignition timing.

The objective of an engine management system is to enable the gas engine to operate at the best combination of fuel efficiency, power output and low exhaust emissions under varying engine speed and load conditions and to retain this performance during the service life of the engine. The system thus automatically corrects for wear of essential moving parts, aging of sensors, etc. The following sections discuss how far this objective can be achieved.

For automotive gas engines analog electronic control systems as well as digital control systems have been developed. Demands on accuracy, response time and versatility can be fulfilled more easily and cheaply by digital control systems. Therefore this chapter only discusses the latter approach.

5.2 DEVELOPMENT BACKGROUNDS

The need for drastic reductions in exhaust emissions from road vehicles has also affected gaseous fuel control systems. Conventional systems, which have been used mostly until now and which are still applied, have conventional hardware. This hardware consists basically of a gas pressure regulator, a gas/air mixer and – at best – some additional mechanical or pneumatic control functions, mostly to match air/fuel ratio to changes in engine speed and load.

Mechanically controlled gas systems are characterized by certain imperfections, which can be described as follows.

The best Venturi gas/air mixers combine high homogeneity and good mixture distribution with high volumetric efficiency [1]. Furthermore, a very high linearity of air/fuel ratio is achieved and such mixers are virtually insensitive to the shape of the airflow duct before the mixer. However, for good transient response, especially at low airflows, additional control equipment is needed and then a good Venturi mixer provides excellent results. In cases where gas/air mixers have not been designed properly they may suffer from one or more of the following weaknesses:

- poor mixture homogeneity causing an inhomogeneous in-cylinder mixture as well as air/fuel ratio differences between engine cylinders (poor mixture distributions).
- limited gas suction (depression) at low airflow rates, as in idle or low part load conditions;
- sensitivity to the shape of the airflow duct before the mixer;
- loss of volumetric efficiency at full engine load;
- non-linear air/fuel ratio over engine load.

These effects, their causes and possible solutions, are described in more detail in Tiedema and Wolters [1] and Klimstra [2]. In most cases, solving one of the problems means worsening one or more of the others.

The gas pressure regulator usually has two or three stages which reduce the gas pressure to a value just below ambient pressure.

In the case of LPG the pressure regulator is combined with an evaporator. The pressure-reducing stages are of the diaphragm/spring type in most cases. This means a relatively slow response to changes in gas flow demand, hysteresis and relatively poor metering accuracy.

In order to match with a Venturi-type gas/air mixer, the gas pressure regulator should act like a perfect zero-pressure regulator. As mentioned already, most of the conventional regulators show deviations from perfect zero pressure and exhibit:

- non-linearity, particularly at low depressions;
- reproducibility problems due to evaporator (in case of liquid gas like LPG) and/or valve and diaphragm problems;
- problems in taking over from idle gas supply to main gas supply (transient behaviour);
- sensitivity to pulsations in the admission system at high load/low engine speed.

The combination of gas/air mixer with gas pressure regulator in a conventional gas system therefore results in insufficient performance in terms of metering accuracy, response under transient conditions and reproducibility, especially if stringent low exhaust emission levels have to be met. These imperfections can be overcome by the application of microprocessor technology, which at the same time offers other engine control functions.

The efforts for cleaner engines have resulted in two main methods of engine operation, indicated by the air/fuel ratio of the mixture supplied to the engine cylinders.

First, engines running at stoichiometric air/fuel ratios (designated by the excess air factor λ, with $\lambda = 1$) enable the application of a three-way catalytic converter. Such systems are applied on most modern petrol vehicles.

The second approach applies a certain amount of excess air in the combustion process, according to the so-called lean burn principle (designated by the excess air factor λ, with $\lambda > 1$).

These combustion principles both have their advantages and disadvantages, the discussion of which is beyond the scope of this chapter. Engine management systems for gas engines have been developed for both approaches, and both types have been applied.

In passenger car applications the use of gaseous fuels usually requires a retrofit conversion, since almost no 'dedicated' gas vehicles are manufactured by the car industry. This results in a duplication of the microprocessor fuel control system, because in many cases it is not possible or not permitted by the car manufacturer for the gasoline engine management system to be used or interfered with when running on gaseous fuel.

In the case of heavy duty engines the application of gaseous fuel normally involves the development of a dedicated gas engine with direct commitment/involvement of the engine manufacturer. In this case only one control system is needed.

Electronic control systems are not only applied in conjunction with carburettors. The newest developments include injections of the gaseous fuel. This can be single-point or multipoint injection and the fuel is injected in the gaseous phase or in the liquid phase (in the case of some LPG injection systems).

The advantages are more accurate metering of the fuel and better response under transient conditions. Moreover, the risk of backfires decreases, certainly in the case of injection close to the cylinders. However, mixture homogeneity, a very important factor for low exhaust emissions, will be not as good as with an optimized gas/air mixer.

The electronic control system hardware and software are not, in principle, different from that for carburetion systems, since similar sensors are used for the input signals to the ECM. The output signal of the ECM controls a single- or multipoint injector assembly.

It is expected that in 1992 the first series-produced systems will be available on the market.

5.3 CONTROL SYSTEM HARDWARE AND SOFTWARE

The system discussed in this section is based on a multi-function engine management system developed by Delco Electronics for use on petrol engines. For the application of gaseous fuels part of the standard hardware and software is used. Additional hardware and software, specific for this application, has been developed in The Netherlands.

First, the original petrol engine management system is discussed, based on the literature [3, 4]. It is based on TBI (throttle body injection) with a so-called GMP-4-type computer.

5.3.1 Petrol system description

The TBI assembly incorporates the injector, pressure regulator, throttle position sensor and bypass air control.

The ignition system can be a distributor and coil system or a direct ignition system not using a distributor. Spark timing is controlled by the computer as a function of engine rev/min, manifold absolute pressure, water temperature, altitude, throttle position and other transient parameters for a fully mapped system. The distributor contains an electronic module and magnetic pick-up which senses engine position and speed and sends this information to the computer. The computer processes this information along with data from other sensors and sends a spark timing signal back to the distributor module. The module then controls the current to the coil to produce the high-voltage spark. In addition, the distributor module also provides a 'back-up' system to provide an independent spark advance curve vs rev/min in case of computer failure.

The optional direct ignition system (designated DIS) provides all of the above functions and calibration flexibility. Moreover, the distributor and separate coil are no longer necessary. The DIS consists of one coil assembly for each pair of engine cylinders with an integral electronic module plus a crankshaft sensor. The advantages include installation flexibility (engine or underhood location), overall package size, improved spark timing accuracy and the possibility of eliminating base timing adjustment in the factory and in the field.

The speed-density concept is used on all TBI applications, the basics of which are well documented. The manifold absolute pressure sensor is a solid-state semiconductor, and water temperature, vehicle speed and exhaust oxygen level (for a closed-loop system) are included. The oil pressure switch is a redundant power activation source for the fuel pump, in case of a relay activation problem.

The computer is designated as a GMP-4 type, first introduced in 1986, and it will be described in more detail. An octane selector allows the customer to use either 91 or 95 octane fuel. Also, in view of developing open-loop systems without the advantage of closed-loop learning, a potentiometer is incorporated to allow adjustment (if required) of idle air/fuel ratio in the factory or in the field. This compensates for the tolerances in the fuel system and sensors plus engine-to-engine variations in volumentric efficiency.

An example of a typical mechanization is shown in Figure 5.1 as inputs and outputs to the electronic control module of a closed-loop system. The required inputs are illustrated on the left and the basic outputs are shown on the right. The optional inputs and outputs illustrate the flexibility of the ECM.

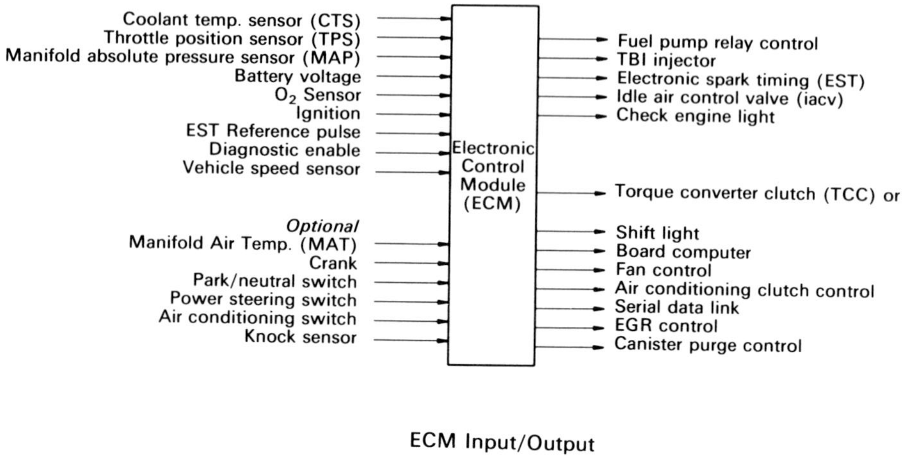

Figure 5.1 ECM inputs and outputs.

The ECM used in this system is the latest in the evolution of multifunction engine management system computers by Delco Electronics. The ECM has been designed with the latest technology, including extensive use of SMD (surface-mounted devices) and a complement of custom integrated circuits (known as the GMP-4 chip set) which maximizes features, functions and performance and minimizes cost and package size. It also provides the flexibility of allowing customer-defined software algorithms and calibration, which are programmed into the plug-in module and uniquely define the customer's application.

One of the main features designed into the ECM's hardware and software is self-diagnostics. Algorithms in the software program executed by the GMP-4 microprocessor continually check the status of the ECM outputs and the validity of sensor inputs. If a fault is detected, then a code is stored in keep-alive memory and the 'check engine' warning lamp is turned on. This code can be read via the 'check engine' warning lamp by a service technician to determine the nature of the problem. A continuous serial data stream of sensor data and other ECM parameters, that can be read by a hand-held scanner, is also provided to assist the service technician in the diagnosis of the fault. The algorithms and hardware are also designed to provide default operation in the event of a component, connector or partial system failure so the driver can drive the vehicle to a service station.

This ECM uses a set of custom-designed integrated circuits, including a custom microprocessor, a co-processor to handle engine timing functions called the ECU (engine control unit), an engine interface circuit called the FMD (fuel moding device), a custom A/D (analog to digital) converter, a

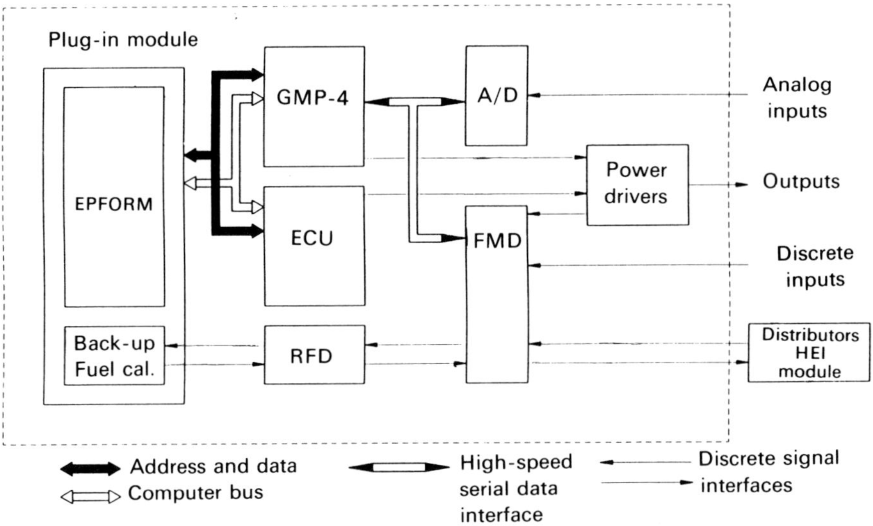

Figure 5.2 ECM block diagram.

custom back-up fuel device called the RFD (redundant fuel device), as well as custom power drivers and power supply circuits. The accompanying ECM block diagram (Figure 5.2) shows how all these various custom parts fit together to form the functional ECM.

The GMP-4 microprocessor is an 8-bit custom microprocessor with many advanced features such as an extended instruction set, error detection and recovery and internal keep-alive RAM (random access memory). The GMP-4 microprocessor sends and receives information, such as the software program and program variables, to and from the ECU chip and the EPROM (erasable programmable read only memory), containing the program and calibration, via a computer bus of address and data lines. To save space, information to and from the other peripheral chips, like the A/D converter and the FMD, is sent via a high-speed serial data link as shown in Figure 5.2.

The software program and the calibration data are programmed into an EPROM which is built into a plug-in module that also contains the resistor networks to calibrate the back-up fuel circuits described hereafter. This plug-in module allows the ECM to be customized with unique software and calibration to meet the customer's requirements. The EPROM in the plug-in module can hold 16 384 bytes of program and calibration data.

For the engineers who need to develop the algorithms and calibrations for particular engine management requirements, the following three systems are most often used for EMS (engine management system) development.

Auxiliary memory

The auxiliary memory allows the use of a production ECM during the development phase of a program. It interfaces to the ECM buffer by way of a ribbon cable. The auxiliary memory provides the following capabilities:

- product operation prior to ROM (read only memory) release;
- expands memory to 20 K capacity using EPROMs;
- provides the proper interconnection to a logic analyser.

Instrumentation interface unit (I^2U)

The instrumentation interface unit is designed to provide a monitor of the engine computer operation. The I^2U system consists of:

- an instrumentation interface unit;
- a head-up display (HUD);
- associated cabling.

The interface to the auxiliary memory is by way of a ribbon cable.

The I^2U system offers various functions during development. Some of these functions are:

- four digital readouts of parameters in engineering units. Two of the four readouts are assignable via 10-position rotary switches;
- eight discrete indicators for display of 16 status indications;
- a duty cycle meter for output monitoring;
- six analog outputs for use with strip chart recorders;
- capability to modify up to five operational parameters.

Calibration development system (CDS)

The CDS is designed to allow 'real-time' ECM calibration development in a vehicle. The CDS gives the user the capability to modify program contents by way of a keyboard. The CDS system consists of:

- a remote interface unit;
- a keyboard;
- associated cabling.

The CDS interfaces to the auxiliary memory unit by a ribbon cable.

The CDS system offers various functions during development by replacing the product's calibration PROM with the non-volatile CDS RAM. Some of these functions are:

- loads the EPROM contents into the CDS RAM;
- loads the CDS RAM with the contents of the product calibration PROM;
- allows 'on-line' change of memory contents via keyboard,
- programs an EPROM with the contents of the CDS RAM.

5.3.2 Software and hardware for gaseous fuels

The system has been modified, especially as far as software is concerned, to enable its use for gaseous fuels. The specially developed software package is called AGEMS [5]. It performs all relevant engine management tasks, such as:

- open- and/or closed-loop (air/fuel) (A/F) control, both on stoichiometric and on lean burn engines;
- open and/or closed loop A/F control in idle and low-load transient conditions:
- mapped spark timing control;
- closed-loop idle rev/min control;
- wastegate control;
- some engine-specific control functions;
- diagnostics.

The AGEMS gas engine management system has been built around automotive components from several General Motors divisions and is completed with automotive components from other suppliers.

The heart of the system is the Delco Electronics EFI-4 engine control module as decribed above. Inputs used are manifold absolute pressure (MAP) sensor, inductive crankshaft sensor (pick-up), manifold air temperature (MAT) sensor, coolant temperature sensor (CTS), throttle position sensor (TPS), park/neutral and other relevant switches, vehicle speed sensor (VSS), battery voltage and exhaust gas oxygen (EGO) or universal exhaust gas oxygen (UEGO) signal. Outputs used are gas control stepper motor, main fuel solenoid valve, deceleration fuel cut-off (DFCO) valve, idle air control valve, idle fuel control valve, electronic spark timing (EST), 'check engine' light and wastegate control valve.

AGEMS software features an advanced A/F control algorithm, using the oxygen sensor signal. This algorithm allows for fully exploiting three-way catalyst capabilities for reducing CO, HC and NO_x emissions at the same time. Furthermore, a powerful learning strategy is incorporated, allowing the system to self-adjust for (limited) adjustment errors or for varying gas compositions. This can be very valuable with regard to conformity of production tests etc.

An important feature of AGEMS is its diagnostics capability. The GMP-4 microprocesor used in the EFI-4 ECM has an internal keep-alive RAM and error detection and recovery. Algorithms in the software continuously check the status of the ECM outputs and the validity of the sensor inputs. If a fault is detected, corrective actions are taken, when necessary and possible. A malfunction code is stored in the keep-alive memory and the 'check engine' warning light is switched on.

By later putting the ECM in the diagnostics mode, the stored malfunction codes can be read from the 'check engine' light by means of a flashing code.

A technician can thus determine the nature of the malfunction. During normal operation the ECM sends a continuous data stream of sensor data and other ECM parameters like stored malfunction codes, stepper motor position and status words. This data stream can be read by a 'TECH-1' hand scanner or by a personal computer (PC).

The diagnostics software runs on most IBM-PC-compatible laptop or desktop computers using MS-DOS. It allows the PC to communicate with the ECM through the PC's serial communications interface. During its main operating mode, the parameters in the serial data stream are continuously updated on the PC screen. Furthermore, error codes can be read with explanatory messages. If necessary the user can request on-line help, which explains circumstances of the detected fault together with applicable testing and repair procedures. The on-line help can also be used for reference without the actual presence of malfunctions.

Another feature of the PC diagnostics software is logging engine and ECM data on disk during operation. The transmitted parameters can be read back for analysis.

5.3.3 Special hardware

So far, common automotive components have been mentioned. In the application examples 1, 2 and 3 the generally used hardware for gaseous fuels is described.

Examples 1 and 2 deal with stoichiometric engine concepts. However,

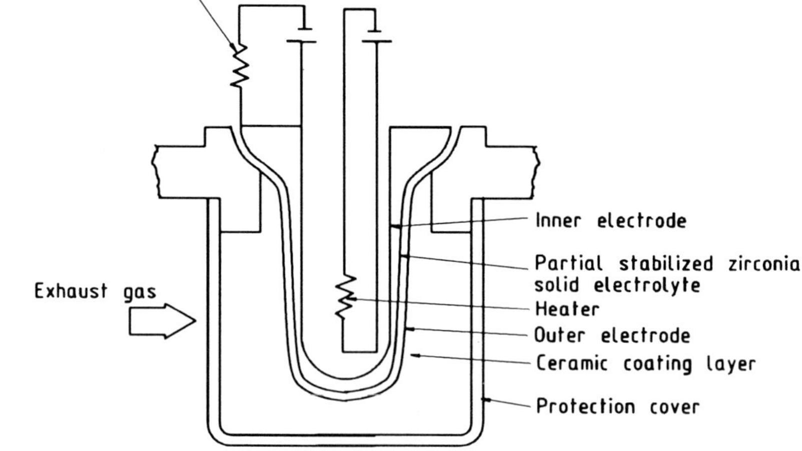

Figure 5.3 Oxygen sensor of NGK Spark Plug Company for lean burn closed-loop control.

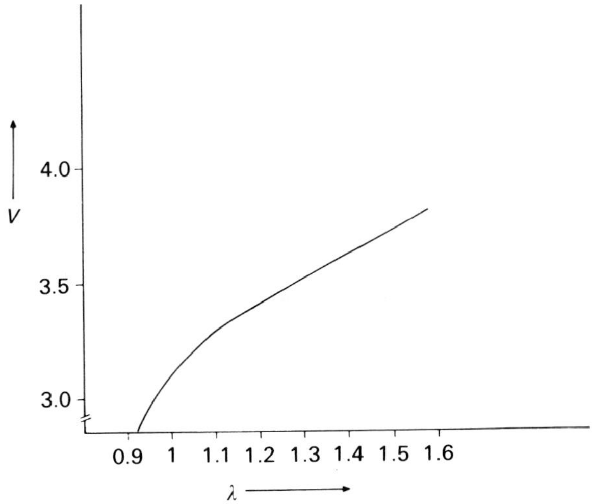

Figure 5.4 Output characteristic of lean burn oxygen sensor.

so-called lean burn concepts for gaseous fuels are also coming into the market. Example 3 in section 5.6 discusses such a system. In lean burn systems the application of an oxygen sensor for closed-loop control is essential. This is because the variation in the fuel composition (which is often substantial in gaseous fuels) results in an A/F ratio change which strongly influences, in particular, NO_x.

An example of such a sensor under development is given in Figure 5.3. Such a sensor is much more complicated than the common '$\lambda = 1$' sensors. Oxygen ions are 'pumped' by voltage to enable detection of high amounts of oxygen.

An output characteristic is given in Figure 5.4.

5.4 APPLICATION EXAMPLE 1: MICROPROCESSOR STOICHIOMETRIC AIR/FUEL-RATIO CONTROL SYSTEM, BASIC LAYOUT

This example describes an application of the microprocessor stoichiometric A/F control system, basic layout, which is used on LPG cars in The Netherlands. These LPG-converted passenger cars are normally dual-fuel petrol-LPG. Under Dutch legislation an LPG-converted car must meet the same exhaust gas limits when running on LPG as were valid for its original petrol type approval. Since the majority of newly sold petrol cars in The Netherlands are now type approved according to US-83 emission standards, LPG-converted cars also have to meet US-83.

The LPG-system [6] is characterized by a gas-metering valve, which is mechanically linked to the mixture throttle valve. Over the whole working range of the engine, this gas-metering valve shows a very predominant role in A/F control. Consequently intervention elsewhere in the system will cause only minor A/F changes, or only in a limited part of the engine's working range. The most obvious thing to do in this case would be to control the gas-metering valve electronically.

Tests, however, have shown that in that set-up it is virtually impossible to meet the engine's response demands. A solution for this problem was found in controlling gas quality rather than gas quantity. As before, the gas-metering valve is mechanically linked to the throttle valve. The gas metering valve is adjusted slightly rich over the whole working range of the engine. By diluting the gas with very accurately metered air quantities, any desired A/F can be obtained. The metering of diluting air is controlled by a small stepper motor controlled valve.

An interesting detail of this system is that the diluting air is mixed into the gas flow by means of a small venturi. This guarantees a constant dilution factor at any engine load, provided that the air-metering valve opening is left unchanged. This way there is no need for the stepper motor to follow sudden engine load variations, thus guaranteeing a very good response behaviour. Figure 5.5 shows the basic system layout. Figure 5.6 gives an example of the stepper motor position at various engine loads and speeds.

Furthermore, the system contains a limp-home function: the ECM A/F control is only a correction on the basic system A/F control hardware. This means that the car will remain driveable if for whatever reason the A/F control fails.

The system runs on AGEMS software, using engine rev/min and MAP as main inputs. Further input signals can be oxygen sensor, engine temperature, throttle position, etc., giving every desired correction possibility. In this application spark timing and idle rev/min control are usually controlled by the car's original (petrol) engine management system.

Three-way catalyst cars equipped with this system quite easily meet US-83 emission limits. With some fine tuning, typical US-83 test results are of the order of 25% of US-83 limits for CO, HC and NO_x. Power loss when compared to the original petrol engine is limited to 2–4%, thanks to the very homogeneous mixture and a very accurate full-load A/F control (Figure 5.7).

Retrofit LPG or CNG systems have a separate microelectronic control system for operation on the gaseous fuel. When switching from petrol to gaseous fuel, another microprocessor system is also switched on. This is because so far there has been little co-operation with the vehicle manufacture (OEM) generally with regard to retrofit systems. This generally means no access to the OEM microprocessor.

There is now a trend towards more OEM involvement. Under development (e.g.) in The Netherlands is such a dual-fuel engine management system, whereby both fuels are controlled from the same processor. The system inputs

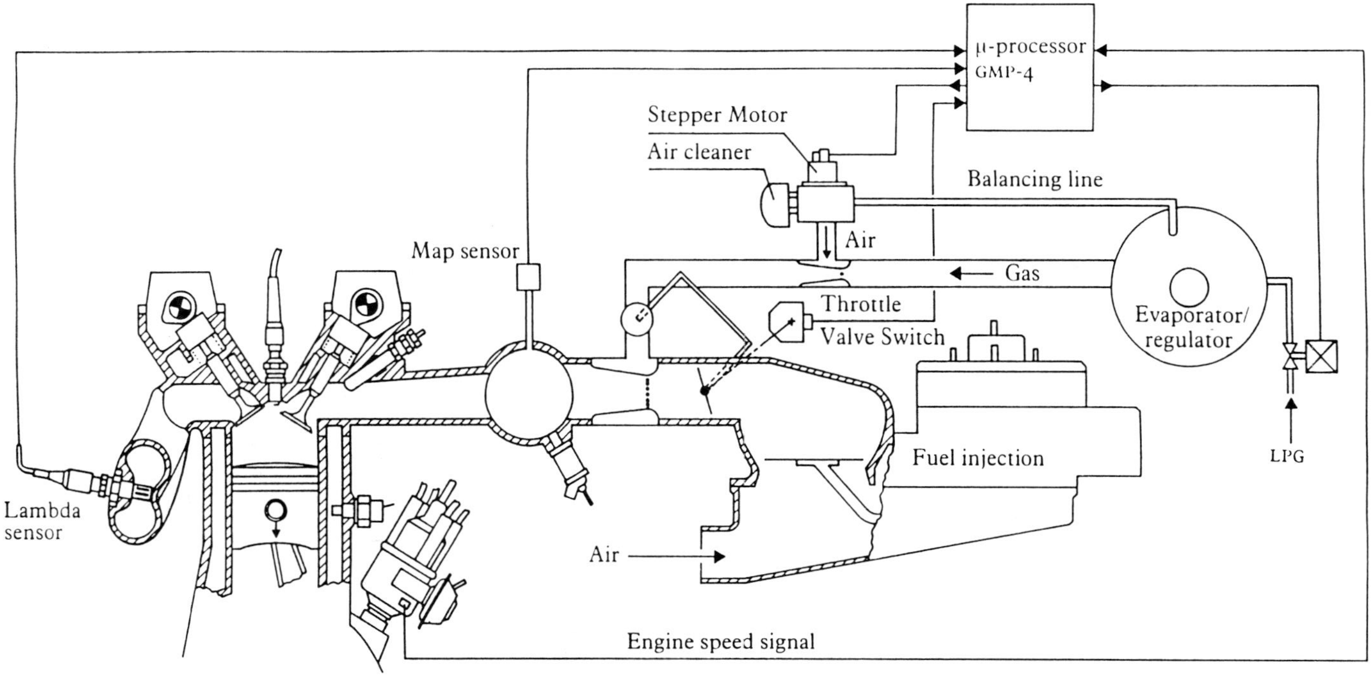

Figure 5.5 Basic system layout of microprocessor-controlled gas system in LPG passenger cars (courtesy of Necam, The Netherlands).

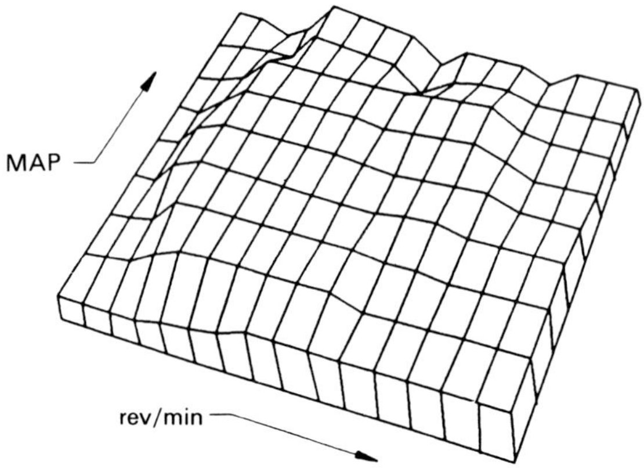

Figure 5.6 Stepper motor position as a function of MAP (manifold absolute pressure) and engine rev/min.

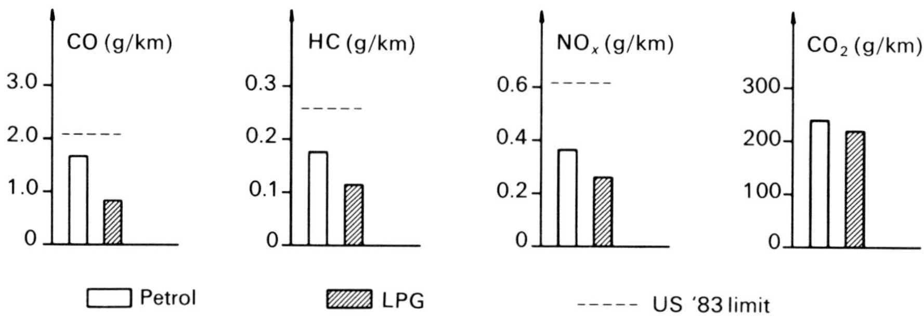

Figure 5.7 Exhaust gas emission with petrol and with LPG; electronic controlled injection engine with three-way catalytic converter; US '83 test procedure.

are:

engine speed pulses	rev/min
manifold absolute pressure	MAP
throttle position	TPS
manifold air temperature	MAT
coolant temperature	CLT
exhaust oxygen	O_2
battery voltage (ignition key)	BAT
fuel switch	G/P
diagnostic mode	DAIG

OUTPUTS \ INPUTS	RPM	MAP	TPS	MAT	CLT	O_2	BAT	G/P	DIAG
GAS VALVE STPPR.	X	X	X			X		X	
HIGH-PRESSURE V.	X						X	X	
PRESS. REG. CHOKE							X	X	
MULTIPOINT FUEL INJ.	X	X	X		X	X	X	X	
FUEL PUMP CONTROL	X						X		
IGNITION TIMING	X	X		X	X			X	
CHECK ENGINE L.								X	X
DIAGNOSTIC DATA L.									X

Figure 5.8 AGEMS dual fuel engine management system; main input/output dependencies.

The system outputs are:

gas value stepper motor control
high-pressure valve control
pressure regulator choke control
multipoint fuel injector control
fuel pump control
ignition timing control
check engine light
diagnostic data link

The essential input/output dependencies are given in Figure 5.8.

5.5 APPLICATION EXAMPLE 2: ENGINE MANAGEMENT SYSTEM ON CNG URBAN BUS ENGINE OPERATING AT STOICHIOMETRIC AIR/FUEL RATIO

This example concerns the application of an engine management system on an IVECO 8469.21 CNG engine for urban buses [5, 7].

The IVECO 8469.21 engine is turbocharged and has a cylinder displacement of 9.5 litre. At stoichiometric operation the power output at rated speed is 163 kW at 2100 rev/min. The engine is equipped with a three-way catalyst.

In this application the engine management software controls the following functions:

- main gas flow to the gas/air mixer;
- gas flow at engine idling;

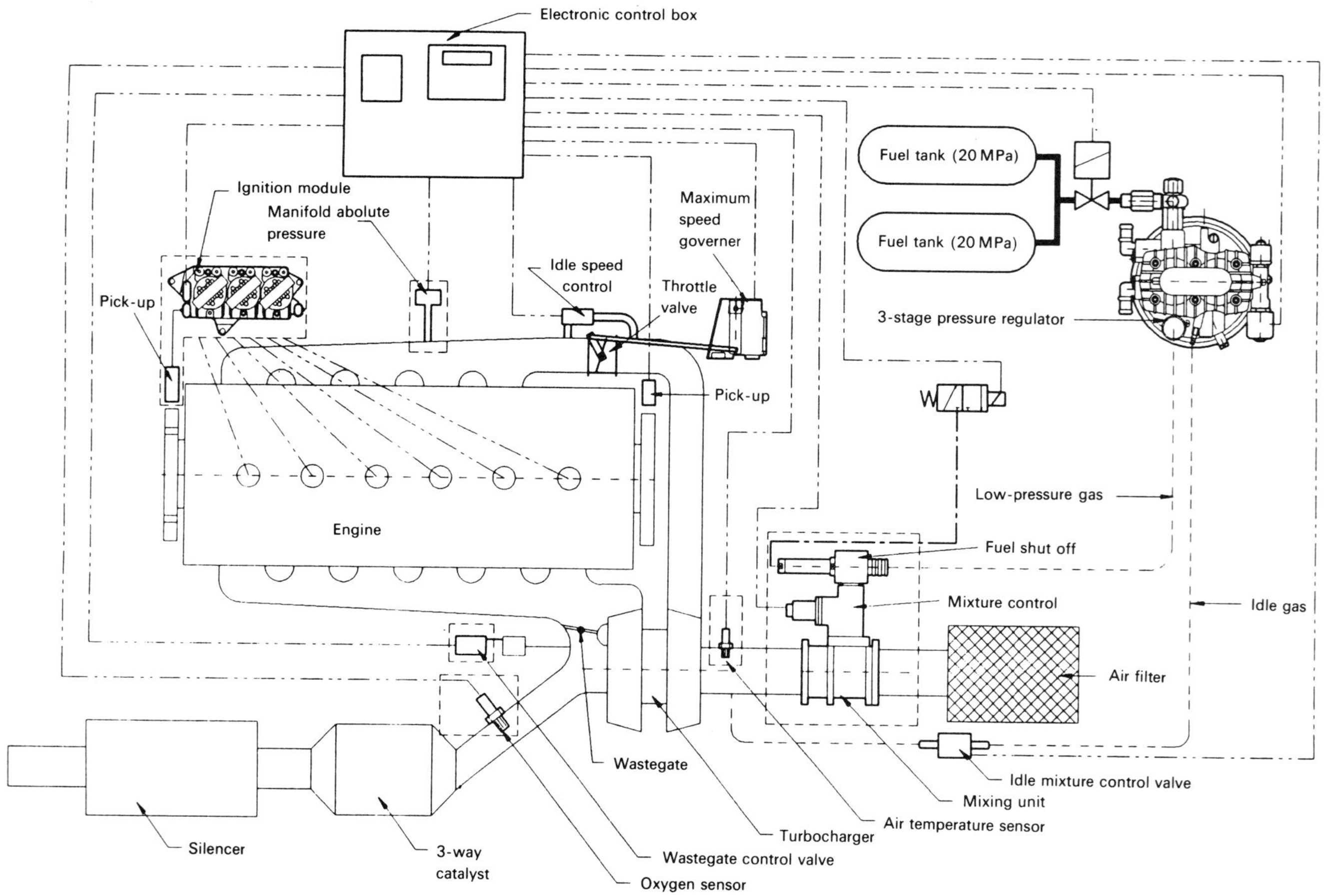

Figure 5.9 System layout IVECO 8469.21 CNG engine.

- engine idling speed;
- gas flow cut-off at deceleration;
- ignition timing;
- boost pressure of the turbocharger.

Figure 5.9 shows the system layout, and Figure 5.10 presents a block diagram of the control system.

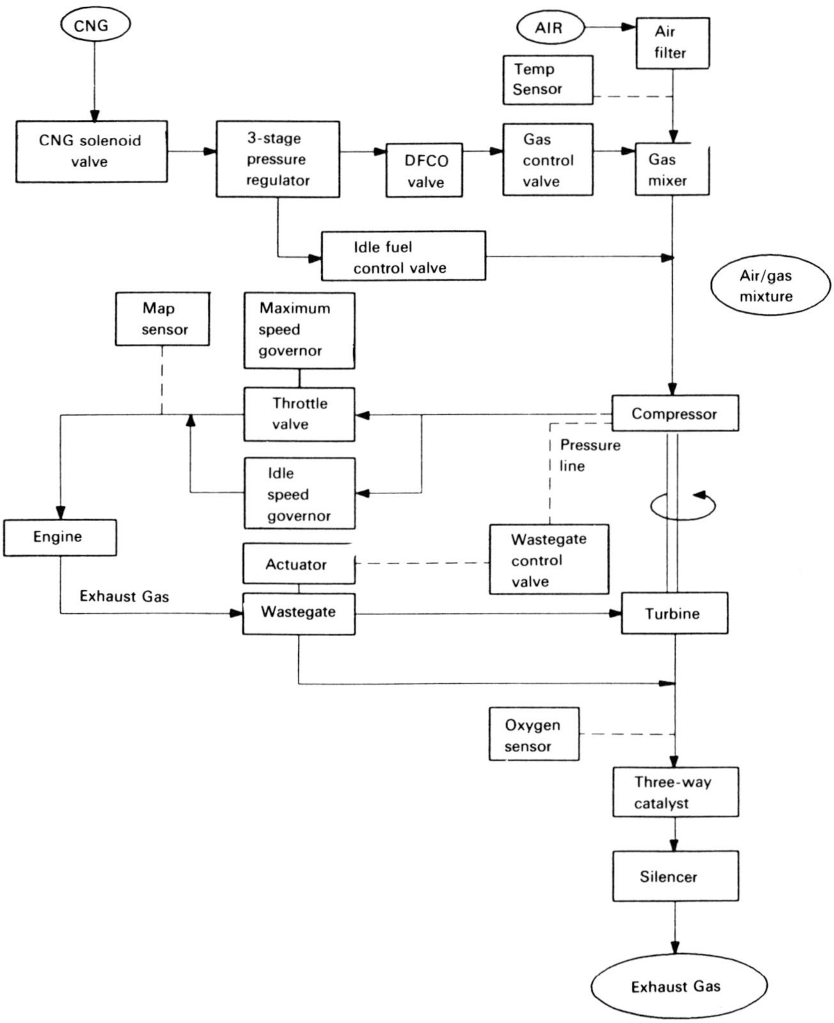

Figure 5.10 Block diagram IVECO 8469.21 CNG engine control system.

5.5.1 Gas system

The gas coming from the high-pressure CNG cylinders enters the CNG pressure regulator, which reduces the gas pressure in three stages to just above atmospheric pressure. The Venturi gas/air mixer creates a depression when the engine is running, which results in a gas flow from the third stage of the pressure regulator to the engine, proportional to the air flow.

The exhaust oxygen sensor measures any deviations from the desired stoichiometric A/F ratio and sends this signal to the ECM. The air/fuel ratio is controlled by a gas flow control valve which receives its input signals from the ECM.

At idle speed and low part load of the engine the depression in the Venturi is too low to obtain a sufficiently high gas supply from the regulator. A part of the flow at idling is therefore supplied under overpressure from the second stage of the pressure regulator by means of a small gas-metering valve, which is duty cycle controlled by the ECM. Also, at idling the A/F ratio is stoichiometric.

As soon as the engine is cranking, the ECM actuates the low-pressure solenoid valve on the pressure regulator so that gas can flow to the gas/air mixer both through the main gas supply line and through the idle gas line. Upstream of the gas/air mixer a deceleration fuel cut-off valve is mounted. Depending on manifold depression and engine speed, this valve is closed during decleration. The valve is actuated by a three-way solenoid valve, which is controlled by the ECM.

5.5.2 Turbocharger system

The compressor of the turbocharger is mounted between the gas/air mixer and the throttle valve.

A wastegate actuator is mounted between exhaust manifold and turbine. As shown in Figure 5.11, the actuator is connected to a capsule which is activated by a wastegate control valve, which is in turn connected to the compressor. The wastegate control valve is actuated by the ECM. The input signals for the ECM are manifold absolute pressure and engine speed.

At a certain boost pressure level the spring in the capsule is pressed in, causing the valve to open in such a way that the boost pressure in the inlet manifold stays nearly constant over the full-load speed range of the engine. The desired engine torque curve can be obtained by software calibration.

If the wastegate opens, part of the exhaust gas bypasses the turbine.

5.5.3 Ignition system

The engine is equipped with a distributorless direct ignition system. Ignition advance (dependent on speed and intake manifold pressure) is controlled by the ECM. In case of a failure the ignition module will function in a bypass mode, with a fixed ignition advance curve, which, however, is not optimum.

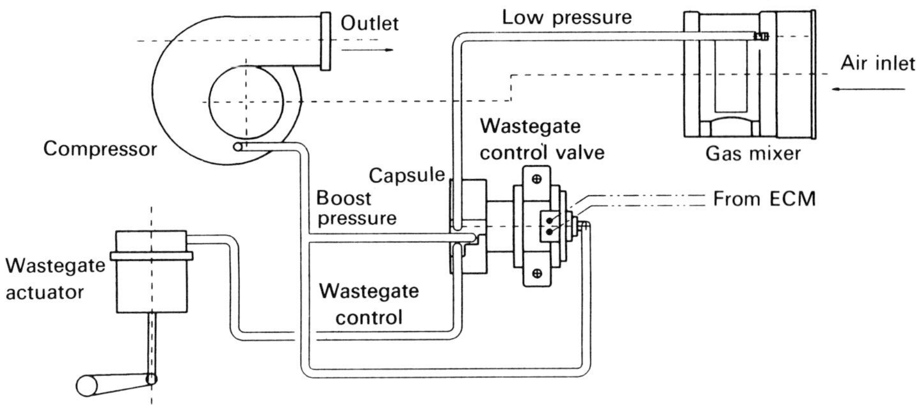

Figure 5.11 Wastegate control system.

5.5.4 Speed-governing system

The engine is equipped with separate electronic governors for idling and for maximum speed. The idle-speed governor is a bypass over the throttle valve. At engine idling the gas/air mixture is flowing directly into the intake manifold, whereby the quantity is controlled by the ECM to give a constant idle speed.

The actuator of the maximum-speed governor is connected to the throttle valve and is controlled by a separate electronic control box.

Engine speed is sensed by a magnetic speed sensor mounted on the flywheel, whereas the other input signal, desired engine load, comes from an electric foot pedal. The movement of the throttle valve is proportional to the stroke of the foot pedal; however, above rated engine speed the throttle valve is closed by the actuator independently of the position of the foot pedal.

Should the maximum-speed governor fail and the engine speed become too high, the deceleration fuel cut-off valve and the idle gas control valve are closed by the ECM (engine overspeed protection).

5.5.5 Engine management system

The engine is equipped with a 12 V DC automotive engine management system, the ECM, which is connected to a DC–DC converter, since standard on-board voltage is 24 V DC.

The ECM controls the gas flow to the gas/air mixer, the gas flow at idling, ignition advance depending on engine speed and load, maximum boost pressure relative to engine speed, the idle-speed governor and the deceleration fuel cut-off valve. The ECM also functions as a safety device in case of engine overspeed. Finally, the ECM switches off the pressure regulator solenoid if the engine is not running.

The ECM makes corrections to deviations from the desired A/F ratio, depending on the air temperature measured upstream of the gas/air mixer. Necessary data are stored in a CALPAC (calibration package), which contains an EPROM.

The gas flow to the gas/air mixer is controlled by a stepper motor valve. Closing the stepper motor valve (less steps) means running at a leaner A/F mixture, while opening of the valve (more steps) means running richer. The basic setting is a stoichiometric A/F ratio over the load and speed range of the engine. This setting is calibrated and stored in the ECM. Stepper motor steps are dependent on the manifold absolute pressure sensor signal, the speed signal from the crankshaft sensor and the intake air temperature.

Corrections in the setting are made by means of the voltage signal of the oxygen sensor which is mounted in the exhaust upstream of the catalyst. The characteristic of the oxygen sensor (λ sensor) is shown in Figure 5.12. When the voltage sent to the ECM is higher than 0.6 V (rich), the stepper motor valve moves towards a leaner mixture position; when the voltage is lower than 0.6 V (lean), a richer mixture is set. The system is continuously switching such that the A/F ratio is always around stoichiometric. The control constants are dependent on engine speed.

The A/F ratio at idling is controlled by the idle gas control valve, operated by pulsewidth modulation (PWM).

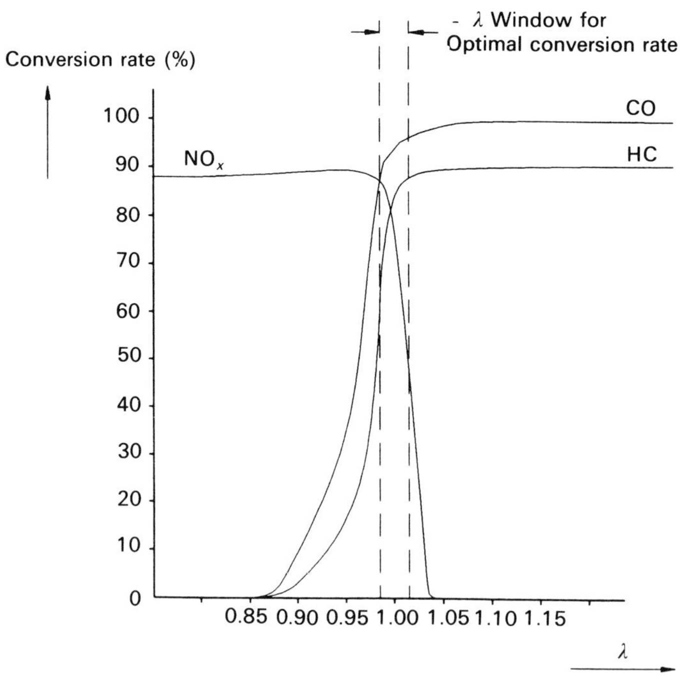

Figure 5.12 Normal operating range between lambda 0.98 and lambda 1.02.

Based on the signal of the oxygen sensor, the mixture is adjusted at stoichiometric; however, the control is made slow to avoid instability due to the time delay at idling. If engine speed increases, the idle valve setting is stored in the memory of the ECM (learning function) and as soon as the engine is idling again, the valve is controlled with the stored PWM value.

The same applies for idling with automatic gearbox engaged; in that case the learned optimum setting for that situation is used, triggered by the neutral/drive switch.

At increased idle speed/low load the idle control valve has a setting which is not influenced by the oxygen sensor and supplies extra gas to the main gas flow; the data are calibrated in a speed/load table in the ECM.

The idle-speed governor, which is a PWM valve too, has two variable settings which are learned by the ECM; one for idling without gearbox engaged and a second setting which is used immediately after engaging the gearbox so that the engine stays constant. To secure speed stability, the learning function is slow.

As soon as the accelerator pedal is pressed, the learning of idle gas control valve and idle-speed governor settings stops and the data are stored to be used as soon as the engine is idling again. At the accelerator pedal, a switch is mounted, which switches on when the accelerator pedal is in idle position.

The wastegate control valve is a PWM valve controlled by the ECM. It converts the pressure signal downstream of the compressor such that the wastegate opens if the boost pressure (measured by the manifold absolute pressure sensor) exceeds a certain value, which depends on engine speed; the data are calibrated such that the engine has the specified torque curve.

In the case of malfunction a 'check engine' light starts to burn. During normal engine running the light is off.

When the engine is not running and the ignition key is off, the output of

Table 5.1 Engine emission performance

European 13-mode emission limits (g/kWh)	*Euro 1 1992/93*	*Euro 2 1995/96*	*Euro 3*	*IVECO CNG engine*
NO_x	8.0/9.0	7.0	5.0	0.7
HC	1.1	1.1	0.6	0.35
CO	1.1	1.1	0.6	0.35
PM	0.35/0.40	0.15	0.15	< 0.05

US Heavy Duty Transient Test: projection based on 8-mode (g/hph)

$NO_x < 1.0$
HC < 0.8
CO < 0.2
PM < 0.05

PM = particulate matter.

the ECM is de-energized; however, the module stays in the standby mode. Several parameters as well as error codes are stored in a memory to be available for later operation. Therefore, the ECM is connected directly with the battery via the DC/DC converter.

If a button near the check engine light is pressed (engine not running) stored diagnostic codes are lit up showing the kind of failure which is occurring or has occurred.

5.5.6 Emission performance

The IVECO CNG engine is equipped with a three-way catalyst. The stoichiometric A/F ratio control system with all the functions as described enables very good emission performance of the engine, as is shown in Table 5.1 [8].

5.6 APPLICATION EXAMPLE 3: MICROPROCESSOR AIR/FUEL RATIO CONTROL SYSTEM ON LEAN BURN CNG URBAN BUS ENGINE

In examples 1 and 2 stoichiometric engine concepts were discussed. The attractive aspect of lean burn systems is the lower fuel consumption (some 10%). However, it is more difficult with lean burn concepts to obtain sufficiently low NO_x emissions, particularly when a relatively high power output is required [8].

Typical emission data are given in the Table 5.2. This gives an overview of different fuels and engine concepts. In the table, $\lambda = 1.5$ can be considered

Table 5.2 Exhaust gas emission levels of several engine types measured using a 13-mode emission test

Engine type	*Fuel*	λ	*Emission (g/kWh)*			
			HC	*CO*	NO_x	*Particulate*
European emission standards (R. 49)			2.40	11.20	14.40	—
US HD transient standards (1991)			1.74	20.79	6.71	0.13
Spark ignited	CNG	1.00	0.35	0.35	0.70	< 0.05*
Spark ignited	CNG	1.00	0.97	6.32	1.19	< 0.05*
Spark ignited	Propane	1.00	1.05	7.31	0.90	< 0.05*
Spark ignited	Propane	1.00	0.20	3.70	1.70	< 0.05*
Spark ignited	CNG	1.44	1.10	0.15	3.50	< 0.05* (cityb)
Spark ignited	CNG	1.48	1.37	0.18	5.70	< 0.05*
Spark ignited	Propane	1.49	1.02	0.28	4.38	< 0.05*
Diesel TC/IC†	Diesel	—	0.50	1.00	8.90	0.35

* with catalyst.

† TC/IC = turbocharged and intercooled.

as lean burn operation. $\lambda = 1.00$ can be considered as stoichiometric operation.

This lean burn example discusses a system applied on a DAF 1160 engine converted to CNG operation according to the lean burn combustion principle. The DAF 1160 engine is a naturally aspirated engine with a cylinder displacement of 11.6 litre. Power output at rated speed is 120 kW at 2000 rev/min. The engine is operating at lean A/F mixture ($\lambda = 1.4$). The bus is equipped with a four-speed automatic gearbox with hydraulic torque converter.

Several components of the engine management system are identical to those of the system on the IVECO CNG engine, described in section 5.5. Also, in this case a venturi gas/air mixer is applied.

Most lean burn gas engine control systems have been open-loop control systems since reliable lean λ sensors have not been available until now. This situation seems to be changing, however. The system discussed in this section applies a UEGO (universal exhaust gas oxygen) sensor, which enables closed-loop control (section 5.3.3). This is important for fast response under transient conditions, as will be shown.

Figure 5.13 shows the components of the system as far as they are of importance for the discussion in this section.

Main A/F ratio control is by the stepper motor controlled gas valve. In open-loop control a three-dimensional map is read from the ECM, giving stepper motor position versus engine rev/min and manifold absolute pressure. In closed loop the stepper motor position is corrected by means of the UEGO sensor signal. This signal is continuously compared to a three-dimensional map, containing the desired A/F ratio.

The system comprises an analog-controlled idle and transient enrichment valve, which is fed by the second stage of the gas pressure regulator with gas at an absolute pressure of 2.3 bar. Like the main gas control valve this valve is also controlled by the ECM in open and in closed loop. As on the IVECO gas engine, the idle rev/min is controlled by a rotating analog valve bypassing the throttle valve.

The idle/transient enrichment valve is necessary to prevent flat spots during acceleration. Such flat spots are a weak point of lean burn engines, and therefore additional measures are required in most cases. The effect of proper idle/transient enrichment control is shown in Figures 5.14–5.17. These figures present actual A/F ratio versus desired air A/F ratio, which have been recorded in a short driving cycle on a chassis dynamometer.

Figure 5.14 shows recorded engine rev/min. The cycle starts with the engine idling, gear in drive. The bus accelerates at wide-open throttle, shifts gear three times and is then brought to a constant speed of 50 km/h. This is followed by a deceleration to 30 km/h, during which the gear shifts back, and finally by a deceleration to 0 km/h. Some 15 s later the gear is put from drive to neutral.

Figure 5.15 shows the resulting A/F ratio versus the desired A/F ratio in

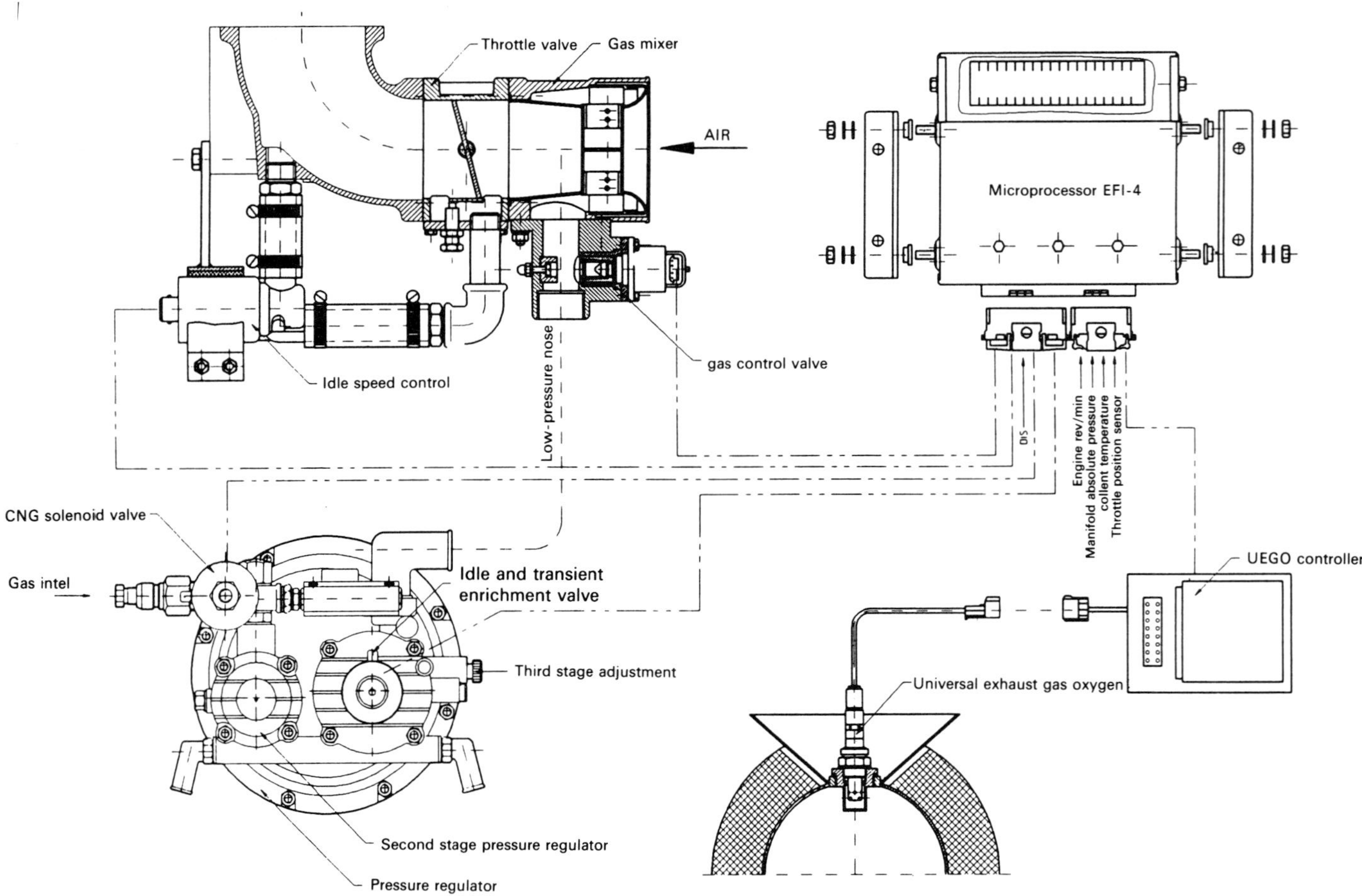

Figure 5.13 System layout DAF 1160 CNG engine.

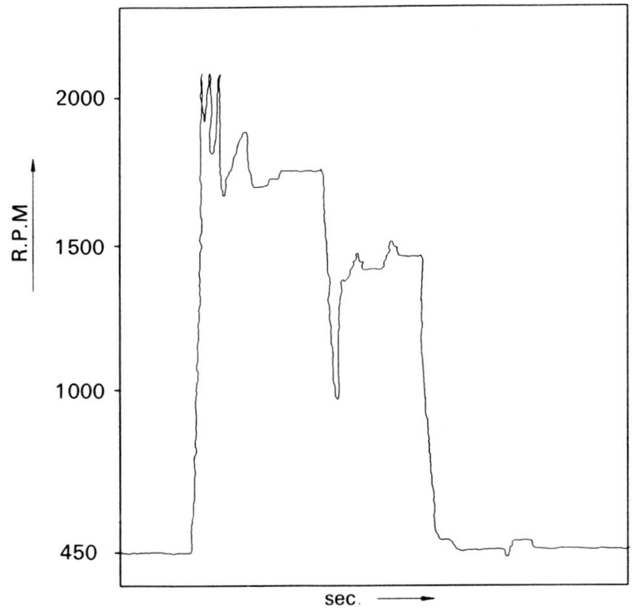

Figure 5.14 Recorded engine rev/min versus time.

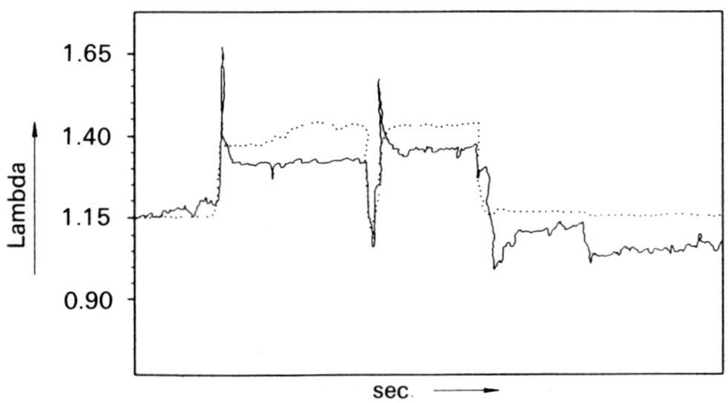

Figure 5.15 Actual air/fuel ratio (drawn line) versus desired air/fuel ratio (dotted line); no idle/transient enrichment.

open-loop control, with the stepper motor main gas control valve but without the idle/transient enrichment valve. The lean peaks (which represent flat spots) at acceleration from closed throttle are clearly shown.

Figure 5.16 shows the same open-loop control system, but in this case the idle/transient enrichment valve is also used. The lean peaks have disappeared and also undesired enrichment during deceleration is limited. A special

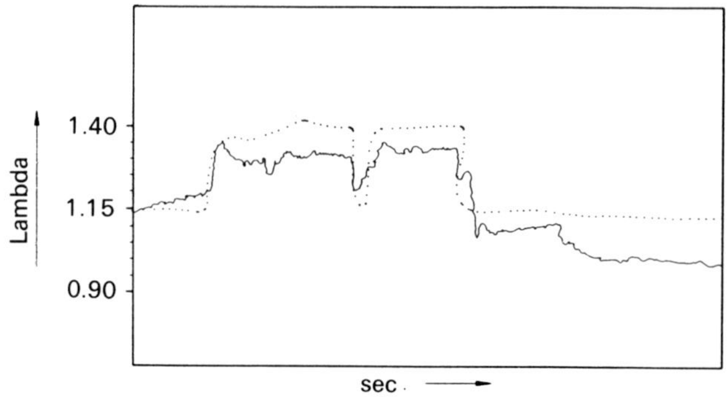

Figure 5.16 Actual air/fuel ratio (drawn line) versus desired air/fuel ratio (dotted line); with idle/transient enrichment.

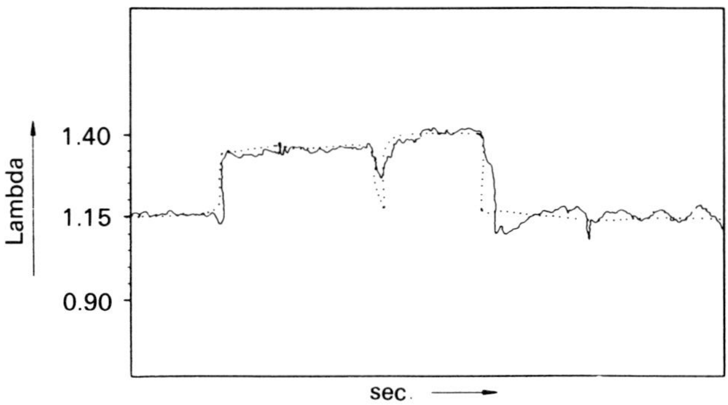

Figure 5.17 Actual air/fuel ratio (drawn line) versus desired air/fuel ratio (dotted line); with idle/transient enrichment and with closed-loop control.

software algorithm is used to control the idle/transient valve. Figure 5.17 finally shows the results when the system operates in closed-loop control with the use of the UEGO sensor.

Since proper transient response of the control system is very important for exhaust emission performance, certainly in urban bus applications, the potential of this type of control is obvious.

5.7 CONCLUSIONS

Electronic control systems such as the ones discussed in this chapter have demonstrated good potential in gas engine applications.

With the help of such systems even conventional gas equipment is capable of fulfilling the requirements of stable open- and closed-loop A/F ratio control, good response under transient conditions and low exhaust emissions. This has been demonstrated on light duty as well as on heavy duty applications.

The software specially developed for gas engine applications incorporates packages which enable a number of engine-specific control functions. Therefore these systems offer a variety of customer-specific options.

REFERENCES

1. Tiedema, P. and Wolters, L. (1990) Recent developments in gas/air mixers and in microprocessor air/fuel ratio control systems. Seminar on Gas Engines and Co-generation, Institution of Mechanical Engineers, Solihull.
2. Klimstra, J. (1989) *Carburettors for Gaseous Fuels – On Air-to-fuel Ratio, Homogeneity and Flow Restriction.* SAE Technical Paper No. 892141.
3. Duncan, C.I. and Tryon, D.C. (1987) *Single Point Fuel Injection for Europe.* ISATA Paper 87011.
4. *Automotive Electronics. Computer Controls for Engine Management Systems.* Delco Electronics Brochure.
5. Van der Weide, J. *et al.* (1991) Engine management systems for lean burn and stoichiometric gaseous fuelled engines. Windsor Workshop on Alternative Fuels, Toronto.
6. *Technische Leergang.* Necam brochure on natural gas systems.
7. Signer, M. and Cornetti, G. (1991) Iveco low emission CNG engine for transit bus. Windsor Workshop on Alternative Fuels, Toronto.
8. Van der Weide, J. *et al.* (1991) Gaseous fuels in heavy duty engines; developments from The Netherlands, in *Symposium on the Use of Compressed Natural Gas (CNG), Liquefied Natural Gas (LNG), and Liquefied Petroleum Gas (LPG) as Fuel for Internal Combustion Engines, Kiev.*

PART THREE

Vehicle Ride and Comfort

CHAPTER 6

Automotive accelerometers for vehicle ride and comfort applications

6.1 INTRODUCTION

Electronics is continuing to replace traditional control technologies in the automobile. With the success of electronic engine control systems, automotive engineers are investigating other functions with an eye towards replacing open-loop, passive systems with closed-loop, active systems. In the area of ride control, fixed damping is being replaced by damping which is programmed based on vehicle parameters such as speed and load, and self-contained damping by the shock absorbers is being replaced by electronically controlled damping to improve performance. The accelerometer is a key component. In antiskid braking systems, sensing deceleration in addition to sensing of wheel speed is being used to improve performance. Accelerometers have long been used in aircraft inertial navigation systems, but their application in automobiles is just beginning. The need for good performance, combined with severe environmental and cost requirements, serve to disqualify many well-developed technologies and are spurring development of new approaches. The more promising technologies are discussed in this chapter and conclusions are drawn regarding which ones are best suited to this application.

6.2 ACCELEROMETER BASICS

The accelerometers discussed measure linear acceleration. The unit of measure of linear acceleration is g, where $1\,g$ is the acceleration produced by earth's gravity, $9.81\,\mathrm{m/s^2}$. Accelerometers all consist of a spring-mass system which moves in response to acceleration and a means of detecting

MICRO SWITCH retains the right to publish material contained in this chapter in conjunction with its advertising, sales promotion, public relations, training and/or technical literature programmes.

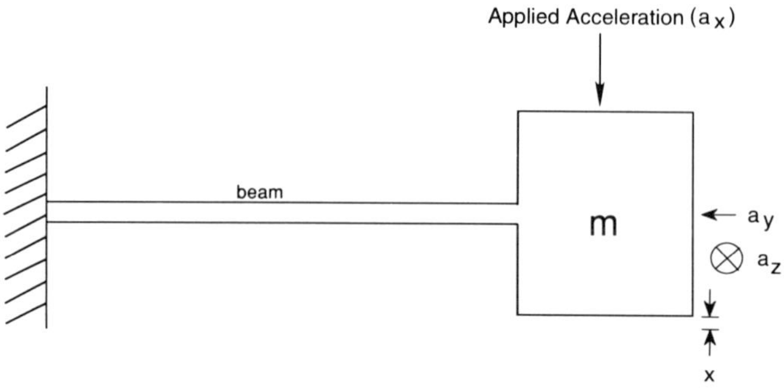

Figure 6.1 Simple accelerometer.

that movement. For example, Figure 6.1 illustrates a cantilever beam with a mass located at the unsupported end. As acceleration is applied in the direction shown, the beam deflects. Add a means of detecting the deflection, and a simple open-loop accelerometer results.

An ideal accelerometer would have infinite frequency response, high sensitivity and no cross-axis response. Basic physics show that this is not achievable. A spring-mass system such as the cantilever beam shown in Figure 6.1 will exhibit resonance at its natural frequency which is given by $f_n = \sqrt{(k/m)}/2\pi$ where k is the spring constant and m is the mass. It is desirable to keep the operating frequency range well below the natural frequency to minimize the sensitivity change as resonance is approached. Displacement of the mass, which is proportional to the sensitivity, is given by $x = -(m/k)\ddot{x}$. Note that the first equation contains $\sqrt{(k/m)}$ whereas the second contains m/k. An inverse relationship exists between the two equations. Therefore, a high-sensitivity accelerometer will have a low natural frequency, and vice versa. The cross-axis sensitivity is also proportional to x, since the applied x-axis acceleration will cause the beam to rotate through an angle θ, with the resultant being $a_x \cos\theta + a_y \sin\theta$, where a_y is the acceleration in the y-axis. For small θ, $a_x \cos\theta \approx a_x$ and $a_y \sin\theta \approx a_y\theta \approx a_y x/l$ where l is the length of the beam. The term $a_y x/l$ is the cross-axis error. A second cross-axis error could result from accelerations in the z-axis, but this effect is typically small, due to flexure stiffness in this direction.

Table 6.1 Inverse relationship between natural frequency, sensitivity and cross-axis error

Accelerometer	f_n	*Sensitivity*	*Cross-axis error*
A	10	k	e
B	50	$k/25$	$e/25$

The following comparison should help to illustrate the above. Take two cantilever beam accelerometers, one with a natural frequency of 10 Hz and the other, 50 Hz. The displacement x and cross-axis error are related as shown in Table 6.1. Accelerometer B, with five times higher natural frequency, has 1/25 the sensitivity and cross-axis error of accelerometer A.

6.2.1 Force rebalanced accelerometer

If a means of applying a restoring force to the beam is incorporated, a closed-loop force rebalanced accelerometer results as shown in Figure 6.2. Closed-loop accelerometers generally have higher accuracy due to the fact that movement of the beam can be reduced, minimizing non-linearity and cross-axis errors. Improved accuracy comes at the expense of increased mechanical and electronic complexity.

Figure 6.3 shows a block diagram equivalent of the force rebalanced accelerometer. Referring to the block diagram, note that the applied acceleration (a) acts on the proof mass (m) to produce a force ($F = ma$). This is opposed by a restoring force ($F_e = k_e x$), which is generated by an electronic circuit that amplifies a signal proportional to the displacement of x and applies it to a means of generating a rebalance force such as a coil. The coil generates a force to minimize the displacement x. The transfer function of the force rebalanced accelerometer is obtained by writing the expression for the displacement (x):

$$x = (ma - k_e x)(-1/k) \tag{6.1}$$

Solving for x yields

$$x = -ma/(k + k_e)$$

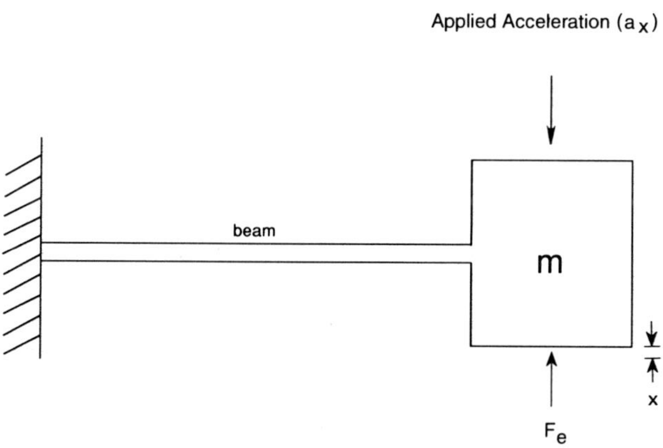

Figure 6.2 Force-rebalance accelerometer.

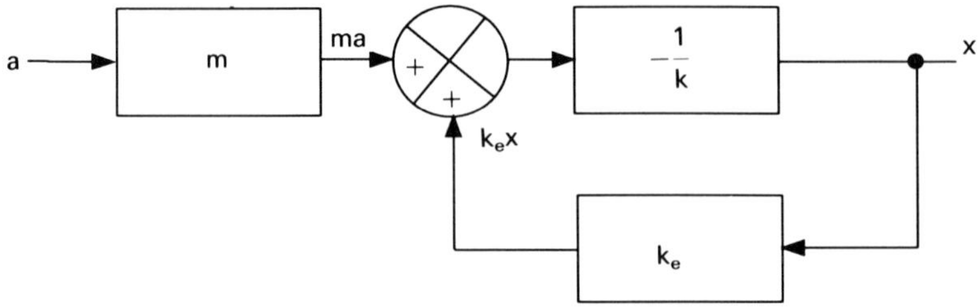

Figure 6.3 Block diagram of force-rebalanced accelerometer.

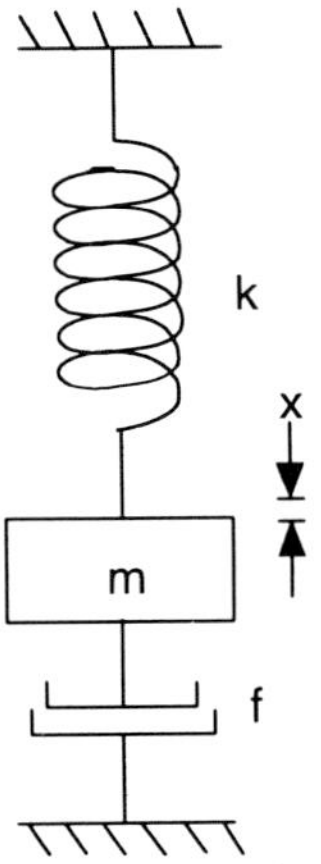

Figure 6.4 Velocity-damped spring-mass system.

As k_e is increased with respect to k, the displacement x is reduced correspondingly.

6.2.2 Damping

A spring-mass system with no damping will oscillate at its natural frequency f_n in response to any disturbance. A practical accelerometer must include a means of damping oscillations. Typically, velocity damping is used since it introduces no steady-state error.

Referring to Figure 6.4, velocity damping has been incorporated in a spring-mass system through the use of a viscous damper. Solving this for the time response yields $x = K_1 e^{-t/\tau} \sin k_2 t$. This response will damp to zero due to the $e^{-t/\tau}$ term. The amount of damping will affect the speed of response and determine whether any overshoot exists, as shown in Figure 6.5.

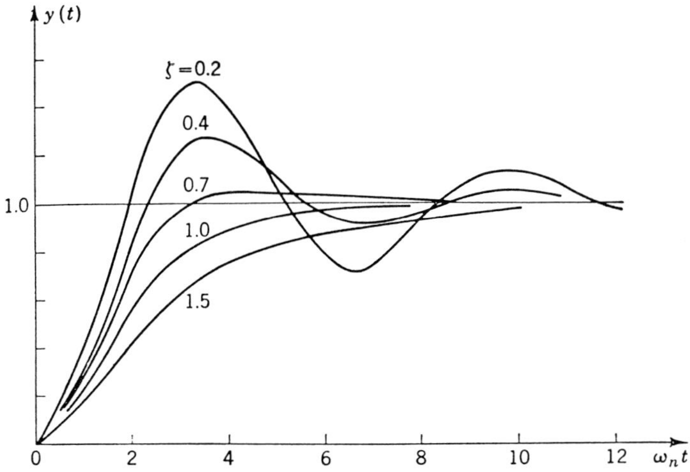

Figure 6.5 Effect of damping on response of spring-mass system.

In the case of the force rebalanced accelerometer, damping may be provided through the electronic feedback circuit, and more sophisticated control algorithms may be implemented.

6.2.3 Overrange protection

Automotive sensors must be able to withstand high g shock impacts which may occur in handling, installation or due to a vehicle being involved in an accident. Physical displacement limiting stops are a necessity to limit the beam (or other movable member) displacement in the event of a shock of several thousand g.

6.2.4 Vibration rectification

In automobiles, various vibratory accelerations are experienced due to road surface irregularities and on-board sources such as unbalanced or out-of-round tyres, etc. If the response of an accelerometer is non-linear, this vibration or oscillatory input will be rectified, resulting in an error signal proportional to the non-linearity and the vibration magnitude. Take an accelerometer with a quadratic transfer function $V_0 = k_1 a + k_2 a^2$ where k_1 and k_2 are the transfer function coefficients and a is the applied acceleration. If an oscillatory input $a(t) = a_0 \cos(wt)$ is applied to this accelerometer, the response is $V_0 = k_1 a_0 \cos(wt) + k_2 a_0^2 \cos^2(wt)$. The time average of the first term is zero but the time average of the second term is $V_0 = (k_2 a_0^2)/2$, the vibration rectification error.

A force rebalanced accelerometer will have a smaller vibration rectification coefficient than an open-loop accelerometer of the same general design and dimensions because its displacement will be smaller for a given acceleration, leading to less non-linearity.

6.3 RIDE CONTROL AND COMFORT APPLICATION REQUIREMENTS

Automotive ride control and comfort accelerometer applications include measurement of vehicle body velocity in electronic suspension control, measurement of lateral acceleration during cornering for ABS proportioning, measurement of longitudinal acceleration for ABS enhancement, and measurement of vertical acceleration for anti-drive. Most of these applications require that the sensor must be able to measure steady-state acceleration. This is because events such as rounding a large-radius curve or accelerating to motorway speed from a dead stop can take of the order of 20–30 s. This has the result of ruling out certain technologies which can only measure dynamic accelerations. Table 6.2 provides a summary of automotive ride control and comfort accelerometer requirements.

The lateral and longitudinal accelerometers measure cornering, acceleration and decelerations, and a 1 g range suffices. The vertical accelerometer, used in ride control, needs a 2 g range. The vertical accelerometer is used to obtain a body velocity signal and therefore does not need to respond to steady-state accelerations. The lateral and longitudinal accelerometers may be mounted in the passenger compartment where the temperature range is more benign, whereas the vertical accelerometer needs to be mounted near the spring/shock absorber mounting to accurately measure accelerations imparted through the suspension. Therefore, it must operate over a wider temperature range.

The combination of good null stability and frequency response to DC (steady state) are the most severe requirements for the accelerometer. The next section will review and compare the technology choices.

Table 6.2 Ride control and comfort accelerometer requirements

Axis	*Range* ($\pm g$)	*Response* (*Hz*)	*Temp. range* (°C)	*Accuracy* (%)	*Transverse sensitivity* (%)	*Null stability over temp.* (*G*)
Lateral	1	DC to 50	– 40 to 85	5	5	0.1
Longitudinal	1	DC to 10	– 40 to 85	5	5	0.1
Vertical	2	0.1 to 50	– 40 to 125	5	5	0.1

6.4 TECHNOLOGY COMPARISON

The more promising technologies in terms of cost versus performance will be described and compared below.

6.4.1 Piezoelectric

Quartz and certain ceramics exhibit piezoelectricity, the ability to convert stress into an electrical output. These materials respond only to dynamic inputs, and typically require very high impedance electronics to achieve low-frequency response. Protecting the interface electronics from the effect of high-g shocks can be difficult because very high voltages result from a large shock impulse. Piezoelectric materials respond equally well to mounting stress so that the packaging must provide good isolation, especially for low-g applications. A low-g piezoelectric accelerometer is shown in Figure 6.6 and a description of its design follows.

Two thin piezoceramic disks are bonded to a thin metal disk to form a symmetrical trilaminar layer construction. Such trilaminar elements are commonly referred to as 'bimorphs'. The metal centre layer provides rigidity for the disk structure and allows electrical contact to be made between the two ceramic disks. Centrally located weights are attached to each face of the bimorph to enhance bending of the piezoceramic elements (sensing elements) when subjected to an acceleration force normal to the plane of the disk structure. The bimorph is mounted near its outer edge using O-rings, which

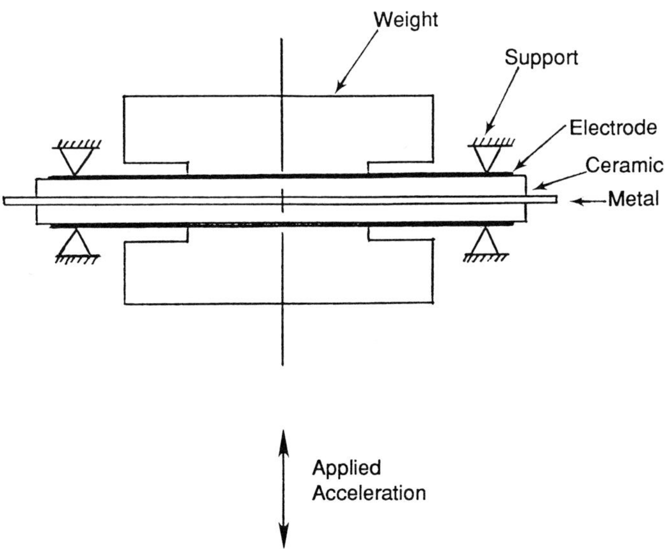

Figure 6.6 Piezoelectric low-g accelerometer.

Table 6.3 Linear Hall IC specifications

Supply V_{DC}	4.5 to 5.5
Response time (μs)	3 typical
Span (gauss)	-500 to 500
Sensitivity (μV/gauss)	5.00 ± 0.10
Null output (V_{DC})	2.50 ± 0.04
Temperature coefficient of null and span (%/°C)	± 0.02 max.

also isolate it from package stress inputs. The signal-conditioning electronics are packaged on a thick film substrate within the accelerometer. Most piezoceramic materials presently available exhibit very large temperature coefficients over the full automotive temperature range. Acceptable performance can be obtained over 0–85 °C, making this technology suitable for location in the passenger compartment with its more benign environment.

6.4.2 Hall effect

The availability of a high-performance linear magnetic sensor using the Hall effect, combined with a cantilever beam that has a magnet located at its unsupported end, resulted in a simple accelerometer capable of satisfying the requirements of automotive applications [1]. The linear Hall-effect sensor used incorporates both thin film on-wafer trim and thick film (application-specific) trim to achieve high sensitivity, low null and sensitivity temperature coefficients, and unit-to-unit interchangeability. A specification summary is given in Table 6.3.

In this design, the response of the Hall sensor to the magnet follows the inverse square law. Therefore, the magnet (and beam) movement must be kept small to approximate a linear response. The accelerometer is shown in cross-section in Figure 6.7.

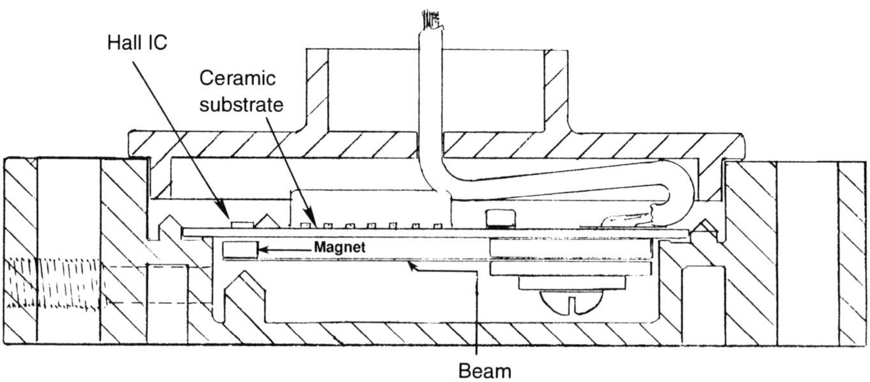

Figure 6.7 Cross-section of Hall-effect accelerometer.

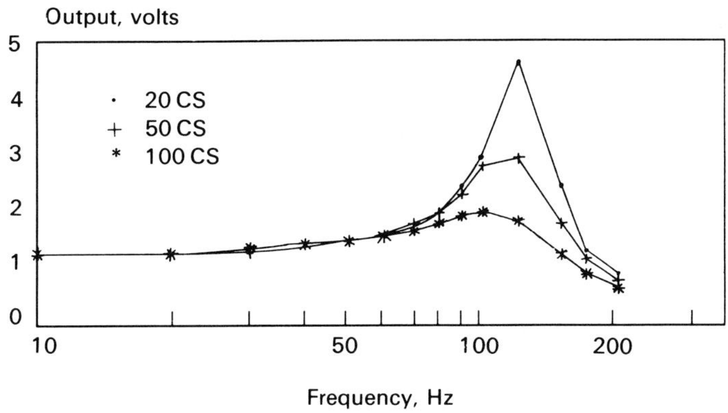

Figure 6.8 Effect of damping fluid viscosity on frquency response.

The ceramic substrate supports the Hall IC, as well as thick film trim resistors and additional circuitry used to amplify and AC couple the output. The beam, formed of a high modulus of elasticity material such as Elgiloy or stainless steel, is clamped to the ceramic. The magnet is located at the end of the beam and acts as the proof mass as well as providing magnetic excitation of the Hall sensor.

One of the potential problems with a spring-mass accelerometer is ringing at the natural frequency due to shock excitation. It is, of course, aggravated by the choice of beam material, which needs a high modulus of elasticity for long-term stability and linearity. Silicone oil damping was incorporated to eliminate any tendency to ring. Both the viscosity and volume of the oil change with temperature. Provided that the device is operated well below the natural frequency, the damping has a negligible effect. The output as a function of damping fluid viscosity is essentially constant as shown in Figure 6.8. This range of viscosity corresponds to operation from $-40\,°C$ to $120\,°C$. Since viscosity change is the major temperature effect in this design, the output should be essentially constant and test data confirm this.

Figure 6.9 shows a prototype device. Scaling this design for various acceleration ranges may be done by changing beam parameters. Table 6.4 illustrates calculated beam dimensions of length l and thickness t and calculated natural frequencies for a constant output of 100 mV and a constant beamwidth of 0.375 inch. One package can easily accommodate all these ranges with minor changes in the beam's dimensioning and sensor IC locations.

6.4.3 Accelerometers using thick film technology

An accelerometer has been reported by Cotignoli [2] which uses a cantilever alumina ceramic beam with a proof mass as shown in Figure 6.10. Readout

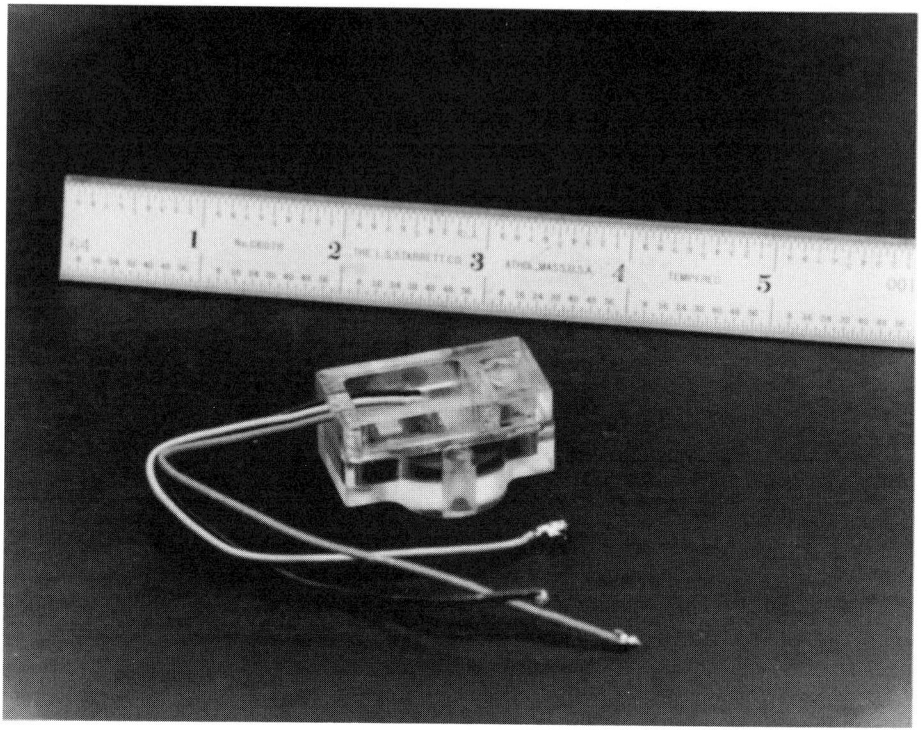

Figure 6.9 Prototype Hall-effect accelerometer.

Table 6.4 Calculated beam design parameters

Range	*l(mm)*	*t(mm)*	$f_n(Hz)$
2 *g*	15.3	1	281
10 *g*	12	1.5	617
50 *g*	10	2.2	1413
100 *g*	10	3	2021

is by a pair of thick film piezoresistors printed on each side of the beam. For a given deflection of the beam, resistors on one side will be in compression and those on the other side will be in tension, causing one pair to increase and the other pair to decrease, maximizing the signal. The remainder of the thick film is used for the signal-conditioning electronics.

6.4.4 Silicon micromachined accelerometers

Silicon is a material with excellent mechanical characteristics and through integration can contain as much electronic circuitry as needed. Although

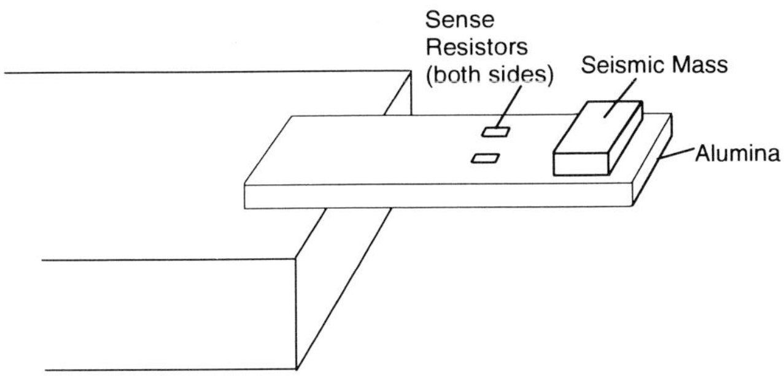

Figure 6.10 Thick film accelerometer (after Cotignoli [2]).

silicon is brittle, and will break like glass when exposed to very high stress rather than deform plastically like most metals, it is almost perfectly elastic below the breaking point. Its elastic behaviour is well suited to sensor applications where physical movement is involved but repeatability must be good and hysteresis very low. Silicon has a Young's modulus comparable to steel, almost twice the yield strength, about half the hardness and thermal conductivity, one-fourth the density and one-sixth the thermal expansion.

With the continued development of micromachining techniques, it is ideally suited for accelerometers, especially for high-g applications. Achieving adequate sensitivity in a low-g silicon accelerometer at low cost poses a significant challenge. A wide variety of silicon accelerometers have been described in a recent survey paper by Yun [3]. Devices have been constructed with piezoresistive, capacitive and resonant detection schemes, with oil, air and electronic damping, using both open-loop and force rebalanced operation. The reason for this variety is that an optimum design approach has not emerged, especially considering application tradeoffs such as size, complexity, power consumption and cost.

Bulk micromachining of single-crystal silicon

There are two basic approaches to building microstructure devices. The first uses single-crystal silicon. This is the starting material used for integrated circuits, and is available for just a few dollars per 10 cm diameter wafer. Upwards of 1000 6 mm^2 microstructure sensors may be obtained from a single wafer. Constructing accelerometers using this material involves selectively etching to form a thin flexure area in the silicon while maintaining the proof mass area at nearly its starting thickness. Next, the moving part must be cut free from the fixed portion. This may be accomplished by a dry-etching technique such as ion milling. A cross-section of a cantilever beam silicon accelerometer is shown in Figure 6.11 and SEM photographs of the front and back side of the chip are shown in Figure 6.12. To protect against high-shock impacts, top and bottom capping wafers are bonded to

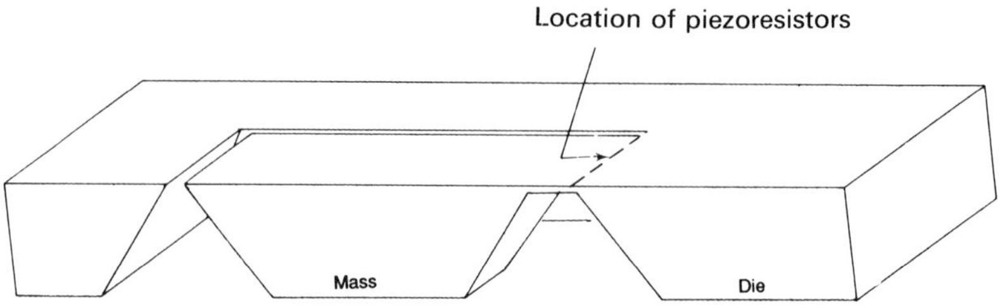

Figure 6.11 Cross-section of silicon piezoresistive accelerometer.

Figure 6.12 SEM photo of silicon accelerometer.

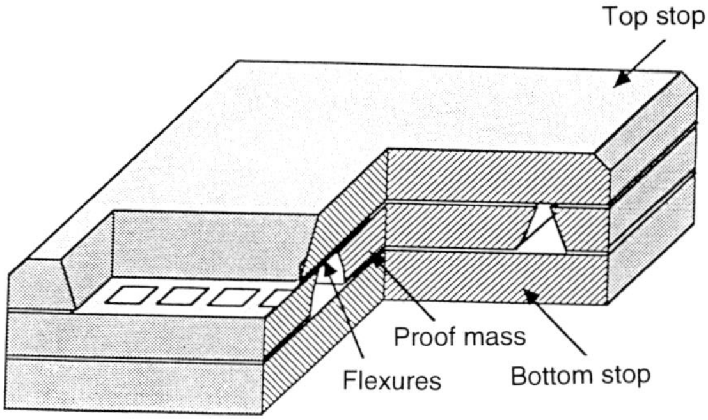

Figure 6.13 Silicon accelerometer with top and bottom caps.

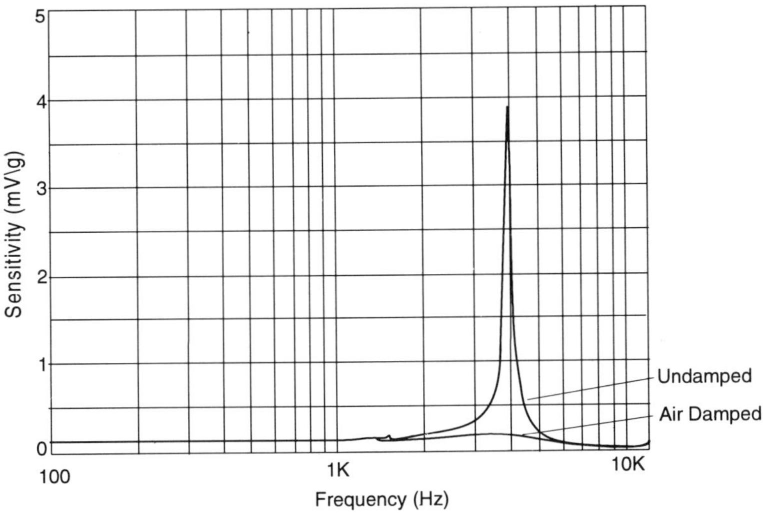

Figure 6.14 Effect of air damping on frequency response.

the sensor wafer. These must be first etched to form a relief to allow for beam deflection during normal operation. An accelerometer using this construction is shown in Figure 6.13. This three-layer sandwich may then be sawn to obtain the individual sensor elements. These overrange stops also provide squeeze-film air damping and effectively damp the accelerometer response as shown in Figure 6.14. One of the potential problems in using silicon for a low-g accelerometer is obtaining sufficient sensitivity. Process control and handling considerations limit the minimum flexure thickness to about 15 μm and, given the relatively low density of silicon, it is difficult to obtain much deflection at 1 or 2 g without making the proof mass large. A

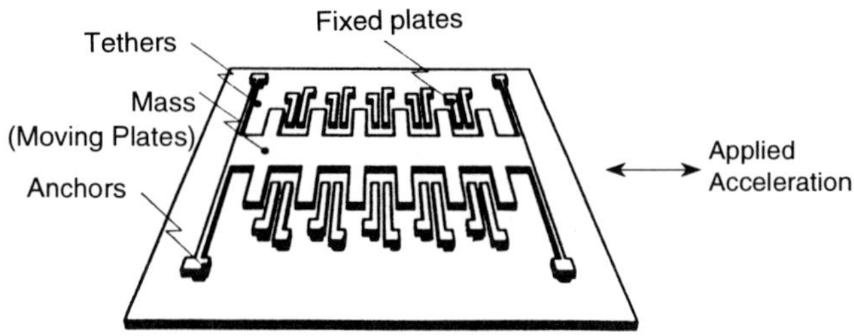

Figure 6.15 Surface micromachined silicon accelerometer (after Yun [3]).

large proof mass is not an attractive option because this directly affects chip size and, thus, cost. Plating a 1–2 μm layer of a high-density material like gold on the proof mass is one approach to increasing the sensor chip mass. This also will serve to put the centre of mass in plane, which simplifies the sensor mounting since it no longer has to be tilted to align the centre of mass properly in the case of a cantilever beam accelerometer.

Surface micromachined accelerometers

Many of the techniques applied to constructing micromachined sensors were first applied to construction of integrated circuits. Polycrystalline silicon (or polysilicon) was first applied to interconnects as the complexity of large-scale integration drove the need for multiple level interconnects. Surface micromachining differs from the bulk process as follows. Rather than etching through the wafer to delineate features, layers of oxide and polysilicon are deposited on the wafer surface. The oxide is then etched away to free the polysilicon sensing structure. This approach leads to a greater degree of miniaturization of the sensing element. These polysilicon structures are limited to several micrometres in thickness and thus have difficulty in achieving adequate mass (and sensitivity) for low-g applications. The proof mass of a surface micromachined accelerometer was given as 0.1 μg whereas a typical value for a bulk micromachined device is 0.1 mg, 10^3 larger. A surface micromachined accelerometer using capacitive sensing electrostatic force rebalance is shown in Figure 6.15.

Electrical readout of silicon micromachined accelerometers

Electrical readout may be performed using piezoresistors located in the flexure area, by measuring the change in capacitance of the moving member with respect to the fixed member(s), or by measurement of the frequency change of a resonant beam. The latter approach is the least developed at this time. Table 6.5 shows major tradeoffs of the three approaches. On the surface, piezoresistance does not appear to be a good choice, since the gauge factor is small and the temperature coefficient of sensitivity is large. However, it does provide a large signal magnitude (ΔR) which is easily read out by

Table 6.5 Electrical readout approaches

	Gauge factor	Signal magnitude	Temp coeff. sensitivity
Piezoresistive	$\frac{\Delta R}{R}=2\text{–}3\%$	$\Delta R = 100\,\Omega$	– 2000 p.p.m./°C
Capacitive	$\frac{\Delta C}{C}=10\%$	$\Delta C = 0.1\,\text{pF}$	– 100 p.p.m./°C
Resonant beam	$\frac{\Delta f}{f}=10\%$	$\Delta f = 20\,\text{kHz}$	– 100 p.p.m./°C

simply operating the piezoresistive bridge with voltage excitation. Simple temperature compensation schemes are capable of reducing the uncompensated error to a few per cent over the automotive temperature range.

Another advantage of piezoresistive and resonant beam readout is that they do not depend on measurements relative to 'fixed' members, as does capacitive sensing. The small signal magnitude (ΔC) of a capacitive sensor means that on-chip electronics are almost mandatory to avoid errors due to stray capacitance variations. The resonant beam approach encompasses both excitation and readout of the beam resonant frequency. The excitation could be thermal or electrostatic and the readout could be capacitive or piezoresistive.

Force rebalance of silicon micromachined accelerometers

Electrostatic force is a simple approach to rebalance a micromachined accelerometer and can be incorporated with minimum structural complexity. This approach is feasible only because of the microscale dimensions of a silicon micromachined accelerometer. The force is given by $F_e = CV^2/2d^2$ where C is the capacitance of the sensor, V is the applied rebalance voltage and d is the spacing between the sensor 'capacitor' plates. Given a square proof mass of 635 μm, and spacing of 5 μm, the capacitance is 1.8 pF. If the proof mass is 0.1 mg, an acceleration of 1.13 g may be rebalanced with 5 V. Note that the electrostatic force is attractive and is proportional to $1/d^2$ so that it is highly non-linear. The servo control electronics used in the force rebalance must take these factors into account.

Cross-axis sensitivity

Cross-axis sensitivity of most silicon micromachined accelerometers is relatively small, irrespective of whether a simple cantilever beam is used or a more complex four-point flexure design. Calculation of maximum deflection of a cantilever beam silicon accelerometer at 2 g shows that it is only 3 μm. Given the beam length of 2000 μm, this is a cross-axis sensitivity of only

Table 6.6 Technology comparison

Technology	*Size* (cm^3)	*Cost* ($)	*Steady-state response*	*Electronics complexity*
Piezoelectric	>1	$10 \approx 20$	No	Medium
Hall effect	>1	$10 \approx 20$	Yes	Simple
Thick film	>1	$10 \approx 20$	Yes	Simple
Bulk silicon				
Piezoresistive	<1	$5 \approx 10$	Yes	Simple
Capacitive	<1	$5 \approx 10$	Yes	High
Resonant beam	<1	$5 \approx 10$	Yes	High
Polysilicon				
Piezoresistive	<1	$5 \approx 10$	Yes	Simple
Capacitive	<1	$5 \approx 10$	Yes	High

0.15%. A more significant error source is in potential misalignments due to the packaging of the basic sensor element or its alignment in the vehicle. This will appear directly as the sine of the misalignment angle so that a 3° misalignment results in a 5% cross-axis sensitivity. This is true for any accelerometer but misalignments are more probable given the small size of a silicon micromachined sensor. Eliminating this error source requires careful attention in packaging and mounting.

6.5 CONCLUSIONS

Several technologies capable of meeting automotive accelerometer performance requirements have been described. Size and cost both favour a silicon micromachined approach. Piezoresistive readout of an open-loop accelerometer will result in the simplest electronics while a resonant beam readout will provide a frequency-modulated signal free from noise corruption and suitable for direct digital interface. Use of a high-density material plated on the proof mass will increase the sensitivity without affecting cost significantly. Table 6.6 summarizes the comparison.

REFERENCES

1. Bicking, R.E. (1989) A Hall-effect accelerometer, in *Proceedings of the 7th International Conference on Automotive Electronics*, Institution of Electronic Engineers.
2. Cotignoli, G. *et al.* (1989) A thick film accelerometer for electronic suspension control. *Sensors and Actuators*, 75–9.
3. Yun, W. and Howe, R. (1991) Silicon microfabricated accelerometers: a perspective on recent developments, in *Proceedings of Sensors Expo*, Chicago, pp. 1–8.

CHAPTER 7

A low-cost multimode fibre optic gyrometer for automotive applications

7.1 INTRODUCTION

Increasing electronic control of vehicle systems is leading to the development of vehicle automation and driver systems. Fully automated driving could be feasible in the future.

New sensors and systems have to monitor the road environment of the vehicle and the driver when conditions deteriorate; to advise the driver how to proceed safely and to provide an input to a co-driver system that could control some vehicle functions automatically, suspension control or guidance for instance. Gradient, change of course, stops and vehicle stability could be managed through gyros. But the aerospatial technologies available for gyros are not adapted to automotive use for cost and maintenance reasons. New technology has to be developed to provide adapted sensors having the advantage of optical sensors but without the inconvenience that can be supported in aerospace industry but not in automotive industry.

Performance has to be lower than in aerospace applications but life duration and condition of use are specific. Sensors will operate in combination on a network; investigations have to be made to evaluate supported driving in safe conditions. But the wide use of optical fibre sensors for automotive monitoring and control is still an innovative and attractive technology, requiring much complementary analysis.

Optical fibre sensing has a 20-year history. Optical fibres were first used for telecommunications because of their low sensitivity to perturbations. But in fact light propagation in a fibre can be modulated in phase, amplitude, polarization and frequency by a large number of physical phenomena.

Finding a mathematical relation between the measurand and the resulting modulation of the light properties is only one side of the problem. The main difficulties arise as one wants to separate the desired effect from the different perturbations. For instance, pressure and temperature variations can both lead to the same light phase modulation at the output of the fibre.

Therefore the main difficulty in the industrialization of optical fibre sensing lies in obtaining effective separation and sufficient signal-to-noise ratio. The extrinsic optical fibre sensor design offers a partial solution by adding a light-modulation device to one end of the fibre, but it has other limitations.

Most of the intrinsic optical fibre sensors were first made from monomode fibres. However, the costs of the components, and signal separation problems, have limited monomode optical fibre sensing to the military market up to now. Recently, some new technological developments have really opened up the civil market.

Intrinsic multimode fibre sensors have proved in recent years that they can compete in performance with monomode optical fibre sensors while being adapted to mass production and actual markets.

Optical fibre sensors are currently used in the following photonics systems:

1. guidance and navigation systems;
2. robotics systems;
3. medical diagnostics systems;
4. sonar systems;
5. power industry systems.

Many other systems which include optical fibre sensors are under development.

Optical fibre sensors can be classified in two main categories, namely the extrinsic optical fibre sensors and the intrinsic optical fibre sensors. Both sensor types operate as follows: a physical parameter to be sensed induces a modulation of the phase, intensity, polarization or frequency of the light. This modulation is then converted into an electrical signal at the output of an optical fibre. In the extrinsic case, this modulation occurs outside the fibre, in an appropriate device connected to it, whereas in the intrinsic case the modulation takes place in the fibre itself, without any interruption of the optical path at the sensing place.

In this latter case, almost any optical fibre can potentially convert a physical parameter such as rotation strain, temperature, magnetic field, acceleration, etc. into a modulation of phase, intensity, polarization, frequency. But the real problem only begins when an accurate and well-separated reading of one particular effect is desired. Specific coding techniques were used to achieve a selective sensor response. The most common one is probably the two wavelengths referencing technique. Combined interferometric–polarimetric methods have also been demonstrated as well as frequency signal-encoding techniques.

Most of the industrial technologies developed up to now rely on such methods for achieving good sensor performances. The most representative advantages of optical fibre sensors are:

- increased sensitivity;
- EMI safety;
- intrinsic safety in hazardous environments;

- large bandwidth and high data rates;
- compatibility with optical fibre telemetry;
- geometrical versatility;
- high temperature and corrosion resistance.

But to fully realize these advantages, technical solutions must be found in order to obtain efficient and reliable sensors at a cost adapted to the aimed-for market. Such solutions partially exist in market segments.

Extrinsic optical fibre sensors are well adapted to certain applications, for instance temperature, chemical or biological sensing. In other cases their apparent simplicity and low costs are counterbalanced by their poor reliability. This can be due to coupling problems between the light-modulating device and the optical fibre that lead to unwanted noise effects, or to application-related limitations, for instance fluid cleanness in the case of flowrate measurements.

Intrinsic sensors are difficult to conceive and to implement, but they could avoid most of these difficulties.

If one considers the light property that is basically modulated at the sensing place, two principal groups of intrinsic optical fibre sensors can be distinguished, namely:

1. those relying on intensity modulation:
   - attenuation/loss
   - scattering
   - micro-bending
   - grating plates
   - frustrated internal reflection
   - impurity diffusion
   - cladding modification
   - fluorescence
   - radiation loss
2. those relying on phase modulation:
   - interferometry
   - polarimetry (differential phase modulation)

The interferometric intrinsic optical fibre sensors are generally modernized versions of classical interferometers (Mach–Zehnder, Michelson, Sagnac, Fabry–Perot) or use mode–mode interferences in multimode optical fibres.

The realization of industrial optical fibre sensor systems will be limited by the need for cost/performance-optimized connectors, light sources, couplers, receivers and optical fibres. Whether this need will be satisfied or not depends strongly on the future technological improvements in these components. It equally depends on how wide the acceptance of this new field will be, from an industrial point of view.

Laboratory results are very promising for a large panel of optical fibre sensors. And yet, the manufacturing of intrinsic optical fibre sensors has up to now been limited only to defence (gyroscope, hydrophones), biomedical, and some very few industrial applications (micro-bending pressure sensors or distributed temperature sensors, for instance). But after 10 years of intensive research, industrial fibre sensors for mass use are now emerging.

In 1989 Frost and Sullivan issued a report that mentioned a world optical fibre market of \$37.4 million and forecast a \$78.2 million market in 1994. It predicted that 36% of the market would be defence oriented and 28% would be for biomedical applications. The European market was forecast for 1994 as \$18.4 million for Germany, \$16.7 million for France and \$15.6 million for Great Britain.

A more recent study issued by Corporate Strategic Intelligence estimates a growth of 38% per year for the optical fibre sensor market in the USA. It would reach a total volume of about \$474 million by 1995. According to this report, the most important growth will occur in the control of industrial processes. More than a hundred firms active in the field of optical fibre sensing have been identified in the USA. The main growth is foreseen for the gyroscopes (31%), for temperature sensors (13%) and for chemical sensors (13%).

Moreover, if one takes the growing interest of the automotive industry for optoelectronics into account, it seems evident that optical fibre sensors will find some uses in this branch of the industry. In a first stage, this will probably concern 'low-level' sensors using plastic fibres.

The range of possibilities offered by distributed fibre sensors and optical fibre sensor networks (especially passive sensor networks) will certainly stimulate in a significant manner new developments as well as new types of applications. In this context, smart structure applications will play an important role.

The world sensor market is expected to grow rapidly in the next 10 years and optical fibre sensors will share a significant amount of this growth, particularly in the domains where they can provide clear improvements either due to their metrological performances or because of specific application-connected advantages.

Lack of experience in the field of optical sensing has induced some misunderstanding and confusion between the already existing optical fibre sensor technologies and those still under development or having only recently been demonstrated to be feasible. The difference between demonstrated feasibility and developing technologies was often underestimated. It took time and money to reach practical conclusions in this new field.

We can mention, for example, the false hopes that were associated with the use of 'exotic' fibres made out of special materials (or with particular design or coatings), twin or more core fibres, etc. Beside the difficulties in producing the fibre itself, other problems emerged. For instance, the injection of light into these fibres required the development of new light sources, new mechanical designs for connectors, coupling devices etc. Very expensive and time-consuming developments have been undertaken, with rather disappointing industrial results, but with interesting potential.

It is now well accepted by many experts that the first mass-produced generation of optical fibre sensors for civil applications will widely use standard optical fibres, connectors, light sources, light receivers, etc. Their performances are well adapted to such civil markets as cars that want, along with the advantages of optical fibre sensors, low costs and reliability.

However, this approach needs a deep theoretical understanding of some light propagation phenomena in an optical fibre that were not of interest for optical communications, but have now become critical, because of the problems mentioned previously. This steadily increasing understanding may lead to unsuspected developments in the field.

Multimode fibres are in this context very interesting because the costs of the fibre itself, as well as the costs of the associated components, are lower than those associated with monomode technology. Moreover, the positioning tolerances are less restrictive, and thus adjusting- and mounting-related costs are also significantly reduced.

At the beginning of optical fibre sensing, interferometric optical fibre sensors were developed using monomode fibre technology. Recently, multimode fibres proved their efficiency and their advantages even for gyroscopes, as we describe below. Besides that, the processes developed for multimode fibres sensing are well adapted for future implementation on integrated optics and further developments in active fibre sensing.

One of the main limitations to the industrial development of optical fibre sensors was the lack of scientists and engineers in this new field able to understand photonics and general metrology, with basic knowledge in electronics and mechanical engineering, and with a good overview of the field in which the sensors will be used (the automotive industry, for instance).

This means that optical fibre sensing educational programmes for scientists and engineers must include basic and practical know-how in photonics and have to be systems-oriented and interdisciplinary. Training at master, or technician, level was initiated in Europe recently. This will soon provide the staff necessary to support mass industrial expansion of optical fibre sensors.

We have modelled, validated and patented a new approach to gyro sensing adapted to the automotive industry. We will now explain the basic principle and the technology of what could be the origin of a new generation of sensors for accurate stabilization and spatial referencing of a car.

Presently, we can observe an expansion of the gyrometer fields towards domains which did not previously require inertial performances, e.g. in the area of civil applications and more particularly by the robotics and automotive industries. These new needs are principally oriented towards low to medium performances and low-cost mass production. The MFOG could represent an original answer to these needs. It allows jointly the use of very low cost components, ease of assembly and low to medium performance.

7.2 SOME SPECIFIC PROBLEMS IN A MFOG FOR AN AUTOMOTIVE

The principal problems are due to modal coupling effects on non-identical counterpropagating modal distributons. The modal distributions are strongly dependent on launching conditions and in the general case on local numerical aperture variations. Therefore, the condition for reciprocity requires the use of perfectly symmetrical couplers, which respect the modal distribution, and ideal fibre. This is, of course, impossible to carry out in practice and we can expect a diminution of the reciprocity, leading to an unbalanced gyro with important zero drift. The couplings between modes are highly sensitive to intrinsic (microdeformations) and extrinsic (thermal and acoustic effects) perturbations with cumulative effects related to the length of the fibre coil. Modal couplings also depend on geometrical and physical parameters of the fibre. The result is a high instability in amplitude and phase of the output signal.

In order to reduce and stabilize the optical bias in the MFOG our first experimental investigations concern the variations of phase and intensity occurring under coupling effects for a given mode.

7.3 AMPLITUDE–PHASE CORRELATIONS OF MODES UNDER COUPLING EFFECTS

For measuring simultaneously the intensity and phase of an interferometric signal we use a single sideband homodyne phase detection method. This method of detection is well knwon in the range of microwave frequencies

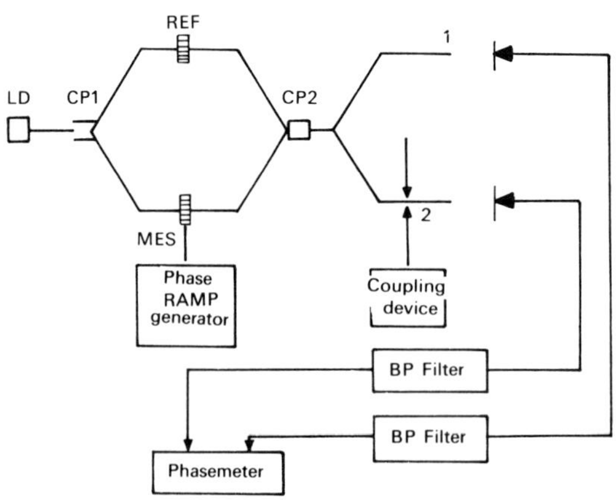

Figure 7.1 Experimental set-up for measuring the amplitude and phase variations occurring under coupling effects.

[1,2] and has been successfully extended to optical frequencies [3]. It uses a linear and odd-phase modulation signal. For appropriate values of the fly-back and the peak phase deviation, it allows a linear transposition of the optical intensity and phase into the amplitude and phase of a low-frequency signal.

7.3.1 Principle of measurements and experimental set-up

The detection principle of phase and amplitude variations induced by modal couplings is shown in Figure 7.1. For more applications it utilizes a Mach–Zehnder interferometer. A multimode cross-coupler provides two outputs. The unperturbed output 1 serves as reference and allows us to neglect bias fluctuations between the two arms of the interferometer. A micro-bending device (Figure 7.2) placed at the output 2 produces phase and amplitude variations by coupling effects. The m-order modes in the information and the reference channel can be respectively expressed as follows:

$$E_{m1} = C_{m1}|E_{m1}|\exp j(\Omega t + \delta_1 + \phi(\omega t))$$
$$E_{m2} = C_{m2}|E_{m2}|\exp j(\Omega t + \delta_2) \quad (7.1)$$

where Ω is the laser frequency, δ_1 and δ_2 are phase terms describing the imbalance of the bridge and $\phi(\omega t)$ is the periodic phase modulating signal. C_{m1} and C_{m2} are complex coefficients featuring amplitude and phase modifications directly due to modal couplings:

$$C_{m1} = |C_{m1}|\exp(j\sigma_1); \quad C_{m2} = |C_{m2}|\exp(j\sigma_2) \quad (7.2)$$

Both signals combine at the second beam splitter and produce mode–mode interference signals which can be written for the m-order:

$$E_m = |E_{m1}C_{m1}|^2 + |E_{m2}C_{m2}|^2 + 2|E_{m1}E_{m2}C_{m1}C_{m2}| \times R_e[\exp\{j(\phi(\omega t + \delta_1 - \delta_2 + \sigma_1 - \sigma_2)\}] \quad (7.3)$$

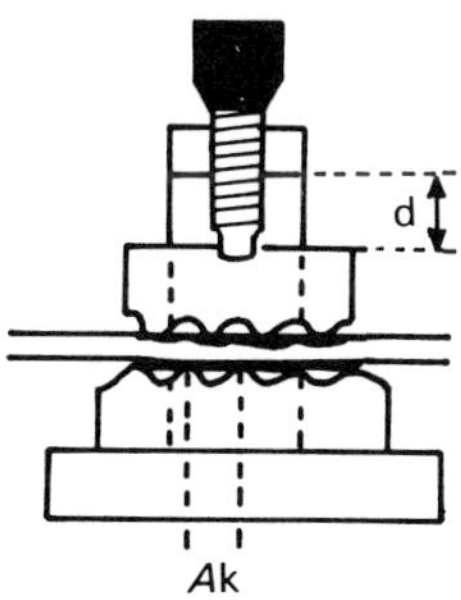

Figure 7.2 Schema of the micro-bending device.

After detection and bandpass filtering at the modulation frequency ω, the first harmonic components issued from the output 1 and 2 can be respectively expressed as follows:

$$i_1(\omega t) = 2\eta |E_{m1} E_{m2} C_{m1} C_{m2}| R_e[\alpha_1 \exp j(\omega t + \psi) + \alpha_{-1} \exp j(\omega t - \psi)]$$

$$i_2(\omega t) = 2\eta |E_{m1} E_{m2}(C_{m1} + \Delta C_{m1})(C_{m2} + \Delta C_{m2})| R_e[\alpha_1 \exp j(\omega t + \psi + \Delta\sigma) + \alpha_{-1} \exp -j(\omega t - \psi - \Delta\sigma)] \quad (7.4)$$

where η includes the quantum efficiency of the diode and $\psi = \delta_1 - \delta_2 + \sigma_1 - \sigma_2$; ΔC_{mi} and $\Delta\sigma$ are amplitude and phase variations due to modal couplings induced by the micro-bending device. The first upper and lower sidebands α_1, α_{-1} of the modulated carrier correspond to the Fourier coefficients:

$$\alpha_{\pm 1} = \frac{1}{2\pi} \int_{-\pi}^{\pi} \exp j[\phi(\omega t) + \omega t] \mathrm{d}(\omega t) \quad (7.5)$$

If one of the two Fourier coefficients α_1, α_{-1} is not exactly equal to zero, we have linearity errors in amplitude ε_A and phase ε_P and we may write:

$$i(\omega t) \sim (1 + \varepsilon_A)\cos(\omega t + \psi + \Delta\sigma + \varepsilon_P) \quad (7.6)$$

The error terms ε_A and ε_P depend on the suppression S of the image sideband defined by the ratio $S = |\alpha_{-1}| / |\alpha_1|$ and their maximum absolute values are:

$$|\varepsilon_A|_{max} = \begin{cases} S \text{ for } S < 1 \\ 1/S \text{ for } S > 1 \end{cases} \qquad |\varepsilon_P|_{max} = \begin{cases} \sin^{-1}(S) \text{ for } S < 1 \\ \sin^{-1}(1/S) \text{ for } S > 1 \end{cases} \quad (7.7)$$

We used a sawtooth phase modulation signal (Figure 7.3) characterized by a peak phase deviation $\Delta\phi$ and a width $2(\pi - \theta)$ of the fly-back. In this case, the first Fourier coefficient $\alpha_{\pm 1}$ has been calculated to be:

$$\alpha_{\pm 1} = \sin(\Delta\phi \pm \theta)\left[\frac{\theta}{\tau(\Delta\phi \pm \theta)} + \frac{\pi - \theta}{\pi[\Delta\phi \pm (\pi - \theta)]}\right] \quad (7.8)$$

When increasing $\Delta\phi$ the lower sideband α_{-1} vanishes for a value of $\Delta\phi = 2\pi - \theta$.

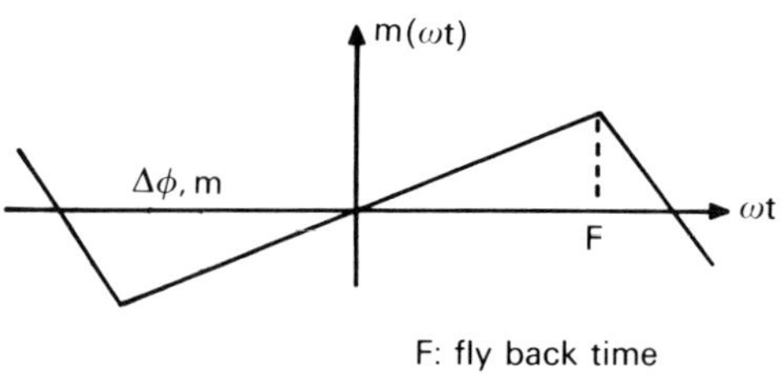

Figure 7.3 Aspect of the linear phase modulating signal.

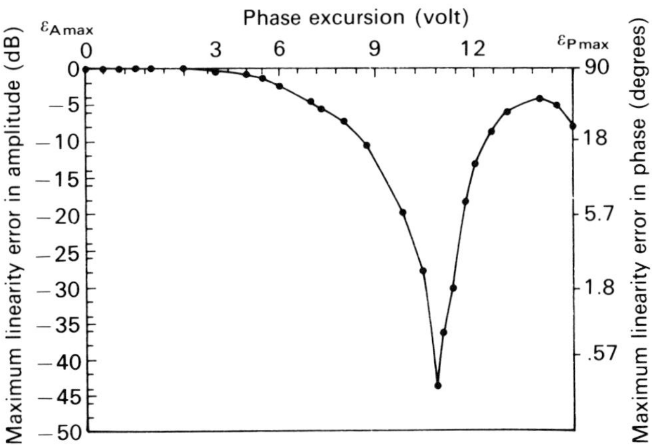

Figure 7.4 Maximum amplitude and phase linearity errors $|\varepsilon_A|_{max}$, $|\varepsilon_P|_{max}$ as a function of the peak phase deviation expressed in voltage unity.

Our interferometer has been realized with 100/140 Si–Si step and graded index fibres. The optical source was 1300 nm wavelength InGaAsP pigtailed laser diode giving 3 mW power. We used as phase modulator a simple 5 cm diameter PZT cylinder with about 30 turns of fibre. The frequency of the phase-modulating signal was 6977 Hz with a fly-back $\theta = 0.863\pi$. The output signals are detected by two germanium photodiodes with active area $\phi = 1\ mm^2$ and placed at a distance $d = 80$ cm from the output end fibres 1 and 2. This gives an angular resolution of 0.07°.

Figure 7.4 shows the variations of the maximum amplitude and phase errors ε_{Amax} and ε_{Pmax} as a function of the peak phase modulation. For a peak value $V_{cc} = 11.7$ V of the modulating signal a suppression $S \sim 47$ dB of the image sideband has been measured from the residual amplitude modulation when varying the DC test signal. This gives a maximum linearity error $\varepsilon_{Amax} \sim 0.7\%$ and $\varepsilon_{Pmax} \sim 0.4°$.

7.3.2 Experimental results and discussion

To produce couplings between guided modes, we used a micro-bending device (Figure 7.3) formed by two gratings with a mechanical wavelength $\Lambda = 1.5$ mm. The displacement of the upper grating which modifies the intensity of couplings, is controlled through a translation stage with 5 μm resolution.

The variations of intensity and phase function of the amplitude of the deformation d have been plotted for different modes in both cases of 100/140 graded and step index fibre. Figure 7.5 shows the evolution of the normalized intensity and the phase cosine for the low-order modes ($\theta = 0.147°$, 0.29°, 1.175°) at the output of the graded index interferometer. For these modes and

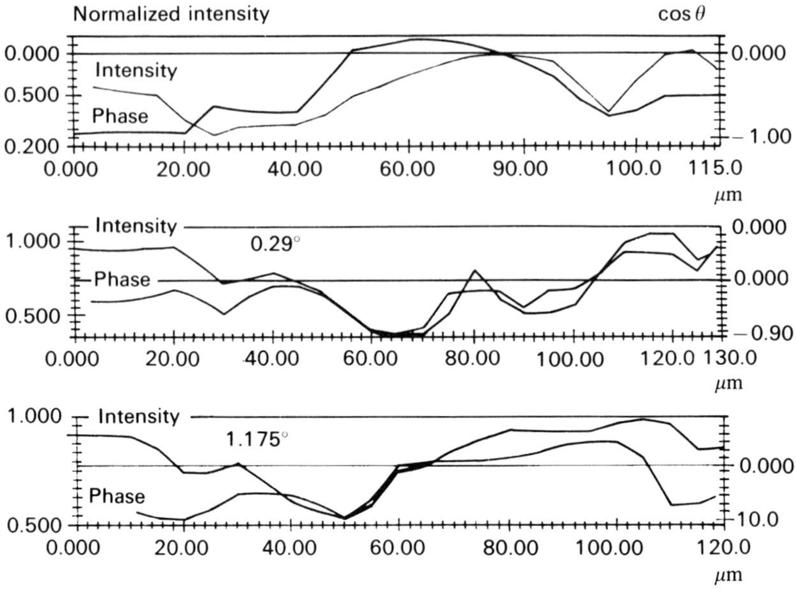

Figure 7.5 Variations of intensity and phase of modes as a function of the deformation depth *d*. Case of modes defined by the angular position $\theta = 0.147°$, $0.29°$, $1.175°$ in 100/140 graded index fibre.

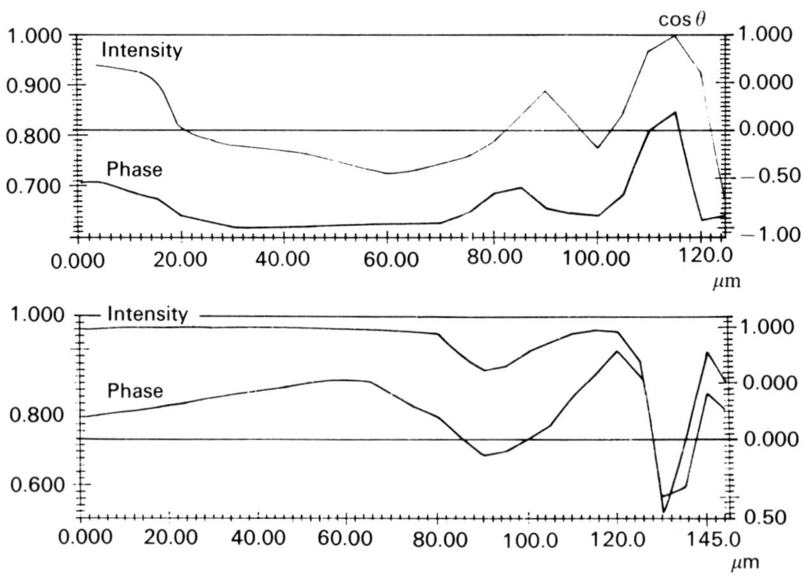

Figure 7.6 Variations of intensity and phase of modes $\theta = 0°$, $1°$ as a function of *d* in 100/140 step index fibre.

others not reported here, we observe similar comportment in the variations of intensity and phases when increasing the depth d. These correlations are also observed for step index fibre (Figure 7.6). Such correlations indicate clearly that only coherent couplings occur when using the micro-bending device.

A simple interpretation of these correlations can be made by comparing the propagation constant and the power distribution. The constant propagation β can be expressed as follows:

$$\beta = kn_1\left[1 - 2\Delta\left(\frac{m}{M}\right)^{2\alpha/\alpha+2}\right]^{1/2} \tag{7.9}$$

where

k = free-space constant of the light
n_1 = refractive index on fibre axis
Δ = relative index difference between axis (n_1) and cladding (n_2): $\Delta \sim (n_1 - n_2)/n_1$
α defines the index profile of the fibre
m/M = normalized order of group of modes

$$\frac{m}{M} = \left[\left(\frac{r}{a}\right)^2 + \left(\frac{\sin\theta_e}{\sin\theta_c}\right)^2\right]^{\alpha+2/2\alpha} \tag{7.10}$$

θ_e = excitation angle
$\sin\theta_c = n_1(2\Delta)^{1/2}$ defines the numerical aperture of the fibre.

In the cases of both graded index ($\alpha = 2$) and step index ($\alpha = \infty$) fibre the expression of the propagation constant for small excitation angles and near the axis ($r \sim 0$) reduces to:

$$\beta \approx kn_1\left[1 - \Delta\left(\frac{\theta_e}{\theta_c}\right)^2\right] \tag{7.11}$$

By considering the power flow equation as derived by Globe [4]:

$$\frac{\partial P(\theta, z)}{\partial z} = -\alpha(\theta)P(\theta, z) + \frac{\partial}{\theta\,\partial\theta}\left[\theta\,\mathrm{d}\theta\frac{\partial P(\theta, z)}{\partial\theta}\right] \tag{7.12}$$

where

$P(\theta, z)$ = optical power distribution as a function of length and axial angle
$\Delta\theta$ = angular space between neighbouring groups of modes
$\mathrm{d}\theta$ = coupling coefficient
$\alpha(\theta)$ = angle-dependent loss

Assuming the presence of an $A\theta^2$ loss law, the best compromise for the power distribution among the guided modes is:

$$P(\theta) = P(0)\mathrm{e}^{-(\theta/\theta_\infty)^2} \sim P(0)[1 - (\theta/\theta_\infty)^2] \tag{7.13}$$

It appears that the power and the propagation constant are well correlated via the deformation function of the fibre z-axis.

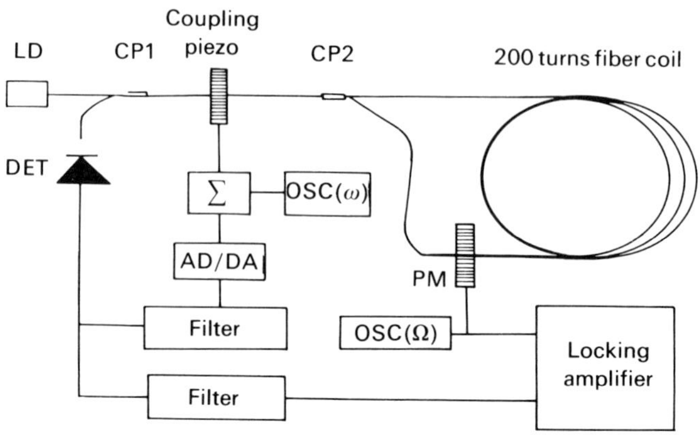

Figure 7.7 Configuration of the stabilized multimode gyrometer.

7.4 APPLICATION TO THE STABILIZATION OF THE ZERO DRIFT IN MFOG FOR AUTOMOTIVE

7.4.1 Principle

The configuration of the MFOG is described in Figure 7.7. It is a closed-loop gyrometer principally characterized by the presence of a stabilization device between the couplers. The principle of the stabilization is based on the properties of coherent couplings between modes. It uses AC and DC voltages applied to a piezoelectric cylinder with 40 turns of fibre. The function of the AC component is to counterbalance the effects of mechanical perturbations. It produces a dynamic mixing between modes which gives at the output a very stable mean modal distribution. It also gives an AM modulation of the DC optical component which will serve as a reference for the stabilization of the slow thermal drifts. This last function is assumed by the DC component and uses the previously described correlations between phase and amplitude variations induced by coherent couplings.

7.4.2 Modes and interference signal expressions

In presence of phase-modulating signal $\Delta\phi(\omega t)$ we may write for homologous modes 1 and 2 at the output ends of the fibre coil (Figure 7.8):

$$I_1 = C_1 I_{01} \exp\{j[\Omega t + \psi_1 + \Delta\phi_R/2 + \Delta\phi(\omega t)]\}$$

$$I_2 = C_2 I_{02} \exp\{j(\Omega t + \psi_2 - \Delta\phi_R/2 + \Delta\phi[\omega(t-\tau)])\} \qquad (7.14)$$

where

$\Delta\phi$ = phase variation due to Sagnac effect
τ = propagation time in the fibre coil

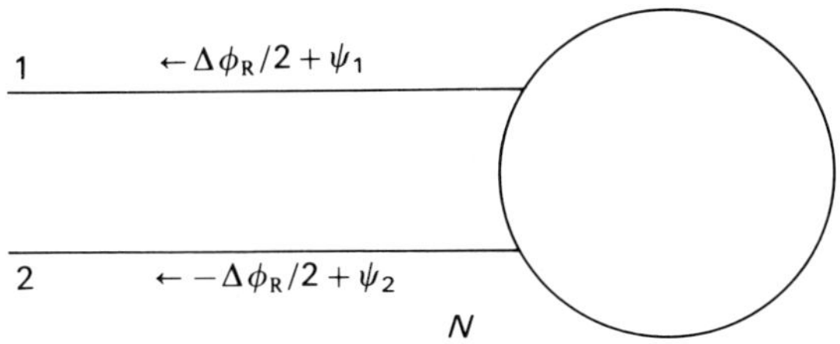

Figure 7.8

I_{o1}, I_{o2} are the powers of the detected modes
ψ_1, ψ_2 are the initial phases of modes at the inputs of the fibre coil
C_1, C_2 are the complex coupling coefficients which may be expressed in the form:

$$C_1 = C_{o1} \exp\{j\delta_1\}; \quad C_2 = C_{o2} \exp\{j\delta_2\}$$

In the case of a sinusoidal phase modulation $\Delta\phi(\omega t) = \Delta\phi \sin \omega t$ we can write for the interference signal:

$$\begin{aligned} I_1 &= C_{o1} I_{o1} + C_{o2} I_{o2} \\ &= 2[C_{o1} C_{o2} I_{o1} I_{o2}]^{1/2} J_o(\eta) \cos[(\psi_2 - \psi_1) + (\delta_2 - \delta_1) + \Delta\phi_R] \\ &\quad + 2[C_{o1} C_{o2} I_{o1} I_{o2}]^{1/2} J_1(\eta) \sin[(\psi_2 - \psi_1) + (\delta_2 - \delta_1) \\ &\quad + \Delta\phi_R] \sin \omega(t + \tau/2) + \cdots \text{higher order terms} \end{aligned} \tag{7.15}$$

where $\eta = 2\ \Delta\phi_m \sin \omega\tau/2$ characterizes the modulation index.

By considering the first harmonic term in Equation 7.15 we can see that the rotation signal may be different from zero even when the gyro is at rest, as has been observed in practice. This offset is firstly described by a constant-phase term (ψ_1, ψ_2) induced by the coupler and the connectors at the input–output of the fibre coil. Its fluctuations are described by the second-phase term $(\delta_1 - \delta_2)$. They are produced by the coherent mode couplings caused by the environmental perturbations. In addition, the sensitivity of the detection depends on modal amplitude modifications, represented here by the coupling coefficients C_{o1}, C_{o2}.

7.4.3 Principle of the stabilization and results

The principle of our zero-drift stabilization is based on the previously shown correlations between the phase and the intensity variations occurring under coherent couplings (section 7.3.2). It consists of using the phase-independent part of the DC component I_o as a reference. For this reason the phase modulation index η must be chosen so that $J_o(\eta) = 0$ and I_o will not depend on rotation phase shift $\Delta\phi_R$:

$$I_o = C_{o1} I_{o1} + C_{o2} I_{o2} \tag{7.16}$$

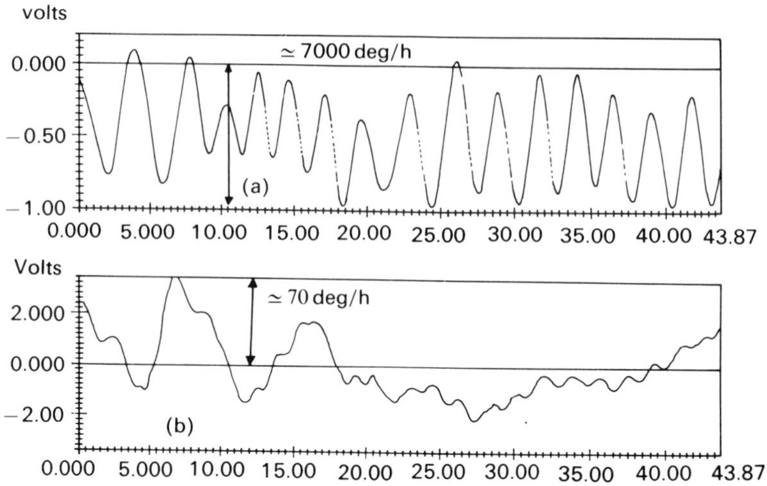

Figure 7.9 Fluctuations of the output signal (a) without stabilization; (b) effect of the AC and DC controlled voltage.

This value of η does not reduce significantly the sensitivity ($J_{1\max}(\eta) = 0.52$ instead of 0.58). In practice I_o is modulated by the AC component applied to the piezocylinder.

When the gyro is at rest ($\Delta\phi_R = 0$) the stabilization consists of applying with the help of the DC component a perturbation P_o is such a way that the amplitude of the first harmonic vanishes ($I_1 = 0$). This condition is realized when:

$$\psi_1 - \psi_2 + \delta_2 - \delta_1 + \phi(P_o) = 0 \tag{7.17}$$

The corresponding value I_o is the reference, which must be maintained constant.

An illustration of this stabilization principle using both effects of static and dynamic coupling is shown in Figure 7.9. We have used for this case an HeNe laser with a microscope objective for focusing the light into the 100/140 graded index fibre gyro. The frequency phase modulation signal was about 90 kHz, while for the dynamic stabilization we used a 16 kHz resonance frequency of the piezocylinder. Figure 7.9a shows the high instability of the open-loop output signal which corresponds to an estimated drift of ±7000 deg/h. The Fourier analysis of this signal shows the periodic contributions of the environmental vibrations around a slow fluctuating offset term due to thermal effects. Figure 7.9b shows the results of the static and dynamic stabilization. We can see that the periodic fluctuations have completely vanished while the thermal offset fluctuations have been significantly bounded to ±30 deg/h. This stabilization could not be maintained for a long time because the DC couplings caused by the piezo are not sufficient for covering the whole range of the thermal perturbations. It is obvious that a more reciprocal configuration of the multimode gyro and particularly an appropriate choice of its components will increase the efficiency of such a stabilization device. Relating to this, previous work [5, 6] had shown that multimode gyros may have intrinsic long-term stability as low as ± 40 deg/h. It appears that the basic conditions for improving the stability of the multimode gyros consist firstly of using an all-fibre technology without any connecting device at the ring input/output. A wide-band and spatially incoherent pigtailed LED seems to be a suitable choice of light source for reducing the effects of coherent cross-couplings. The second coupler must be as symmetrical and reciprocal as possible with respect to the modal distributions.

It seems that the coated micro-optic beamsplitter with fibre pigtails [6, 7] constitutes the more convenient choice. In addition, we had shown previously that the selection of particular modes with the help of a modal filtering device may increase significantly the fringe visibility [8] and also improves the reciprocity of the gyro if only a low-order preponderant group of modes is detected. However, assuming that the photonic noise is equally repartited over all the modes, a compromise must be found for obtaining an acceptable electronic signal-to-noise ratio. Under these conditions this stabilization principle may bound the long-term drift to the 1–50 deg/h range depending essentially on noise performance.

7.5 CONCLUSIONS

Our aim was to describe some of the specific MFOG problems and to demonstrate its feasibility for automotive applications. The basic problems in the MFOG have been assumed to be due to coherent coupling effects and non-identical clockwise and counterclockwise modal distributions. We have proposed a simple method for stabilizing the zero drift based itself on coupling effects induced by DC and AC voltages via a piezocylinder. The role of the AC component is to produce a dynamic mixing of modes and a very stable mean modal distribution with regard to mechanical vibrations, thus allowing the gyro to work in a perturbed environment. Coherent couplings induced by thermal effects which appear to be a large source of drift error may be significantly reduced by using the good correlations found between the phase and amplitude variation. The manufacture of such a device could start soon if the market is open; the production cost in small quantities is estimated to be around \$250. Further implementation in integrated optics, at least partially, is very possible; this will lower the cost for large-volume production.

ACKNOWLEDGEMENT

The author wishes to thank IAR, ANVAR and the CRA for their support.

REFERENCES

1. King, R.J. (1977) *Microwave Homodyne Systems*, Peregrinus.
2. Kaffee, J.S. (1965) Microwave frequency translator. *IEEE Trans Microwave Theory Tech.*, **MTT 13**, 371–8.
3. Voges, E., Ostmald, O., Schiek, B. and Neyer, A. (1982) Optical phase and amplitude measurement by single side band homodyne detection. *IEEE Journal Quantum Electronics*, **QE-18**, no. 1.
4. Globe, D. (1972) Optical power-flow in multimode fibre. *The Bell System Technical Journal*, **51**, no. 8.
5. Bazarov, A.E. and Semenov, A.T. (1984) Non-reciprocal effects in a ring interferometer with a multimode fibre waveguide. *Soviet Journal of Quantum Electroncs*, 522–7.
6. Fredricks, R.J. and Johnson, O.R. (1987) Low cost multimode fibre optic rotation sensor. *Fibre Optic and Laser Sensors V*, Society of Philosophical Instrument Engineers, vol. 838.
7. Alenseev, E.I. (1984) Ring interferometer with a multimode fibre waveguide. *Soviet Journal of Quantum Electronics*, 1436–42.
8. Bouamra, M., Chakari, A. and Meyrueis, P. (1991) Méthode de mesure de déphasage adaptée aux capteurs interférométriques à fibres multimodes, *OPTO-91*, Paris.

PART FOUR

Security and Safety Systems

CHAPTER 8

Development of Toyota airbag sensor and system

8.1 INTRODUCTION

An electrical-type airbag (Air Bag 1) was marketed by Benz in 1980. Initially, the growth rate of worldwide airbag installation was not particularly rapid, but there was a considerable increase in the use of airbags subsequent to the application of relevant US safety standards in 1989. This encouraged Toyota to develop sensors and airbag systems most suitable for all catagories of vehicles, from medium to sub-compact ones, passenger cars etc.

8.2 TOYOTA SUPPLEMENTAL RESTRAINT SYSTEM

Although seat belts have been used in vehicles for passenger protection, the introduction of airbags has gradually been accepted as a supplemental form of passenger protection. In case of a severe collision, the airbag is activated and absorbs the shock before the passenger's head makes any contact with the steering wheel. The airbag is inflated immediately and deflated gradually at the end of operation.

In addition, automobile manufacturers started employing seat belt pre-tensioners to increase seat belt effectiveness. In the case of a severe collision, the seat belt pre-tensioner operates immediately to eliminate looseness in the seat belt.

Figure 8.1 shows an operation diagram of Toyota's supplemental restraint system, composed of a driver's seat airbag and seat belt pre-tensioner for each seat.

8.3 SENSORY SYSTEM

8.3.1 Various types of sensory system

Various types of sensory system are used (Figures 8.2–8.7), since the location of the sensor which determines whether or not the airbag and seat belt pre-tensioners should be activated depends on the construction of the vehicle.

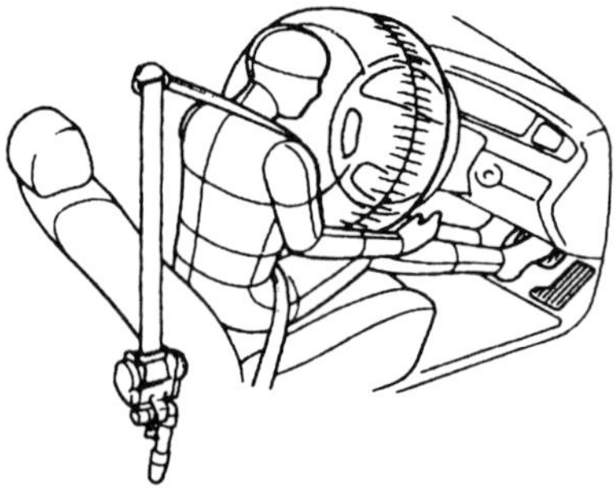

Figure 8.1 Operation diagram of the supplemental restraint system.

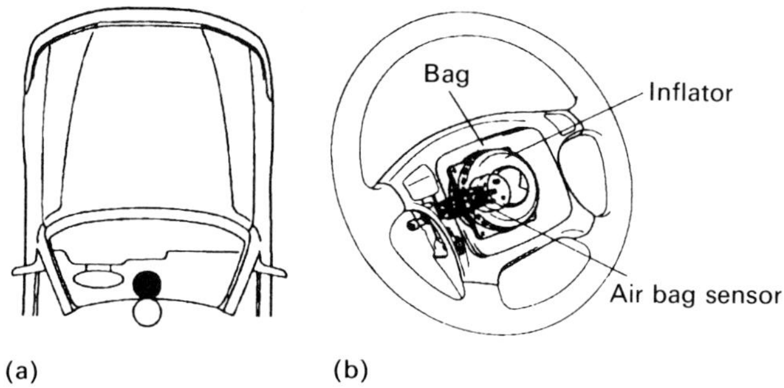

Figure 8.2 One-sensor type. (a) One floor sensor: Benz, Volvo, Audi; (b) One steering sensor: Toyota, Jaguar, Honda.

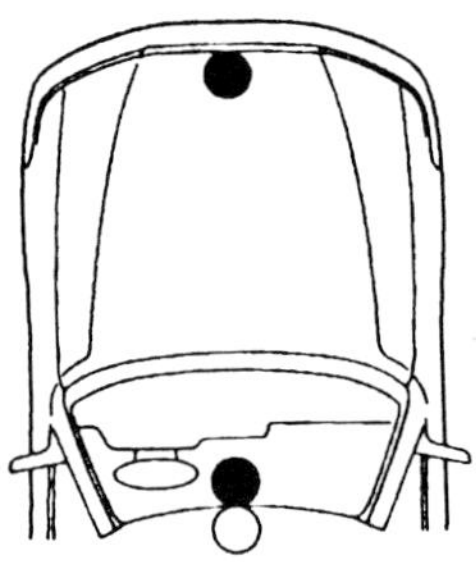

Figure 8.3 Two-sensor type. One front sensor and one floor sensor: GM (Olds 88).

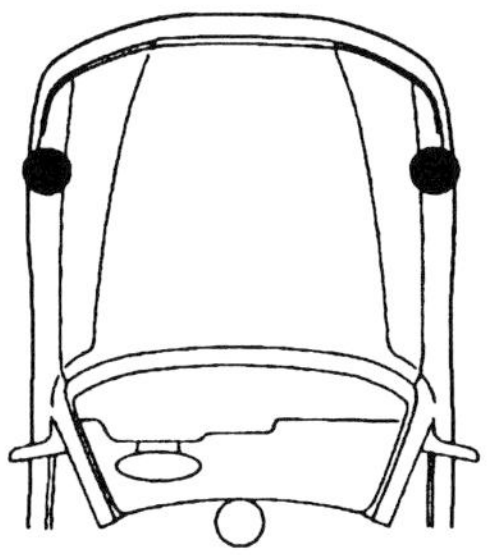

Figure 8.4 Two-sensor type. Two front sensors: BMW, Porsche, GM (Corvette), Chrysler, Honda, Saab, Rolls-Royce, Rover, Mitsubishi, Subaru, Alfa Romeo, VW.

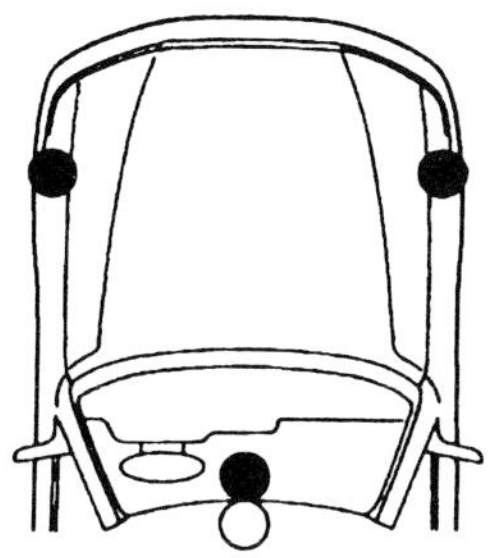

Figure 8.5 Three-sensor type. Two front sensors and one floor sensor: Toyota.

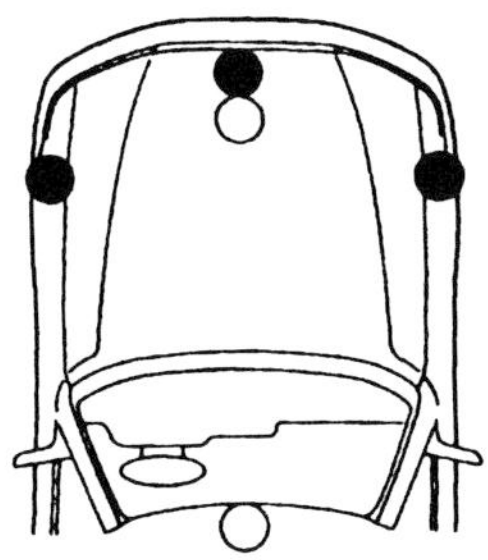

Figure 8.6 Three-sensor type. Three front sensors: Ford.

8.3.2 Toyota sensory system

Toyota uses both mechanical- and electrical-type airbags; the mechanical type is shown in Figure 8.2b and the electrical type in Figure 8.5. The seat belt pre-tensioners are the electrical type and are operated simultaneously by the ignition signal from the electrical-type airbag.

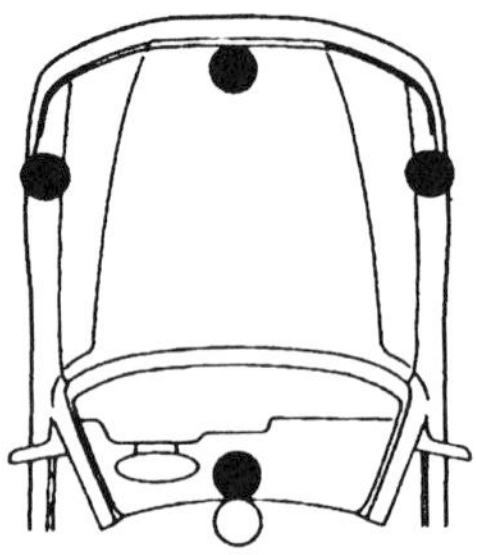

Figure 8.7 Four-sensor type. Three front sensors and one floor sensor: Nissan.

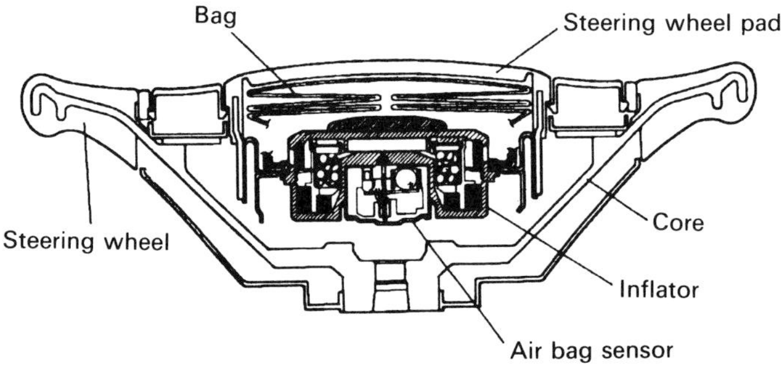

Figure 8.8 Mechanical-type airbag system.

Mechanical-type airbag

With the mechanical-type airbag system, a bag, inflator and sensor are integrated compactly into the steering wheel pad. The features of this type are high reliability and low cost (Figure 8.8).

Electrical-type airbag

With the electrical-type airbag, a bag and inflator are integrated into the steering wheel pad, and sensors are located at suitable places in the vehicle (Figure 8.9). Generally, a front sensor, which decides whether or not the airbag should be inflated in a low-speed collision, is located at the front of the vehicle; whilst a floor sensor, which decides whether or not the bag should be inflated in a high-speed collision, and safing sensor, which judges roughly whether or not the vehicle has collided, are located on the floor inside the vehicle. The inflator is ignited if both front and safing sensors, or both floor and safing sensors, give ignition signals simultaneously.

In the case of a low-speed collision, as shown in Figure 8.10, the onset of

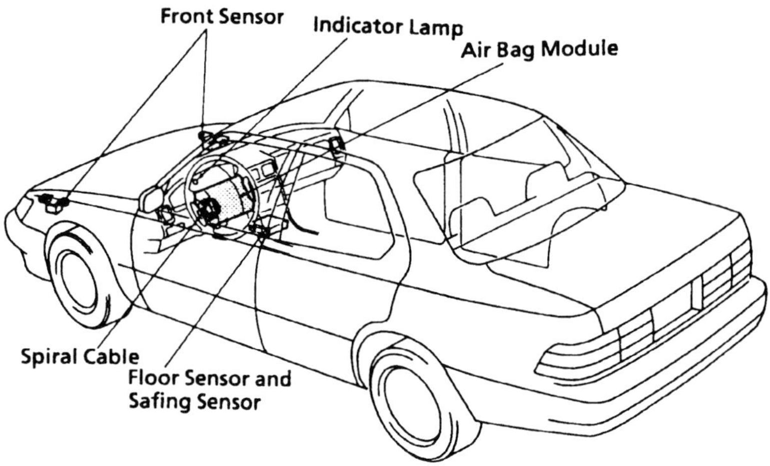

Figure 8.9 Location of components.

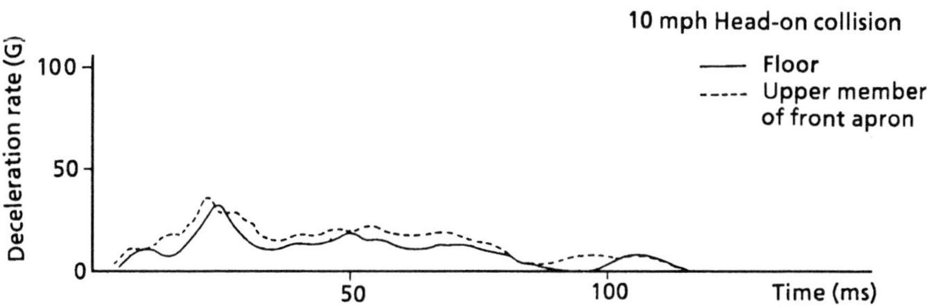

Figure 8.10 Variations of component deceleration for low-speed collisions.

deceleration of the car body occurs at the front of the vehicle earlier than at its floor, and therefore the front sensors reach the TTF (time to fire) earlier than the floor sensors in almost all forms of low-speed collisions.

In high-speed collisions, as shown in Figure 8.11, the onset of deceleration of the car body occurs all but simultaneously at the front and the floor of the vehicle.

8.4 SENSOR

8.4.1 Sensor for mechanical-type airbag

The shock conveyed through the steering shaft causes the ball in the cylinder to move the trigger shaft, whose opposite end is fixed by a bias pin. As a

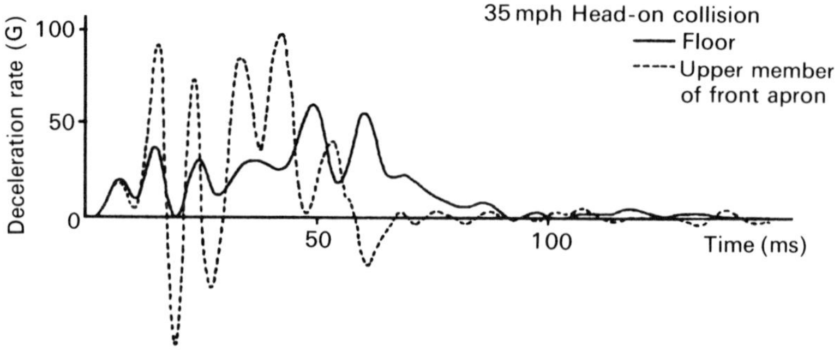

Figure 8.11 Component deceleration varies for high-speed collisions.

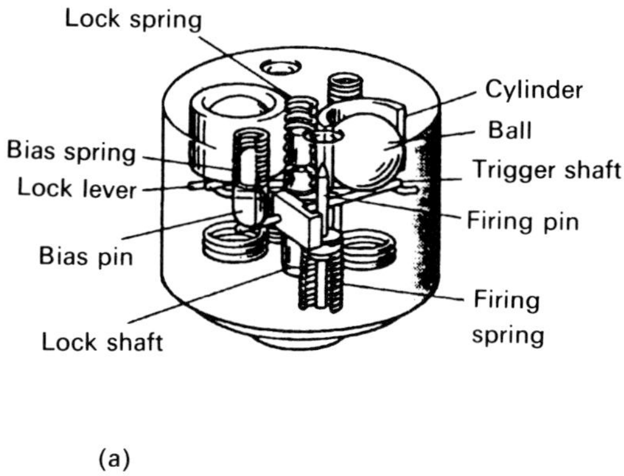

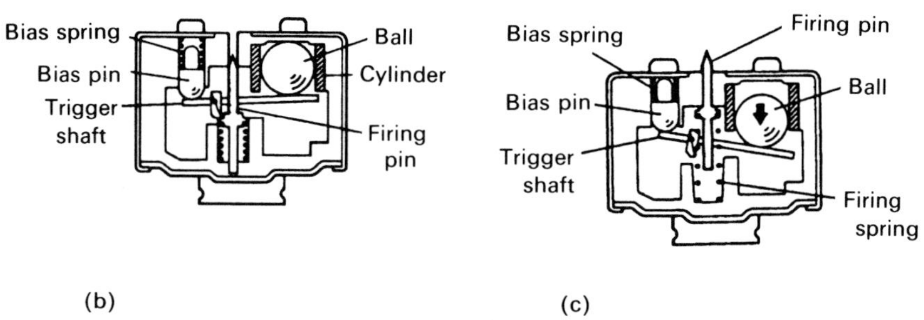

Figure 8.12 Mechanical-type airbag sensor: (a) construction drawing; (b) when not activated; (c) when activated.

result, the firing pin is released to shoot the firing spring in an instant, igniting the ignition agent (Figure 8.12).

8.4.2 Sensor for electrical-type airbag

Sensors for the electrical-type airbag can be classified into front sensor, floor sensor and safing sensor according to their functional aspects, as explained in section 8.3.2. Performances required for each type of sensor and examples of successful sensor systems are given below.

Front sensor

In order to withstand a severe environment, since the front sensor is located at the front of the vehicle, rigid construction, as well as durability and reliability, are required.

Rolamite sensor

The Rolamite's highly responsive frictionless spring-mass system provides rapid sensor closure and airbag deployment. This sensor uses a cylindrical mass which is restrained by a flat spring that is wrapped around the mass (roller), shown in Figure 8.13. The spring is designed to provide unequal forces on the mass, causing it to be restrained against one end of the sensor body. By controlling the size of the mass, the spring rate and the travel distance, the sensor can be designed to match any desired acceleration time envelope characteristic of a particular vehicle's crash response pattern.

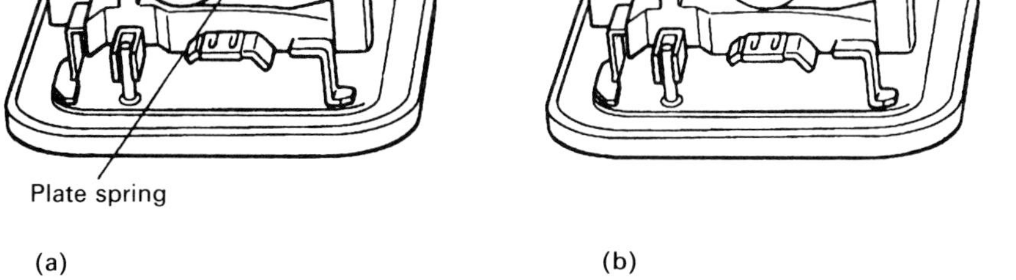

Figure 8.13 Rolamite sensor: (a) when not activated; (b) when activated.

The equation of motion for a crash sensor can be expressed as follows:

$$\ddot{s} = -\frac{M}{I} f_1(s)\dot{s}^2 + f_2[g(l), g(h)] - f_3[g(s)] \qquad (8.1)$$

where

$\ddot{s}$ = acceleration of the mass
$\dot{s}$ = velocity of the mass
s = position of the mass
M = weight of the mass
I = moment of inertia
$f_1(s)$ = coefficient of the mass
$f_2[g(l), g(h)]$ = function of $g(l)$ and $g(h)$
$g(l)$ = longitudinal component of forcing function
$g(h)$ = horizontal component of forcing function
$f_3[g(s)]$ = function of band force on mass as a function of position

In the event of vehicle deceleration due to a severe crash, the mass will move forward and the contact point will close. The Rolamite can be calibrated to specific vehicle models by adjusting the above sensor parameters to meet the exact crash requirements. This can be accomplished by designing the band most favourable for specific crash requirements, and adjusting the mass position at a specific location within the sensor during the manufacturing process.

Figure 8.14 shows simulation results in a high-speed collision using the mathematical model described above.

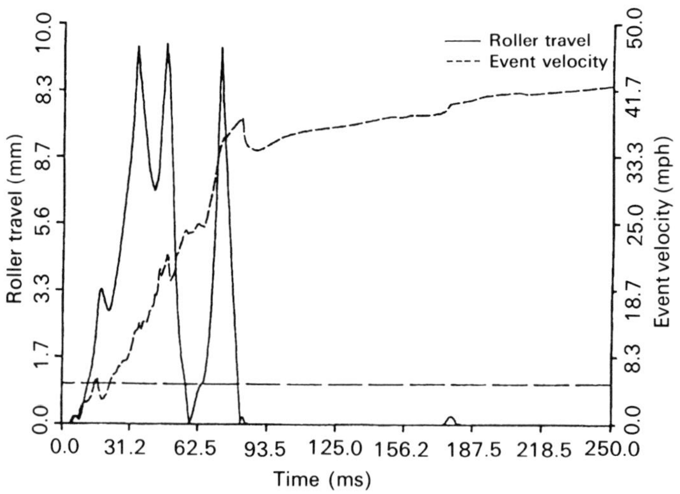

Figure 8.14 Simulation for high-speed collision.

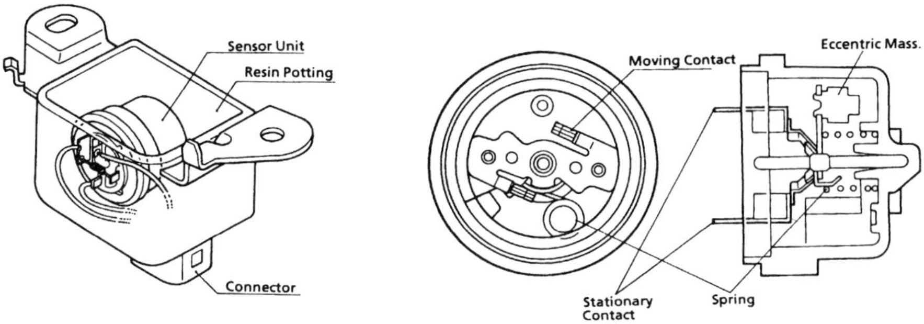

Figure 8.15 Rotary sensor.

Rotary sensor

An axial rotor with an eccentric mass is set into an initial position by a spring with an initial spring force and a spring constant. The rotational angle can be expressed in the following formula:

$$\ddot{\theta} = \frac{MR}{I} f[G(f), G'(f)] - \frac{K}{I} f(\theta) - C \tag{8.2}$$

where

$G(f)$ = deceleration (longitudinal)
$G'(f)$ = deceleration (vertical)
I = inertia moment
θ = rotational angle
C = constant
MR = eccentric mass moment
K = spring constant

When θ reaches θ_{on}, the contact points are closed. The aforesaid parameters are optimally tuned for each car model. Ensuring the reliability of current flow between the contact points, as well as an appropriate bearing structure, are important design requirements with respect to the long-term maintenance of sensor reliability.

Figure 8.16 shows simulation results in a high-speed collision using the mathematical model described above.

Floor sensor

Various types of sensors, including an electrical type, can be used because they are normally installed inside the vehicle and not within the crush zone.

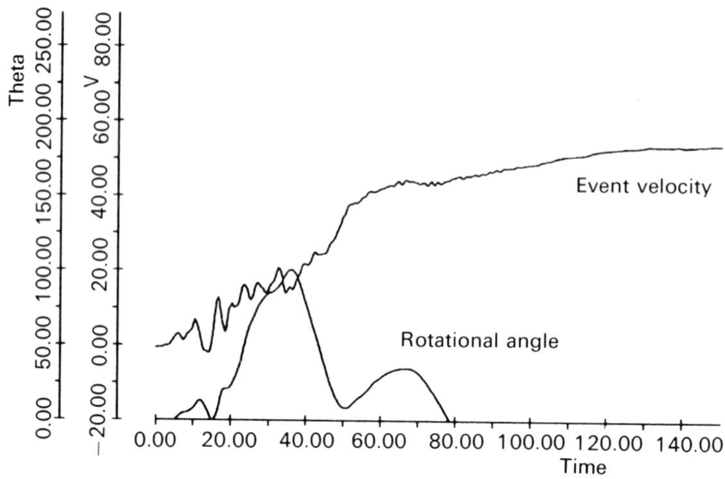

Figure 8.16 Simulation for high-speed collision.

Table 8.1 Electric *G*-sensors

Method	*Item*				
	Sensitivity	*Frequency*	*Temperature*	*Matching of circuits*	*Reliability*
Piezoresistive	Good	Good	Good	Good	Good
Metal	Fair	Good	Good	Fair	Fair
Piezoelectric	Excellent	Fair	Fair	Fair	Good

Electrical type

There are three main types of electrical *G*-sensor; piezoresistive, piezoelectric and metal strain. Each type is characterized in Table 8.1 and Figure 8.17.

Toyota uses the piezoresistive type in its airbag system for the following reasons: the temperature characteristic is especially good; changes in sensitivity due to plastic deformation rarely appear; a silicon type can be produced through IC technology; and its mass productivity is high.

Impacts caused by collision are detected by the cantilever of the sensing head of the sensor. The fabrication process of the cantilever is shown in Figure 8.18. First, an amplification circuit (which will be discussed later) is formed on silicon using bipolar IC technology, and then the thin parts and outer surface of the cantilever are etched from the rear side using anisotropic etching, and, finally, the outer surface of the cantilever is etched from the front side by isotropic etching. This overall etching can eliminate any micro-cracks

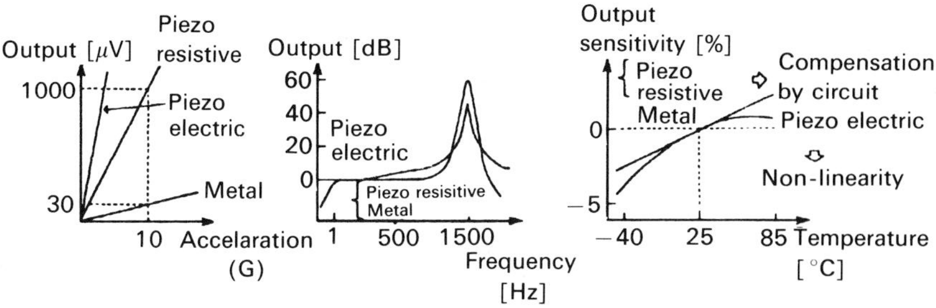

Figure 8.17 Electric *G*-sensors: (a) sensitivity; (b) frequency; (c) temperature.

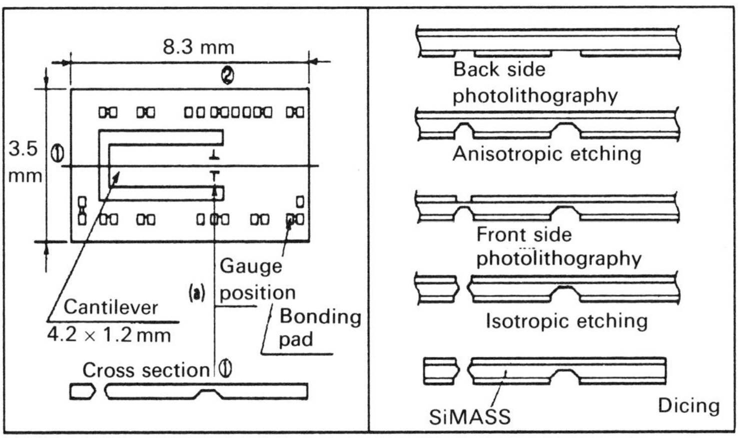

Figure 8.18 Fabrication of cantilever.

that may lead to breakage of the cantilever. A photograph showing the thin part of the cantilever after the etching process is given in Figure 8.19.

P-type diffused resistors are arranged in $\langle 1\,1\,0\rangle$ and $\langle 1\,\bar{1}\,0\rangle$ directions on the n-type silicon [1 0 0] on the thin part of the cantilever to form a bridge circuit which will act as a sensor. The equivalent circuit is shown in Figure 8.20, and the piezoresistance coefficient for each crystal orientation is given in Table 8.2.

When deceleration generated by a vehicle collision is exerted on the cantilever, the bridge resistance varies due to the differences in the piezoresistance coefficient. This variation can be expressed in the formula:

$$\Delta R_c/R_c = -\Delta R_s/R_s$$

Therefore, the output voltage (V_{out}) can be obtained from the following

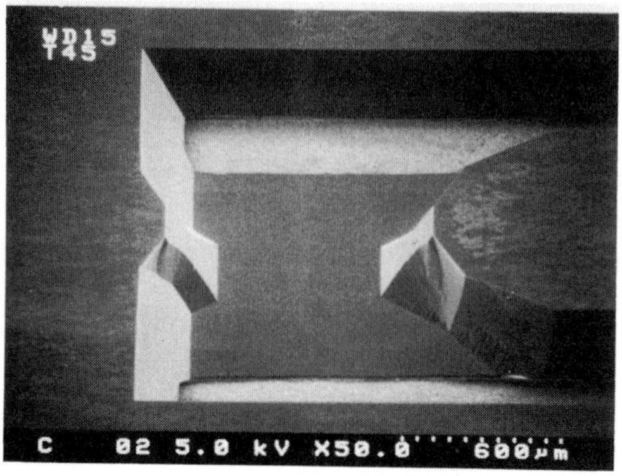

Figure 8.19 Bottom view of cantilever.

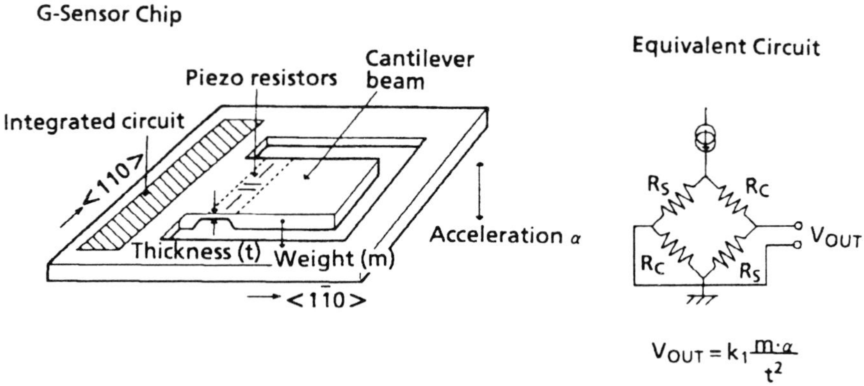

Figure 8.20 Cantilever mechanism of *G*-sensor chip and equivalent circuit.

Table 8.2 Piezoresistance coefficients of crystal orientations

	Crystal orientation	*Piezoresistance coefficient*
R_s	$\langle 110 \rangle$	$\pi = 1/2\,\pi_{44}$
R_c	$\langle 1\bar{1}0 \rangle$	$\pi = -1/2\,\pi_{44}$

theoretical formula:

$$V_{out} = K_1 m\alpha/t^2$$

where

$$K_1 = 0{\cdot}61 \times 10^6\ \mathrm{Vcm}^2/gG$$
$$m = \text{mass of the cantilever}$$

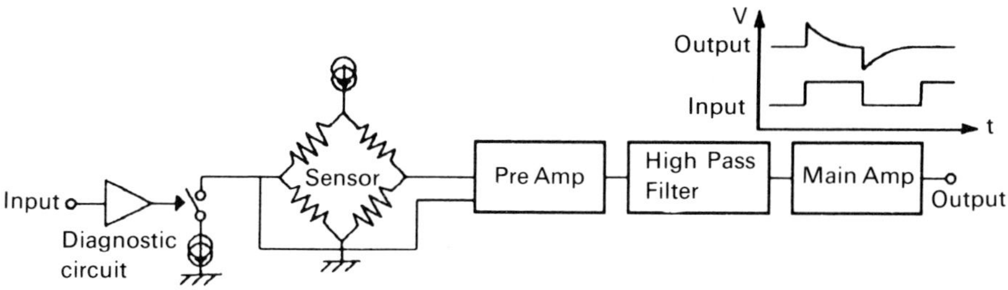

Figure 8.21 Block circuit diagram of sensor chip.

$$\alpha = \text{acceleration (deceleration)}$$
$$t = \text{thickness of the main part}$$
$$gG = \text{forcing fraction } g \text{ causing deceleration } G$$

This formula indicates that the output voltage is proportional to the deceleration exerted. The theoretical formula for rupture strength of the cantilever is expressed as follows:

$$\alpha_{\text{max}} = t^2 K_2 / m$$

where

$$K_2 = 0.87 \times 10^{-6}\, gG/\text{cm}^2$$

Since sensitivity and rupture strength contradict each other, the most suitable m and t are selected to satisfy the relation of these conflicting parameters.

Since the output voltage from this bridge circuit is normally very low, it must be amplified. Although such low signals are difficult to handle in an automobile environment, this problem is solved by forming an amplification circuit on the same silicon chip where the cantilever is fabricated. The block circuit diagram of the integrated G-sensor chip is given in Figure 8.21 and an overall photograph in Figure 8.22.

The output signal based on the deceleration signal detected by the bridge circuit is first amplified by the preamplifier, and then passes the high-pass filter, which eliminates errors and drift generated at the bridge section and the preamplifier, and finally the signal is amplified by the main amplifier so

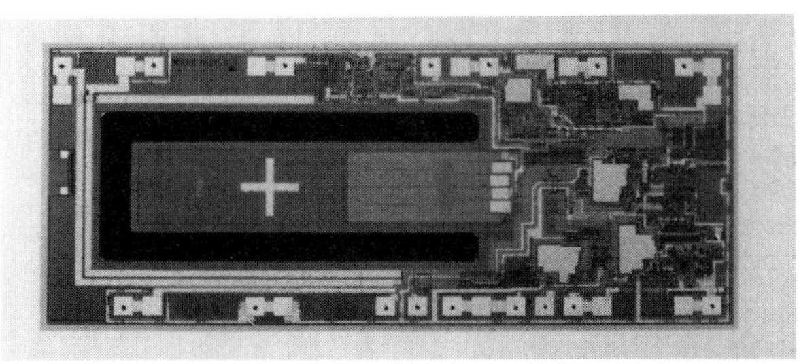

Figure 8.22 Sensor chip.

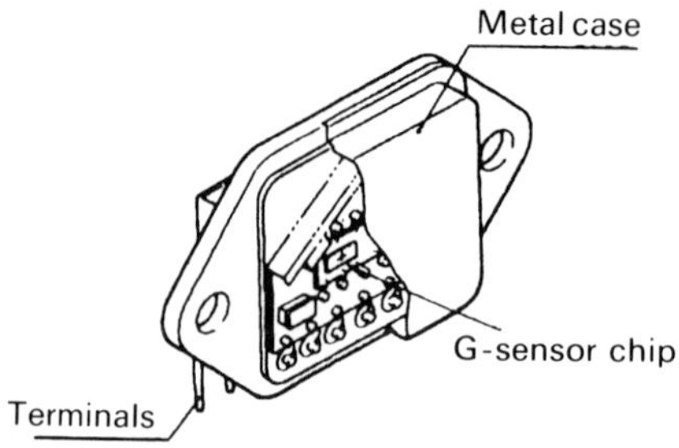

Figure 8.23 Perspective view of *G*-sensor unit.

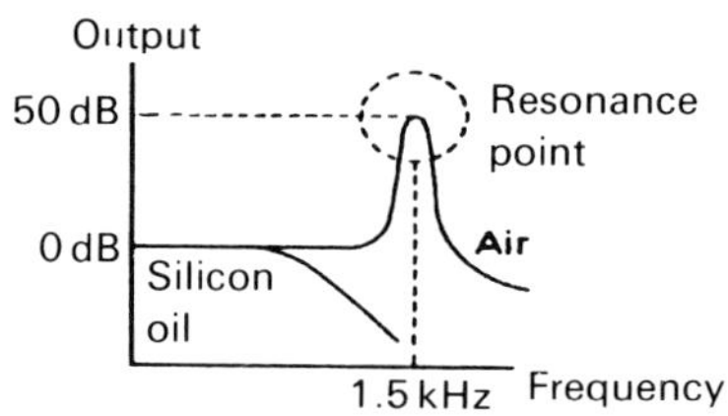

Figure 8.24 Frequency characteristic of *G*-sensor unit.

that the sensor can be tuned to the required sensitivity and '0 *G*' operating point. The integrated circuit has a diagnostic circuit that checks whether the sensor, including the bridge circuit, is operating correctly, Electrical imbalance is forcibly caused in the bridge circuit by the external diagnostic control signal to judge whether the circuits from the preamplifier to the output terminal function correctly. This control and judgement are operated by the diagnostic circuit located outside the *G*-sensor, and carried out each time the ignition is switched on.

Table 8.3 Maximum Ratings of *G*-sensor*

Item	*Value*
Supply voltage	7 V max.
Recommended operating voltage	5 ± 0.25 V
Operating temperature	-40 to $+85$ °C
Storage temperature	-40 to $+105$ °C
Acceleration limit	1 metre drop within 6000 G, 0.1 ms

* $T_a = 25$ °C

The cantilever of the *G*-sensor chip resonates at approximately 1.5 kHz. This is not desirable when taking into account the performances expected from a sensor and durability against excessive impact. To solve this problem, the entire *G*-sensor chip is installed in a hermetic sealing case filled with silicon oil, which not only improves reliability but also reduces resonance.

An example showing the specifications of the *G*-sensor designed, developed and produced with the above points in mind is given in Table 8.3 and Figure 8.25. The sensitivity, '0 *G*' operating point and filter characteristics can be determined freely. Furthermore, because of the silicon oil the performances of the sensor do not deteriorate even if it is dropped from 1 m.

The signal from the *G*-sensor passes the anti-aliasing filter and then enters the microprocessor which judges whether or not a collision has taken place. The signal is then processed by the microprocessor to decide whether or not the airbag should be operated. First, the signal passes the high-pass filter used to eliminate drift, and then it is divided in time window and an integral process is carried out. If the equivalent calculated velocity exceeds the constant value (V_{th}), an airbag-activating signal is output. The V_{th} is determined from

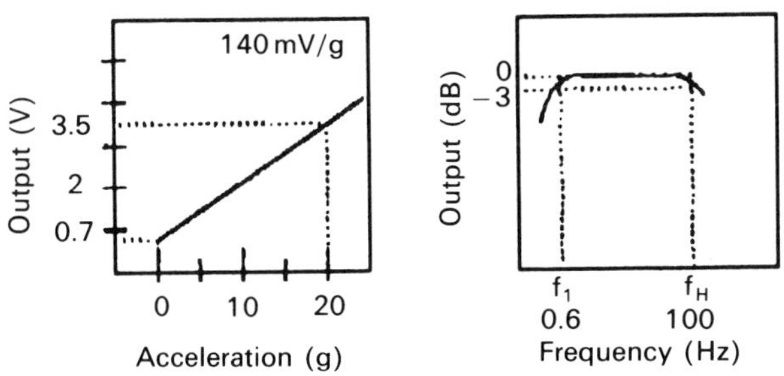

Figure 8.25 Output of *G*-sensor unit (a) versus acceleration and (b) versus frequency.

Table 8.4 Electrical characteristics of *G*-sensor*

Item	*Value*
Acceleration range	$-2 \sim +20\,g$
Sensitivity (1 *g*)	140 mV ± 5%
Output at 0 *g*(1 *g*)	0.7 V ± 42 mV
Nonlinearity	Less than 1%
Frequency response range	0.6 ~ 100 Hz
Directional sensitivity	$V_{0\cos}(30 \pm 5°)$
Transverse and vertical sensitivity	Less than 10%

* $V_{cc} = 5$ V, $T_a = -40 \sim +85$ °C

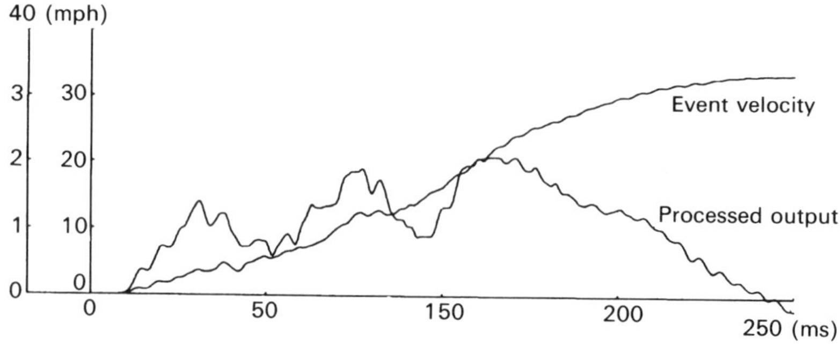

Figure 8.26 Simulation for high-speed collision.

the collision waveform, the G waveform at various running pattern and the construction of the vehicle.

Figure 8.26 shows simulation result for high-speed collisions.

Electromechanical type

The gas damping system composed of the damping disk, damping spring, damping cup and damping slide makes the sensor resilient to short-duration, high-acceleration pulses. The seal around the perimeter of the damping system compartment restricts movement of air into the space when accelerating force is applied to the mass. The air leakage into the space allows movement of mass only if the force is applied over a period of time. When the damping disk exceeds the predetermined distance the seal is broken, equalizing the pressure system with ambient pressure, and thus allowing the mass to accelerate and activate the switch. The equation of motion for the gas-damped sensor can be expressed as follows:

$$\ddot{s} = f_{\mathrm{E}}[G(l), G(r)] - f_{\mathrm{D}}(s) - f_{\mathrm{k}}(s) - f_{\mathrm{M}}(s) \tag{8.3}$$

where

$\ddot{s}$ = acceleration of the mass

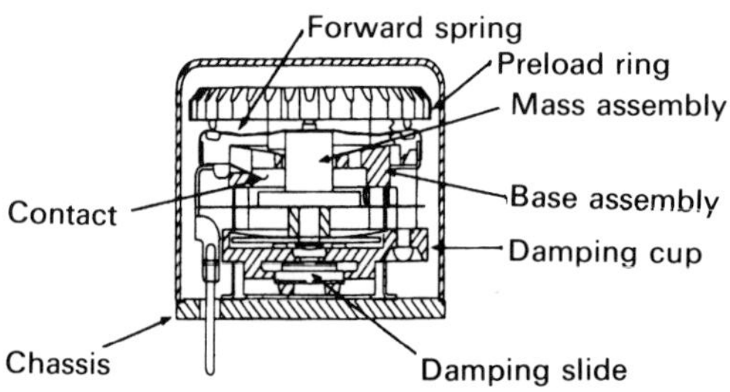

Figure 8.27 Gas-damped sensor.

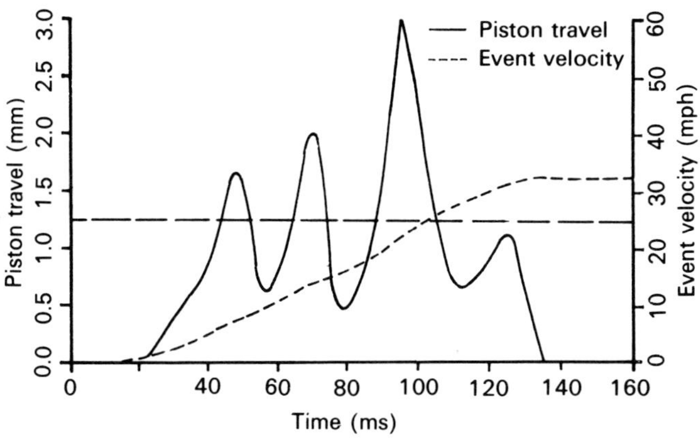

Figure 8.28 Simulation for high-speed collision.

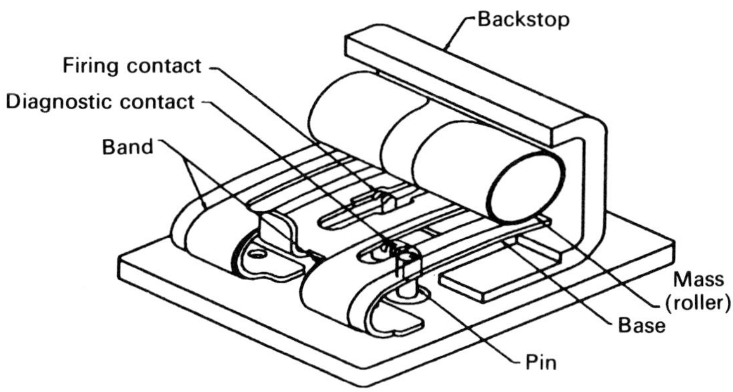

Figure 8.29 Downsized Rolamite sensor.

$f_E[G(l), G(r)]$ = acceleration function of external force (longitudinal and radial)
$f_D(s)$ = acceleration due to damping effects as a function of position
$f_k(s)$ = acceleration due to spring effects as a function of position
$f_M(s)$ = acceleration due to other moving components as a function of position

In the event of a severe crash, the deceleration of the vehicle will cause the mass to move forward from the rest to the actuated position, closing the firing circuit and activating the vehicle occupant safety device.

The gas-damped sensor can have both spring-mass and damping performance characteristics calibrated to meet different performance specification requirements for all vehicles.

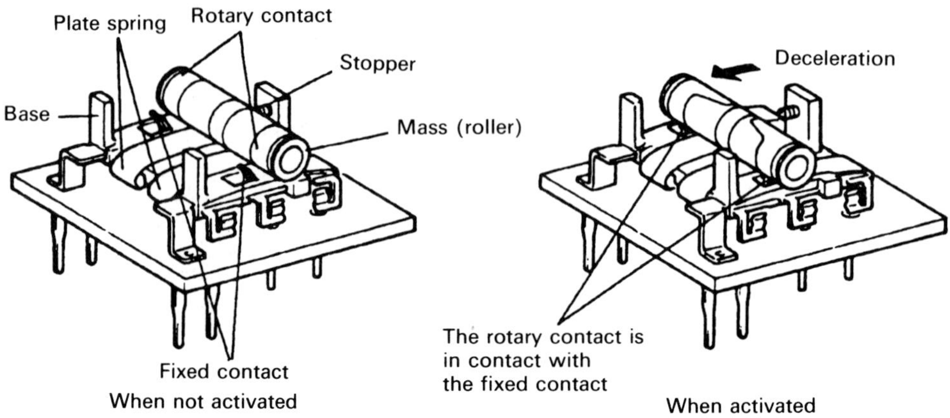

Figure 8.30 Dual-pole Rolamite sensor.

Safing sensor

Like floor sensors, various types of safing sensors can be used because they are normally installed inside vehicle and not within the crash zone. However, the electromechanical type is used since the safing sensor acts not only as a sensor to judge whether a collision has taken place but also as an inhibitor, to prevent the inflator from igniting immediately in case of an abnormality within the electric circuit.

Downsized Rolamite sensor

The structure is basically the same as that of the Rolamite used in floor sensors. However, this sensor is built into the electronic module, and an attempt has been made to lower the cost through reduction of the size of the sensor and the number of components, as well as the introduction of fully automated assembly operation.

Dual-pole Rolamite sensor

Although the structure is basically the same as that of the Rolamite used in floor sensors, the sensor is divided into two sections: the airbag and the seat belt pre-tensioner ignition circuit section, including the power source. This separation is made possible by dividing the plate spring into two and using the insulated mass and base.

8.5 ACTUATOR

Whether the mechanical type or electrical type of airbag system is used, the inflator basically has an ignition mechanism, enhancer and gas generator

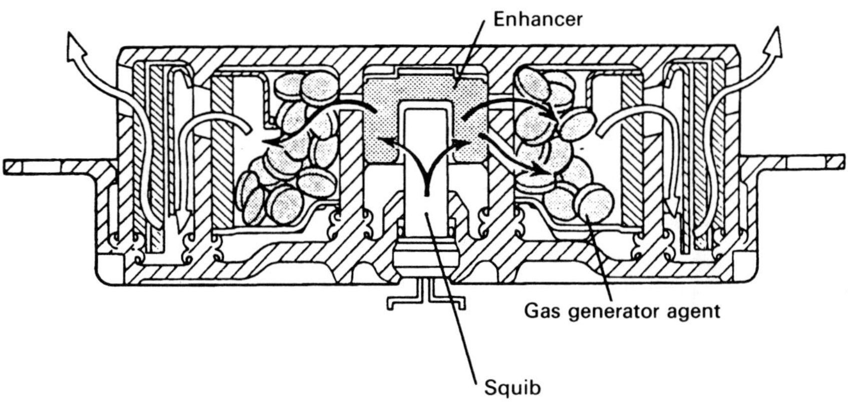

Figure 8.31 Inflator.

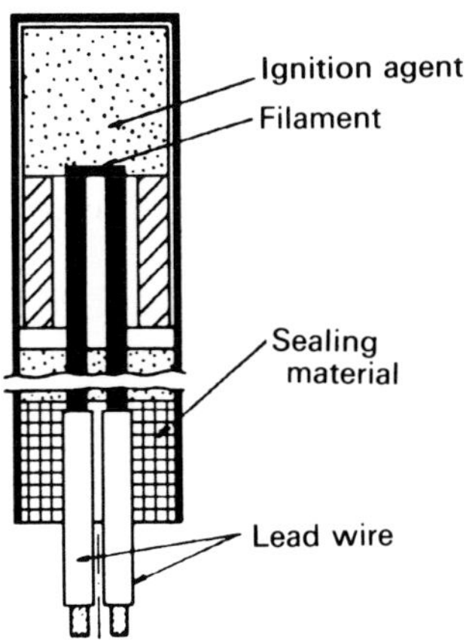

Figure 8.32 Squib.

agent. In this section, the inflator used in the electrical-type airbag system is explained.

8.5.1 Inflator

Once the squib is ignited, both enhancer and gas generator agent are ignited momentarily to produce sufficient gas to inflate the airbag.

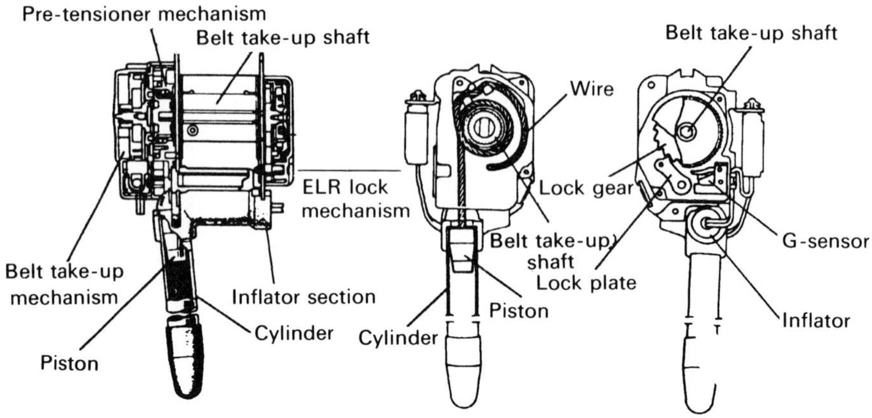

Figure 8.33 Seat belt pre-tensioner: (a) retractor; (b) pre-tensioner mechanism; (c) ELR lock mechanism.

8.5.2 Squib

When electrical energy conveyed to the squib via the sensor exceeds a certain level, the squib is ignited, though this depends on the efficiency of the electrical and heat conversion sections of the system.

8.5.3 Seat belt pre-tensioner

Although the basic function is the same as that of the inflator, the pre-tensioner operates the piston to absorb the slack in the seat belt, whilst the inflator inflates the airbag using generated gas.

8.6 CONTROL CIRCUIT

8.6.1 Block diagram

Figure 8.34 is an example of the entire block diagram, including the control unit (ECU) for the airbag system, which uses an airbag for the driver's seat and seat belt pre-tensioner for both driver and passenger seats.

The front sensor located at the front of the vehicle, squib for inflating the airbag, squib for seat belt pre-tensioners, indicator lamp for the driver etc. are all connected to the EUC and controlled.

The ECU consists of an electric floor sensor, safing sensor, a back-up power source which ensures operation of the squib, and a diagnostic circuit which detects abnormalities in these components and gives a warning to the driver.

The electric floor sensor located inside the EUC and two front sensors located at the front of the vehicle are all connected in parallel. Additionally

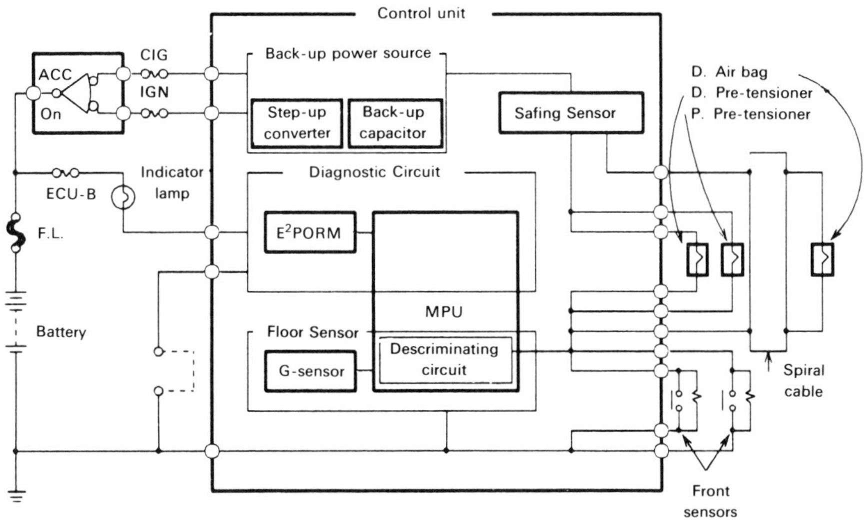

Figure 8.34 Block diagram.

these three sensors, squib and safing sensor are connected in series. This means that when one of these three sensors and the safing sensor output the ignition signal at the same time, the airbag and seat belt pre-tensioners are activated simultaneously.

8.6.2 Back-up power source

Back-up capacitor

In case of a collision, the airbag etc. should be activated even if an abnormality occurs at the same time in the power source. For this purpose, this system

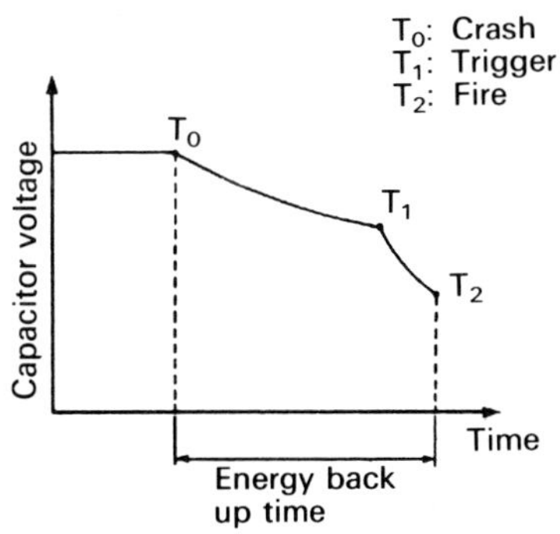

Figure 8.35 Energy characteristic.

employs a back-up capacitor inside the EUC to ensure the operation of the airbag.

The duration (back-up time) allowing the activation of the airbag by the back-up capacitor, in case of an abonormality in the power source, is determined from the time required for the specific collision and the standby time required before discharge of the back-up capacitor, in order to ensure safety during the rescue of passengers from the crashed vehicle.

Not only should the capacity of the back-up capacitor be sufficient to hold energy to ignite the squib, but also the internal impedance of the capacitor should be low enough for the required current and the capacitor itself should have a long life.

Step-up converter

The airbag etc. should be capable of being activated properly and constantly within the voltage range of the electrical power of the vehicle under normal conditions. Therefore, to maintain sufficient energy in the back-up capacitor, the ECU has a step converter which can operate with a wide range of input voltages, as well as supplying efficient output voltage.

8.6.3 Diagnostic circuit

A high level of reliability is naturally required for the airbag system. In addition, should any abnormality occur regarding the operation of the airbag, it should be detected quickly and a warning given to the driver. The ECU used in Toyota's vehicles also employs this type of diagnostic circuit. Breakdowns, such as those which may cause malfunction of the airbag at the time of a collision or one which eventually activate the airbag by mistake, due to accumulated breakdowns during normal driving, are detected in the first stage of breakdown and an indicator lamp is lit to inform the driver.

Details of the breakdowns detected by diagnosis are coded and memorized by the internal electrical circuit. These data are analysed at the repair garage to enable appropriate and quick repair.

8.7 CONCLUSIONS

In this chapter the supplemental restraint system has been explained with the main focus on various sensors. We believe that the required specifications with respect to construction of the vehicle, location of the sensors, vehicle destinations, firing electric current, environmental temperature and recycling aspects are satisfied.

For the mechanical-type airbag system, sensors compact enough to be installed in the steering wheel pad have been developed.

Since front sensors used in the electrical-type airbag system are installed in the vicinity of the engine housing, which is located in the crash zone and which generates electrical noise, electromechanical types such as the Rolamite or rotary type have been developed.

The electromechanical type has been developed for safing sensors to prevent the inflator from igniting by mistake due to a breakdown in the electric circuit. Since the safing sensor is to be installed inside the control circuit (ECU), an attempt to make it compact has been made.

Various types of floor sensor, whether of the electronic or electromechanical type, can be used, because they are normally installed inside the vehicle and not within the crash zone. However, as they are installed in the same location as the safing sensor, special care has been taken to prevent the inflator igniting by pulse impact.

PART FIVE

Driver and Occupant Information Systems

CHAPTER 9

Sensor techniques using image processing for driver behaviour study

9.1 INTRODUCTION

During recent years there has been an increasing amount of interest in the possibilities offered by information technology applied to the automobile industry. Concepts such as intelligent vehicle highway systems (IVHS), including driver support and information systems, advanced vehicle control systems, etc. will have strong effects on driver security and on highway infrastructure usage optimization [1].

To determine the effectiveness of these developments, it is necessary to study their influence on driver behaviour, for which field trials are essential [2]. Field trials, in turn, require the availability of systems and tools for studying driver reactions to a variety of real and simulated situations.

Two approaches can be taken for obtaining this kind of information: the use of vehicle simulators and the adaptation of standard vehicles for gathering the most significant information while being driven. This last option was chosen for the Argos project, in which a medium-sized car was adapted for recording driver, vehicle and environmental information during real driving sessions. Argos was developed at the Computer Science and Engineering School of the Polytechnical University of Madrid (Facultad de Informática – UPM) for the Spanish Traffic Department (Dirección General de Tráfico).

The objective of the Argos project was to develop a tool that would permit the registration of selected information and the laboratory reproduction of driving sessions under real road conditions. For that purpose, it was necessary to develop a sensory and data acquisition system able to collect the required information, and a laboratory reproduction system able to present and analyse previously recorded sessions.

The system developed in the Argos project has an open and modular structure, in order to allow easy system upgrades and modifications. The data acquired in each driving session can be grouped as follows:

- driver actions on the vehicle, such as lights, horn, blinkers, pedals and steering wheel positions, etc.

- driver data, such as eye point-of-gaze, skin resistivity, electromyogram, etc.
- vehicle conditions, such as speed, engine turning rate, vehicle lateral position on the road, etc.
- environmental conditions, such as environmental illumination level, temperature, humidity, etc.

The data acquired are presented interactively at session reconstruction time on a high-resolution graphics terminal. Video data recorded while driving can also be displayed synchronously with the rest of the data on an additional video monitor.

The rest of the chapter is organized as follows. Section 9.2 describes the global system structure, decomposed into its mobile and laboratory subsystems. Sections 9.3 and 9.4 focus on two sensors based on image-processing techniques: SAISMI, devoted to the acquisition of driver eye point-of-gaze, and SAIDDL, responsible for measuring lateral road vehicle position.

Sections 9.5 and 9.6 focus on system use aspects: the integration of the information supplied by each of the sensors, and the laboratory subsystem exploitation, from the user's point of view. Sections 9.7 and 9.8 describe future improvements under way and the conclusions.

9.2 SYSTEM STRUCTURE

Argos is composed of two independent subsystems, each of which performs one of the two major functions of the system. The first subsystem, the Signal Acquisition and Recording System (SARS), is responsible for gathering and recording the information required to characterize the driver behaviour. SARS is mounted on a medium-size car with double pedals. The second subsystem, the Signal Reproduction and Analysis Laboratory System (SRALS), processes all the information recorded by SARS, allowing driving sessions to be reproduced in a fully user-controlled way [3].

The following paragraphs provide an overview of both subsystems, and their hardware and software structure.

9.2.1 SARS subsystem

In order to achieve a complete characterization of driver behaviour, SARS is provided with a wide range of sensors. They allow three types of measurements to be recorded:

- driver operations when driving and vehicle status, including vehicle speed, pedal movements, and use of lights;
- driver's physical conditions, monitored by means of several specialized biomedical sensors;
- environmental conditions of the driving session, such as temperature, humidity and luminosity.

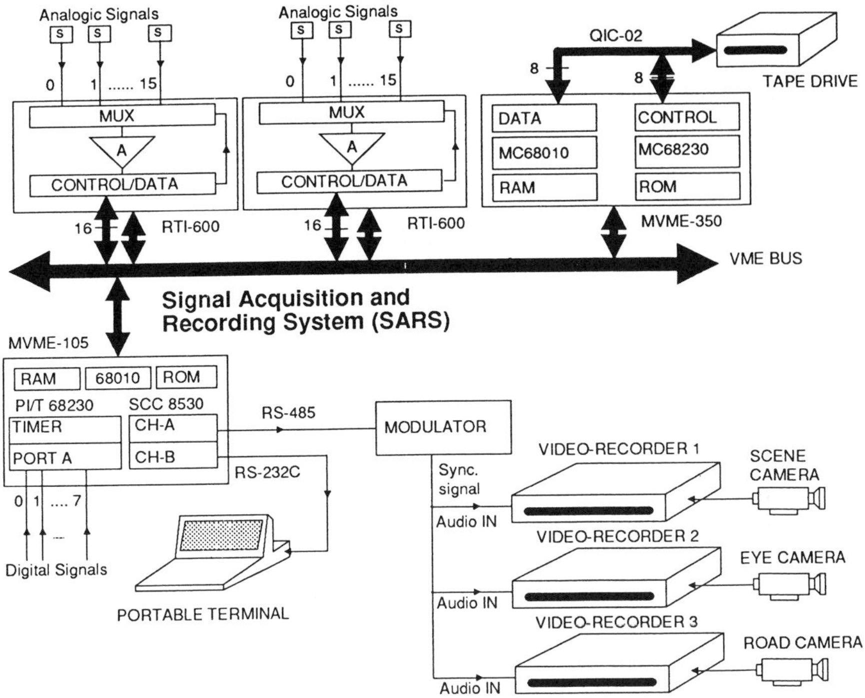

Figure 9.1 Structure of SARS subsystem.

In addition to these sensors, SARS is also equipped with three CCD microcameras. They allow the determination of the vehicle lateral road position and the driver's point-of-gaze (where the driver's eyes are fixed), by applying image-processing techniques [4]. Two cameras are attached to the driver's head with the help of a lightweight helmet, and they respectively capture the driver's field of vision and the pupil position in the driver's eye. The third camera is attached to the top of the car and captures a frontal panoramic view of the road in order to define the position of the vehicle on it.

The heart of the SARS subsystem is a data acquisition embedded system, composed of several VME boards [5] (Figure 9.1). It periodically samples each sensor and records sampled data on a streaming magnetic tape to be processed by SRALS. Likewise, images provided by the microcameras are stored on three portable videorecorders. Synchronization of both data and video images is achieved by adding some timing information to every data record and every stored image.

In order to allow the system to be monitored, an intelligent terminal is attached to the data acquisition system. This terminal can also be used to enter some special codes for miscellaneous noteworthy events and to note any useful comment that may be helpful for the analysis and evaluation of

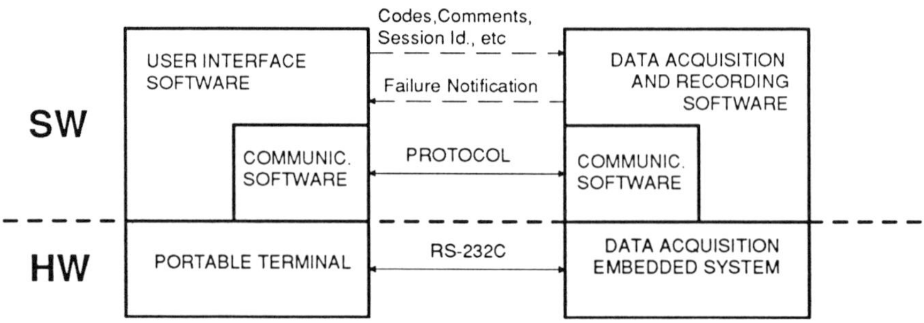

Figure 9.2 SARS software subsystem: HW = hardware; SW = software.

SARS recorded information. Both codes and comments are stored on the streaming tape together with the information provided by the sensors. The terminal is used by the passenger who must go with the driver to introduce comments, pauses or events, to set up the testing equipment, and to control driver's actions by means of dual controls.

The system is powered by five rechargeable lead batteries, in order to obtain the required voltages. This allows five hours of continuous operation.

The software structure of SARS is closely tied to its physical structure, as shown in Figure 9.2. It is made up of three major components, referred to as Data Acquisition and Recording Software, User Interface Software, and Communication Software.

The Data Acquisition and Recording Software runs on the data acquisition system's microcomputer and is responsible for sensor sampling, sample recording, timing information updating, and failure detection. The core of this software component is a non-pre-emptive priority-based scheduler, which allows all system tasks to be carried out without losing any relevant information. Highest priorities are given to the most critical tasks, such as sensor sampling and timing information recording.

The User Interface Software runs on the intelligent terminal, and provides monitoring and dialogue options such as signal tuning, biomedical signal real-time displaying, session identification, codes and comments input, guide screen displaying, and failure notification.

The Communication Software runs on both the intelligent terminal and the microcomputer in charge of data acquisition and provides a reliable communication link between them.

9.2.2 SRALS subsystem

Driving sessions are laboratory reproduced on the SRALS subsystem, which is composed of the following elements (Figure 9.3):

- A high-performance MASSCOMP 6400 biprocessor computer with

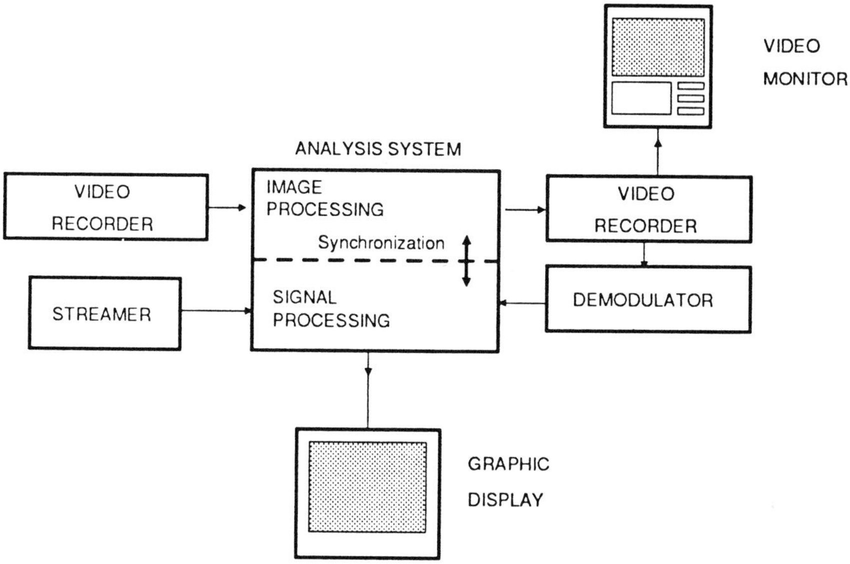

Figure 9.3 SRALS general structure.

graphics hardware, vector facilities, and floating point accelerator. The computer is used to analyse the signals acquired by SARS. A demodulator, attached to the computer, uses the timing information recorded in the videotape audio channel to synchronize data and images.
- A videorecorder and player, which is connected to the computer via a serial interface. It allows the user to control the driving session reproduction through the computer.
- Four image-processing boards to analyse the images acquired with the helmet and panoramic cameras.

SRALS software is structured using the following modules, depending on the performed functions (Figure 9.4):

- A user-friendly Interface Module allows the parameters set up to analyse the session.
- A Representation Module displays the signals synchronized with the video images.
- A Statistic Module provides a set of functions for automatic marking and statistical studies.
- A Utility Module allows the creation of a set of files that will be used in the representation in order to get synchronization.
- A Video Module performs the more common video activities (Play, Rewind, Search, etc.).
- A File Manager allows the other modules to access the session information.

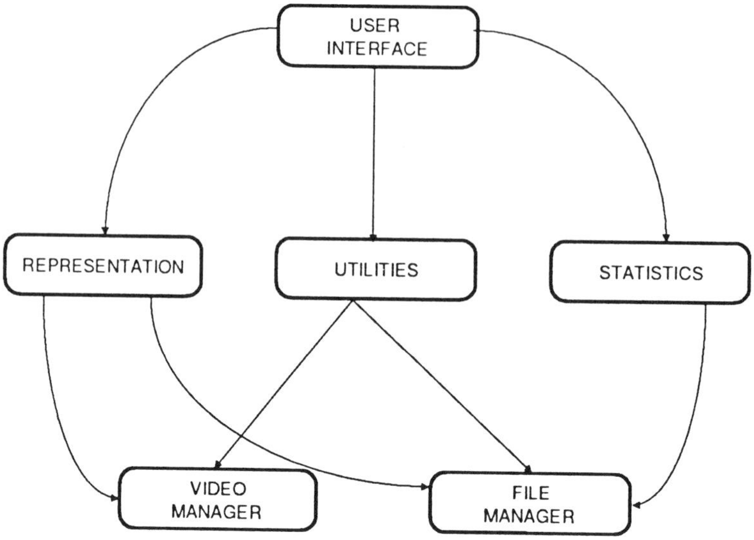

Figure 9.4 SRALS software structure.

9.3 SAISMI MOBILE SUBSYSTEM

9.3.1 General description

The problem of point-of-gaze determination has been addressed with different approaches. The SAISMI module is based on the analysis of two sets of images to determine the actual point at which the driver is looking, since the point-of-gaze depends both on the head orientation and on the pupil position within the eye. Therefore, SAISMI's mobile subsystem takes panoramic and eye images that indicate the scene at which the head is pointing and the point at which the pupil is pointing. These two image sequences are also taped on portable VCRs, with synchronization signals added during the taping process.

The cameras and their associated equipment are mounted on a helmet, as shown in Figure 9.5. The system's major components are:

- *Scene camera.* This is rigidly attached to the side of the helmet. This configuration was preferred to others because it is less cumbersome to the driver, even though it will introduce parallax errors. Other configurations required the placement of objects in front of the driver. An automatic iris adapts the camera to the light changes occurring during the driving session.
- *Pupil camera.* To be able to take eye images from a frontal direction, a plate, which is transparent to visible light but reflective to infrared, is placed

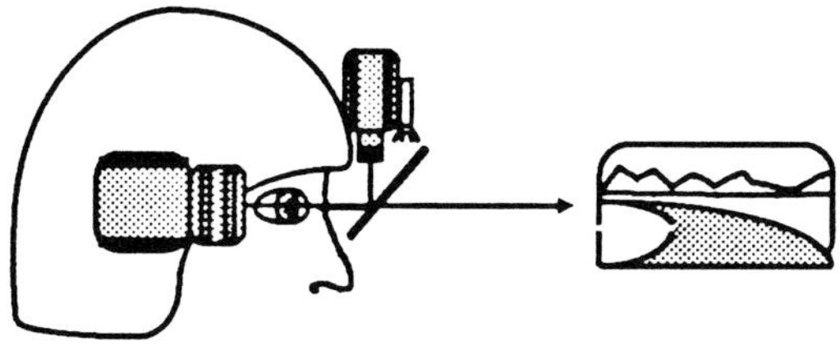

Figure 9.5 SAISMI image acquisition subsystem.

at 45° between the pupil camera and the driver's eye. The eye is illuminated with infrared radiation, and a filter blocks out visible light from the camera sensor.

- *Pupil camera plate.* Designed on surface-treated polycarbonate, the plate allows the obtention of eye images without noticeably limiting the driver's vision.
- *Pupil camera filters.* To minimize environmental light interference on the eye images, only radiation of the near infrared spectrum (of the wavelength of the illumination device) is used for imaging. Of the filters tested, a combination of a narrowband interference and gelatine wratten filters gave best results.
- *Eye-illuminating system.* To minimize lighting changes due to environmental changes, a constant level of infrared ($\sim$ 850 nm) is projected on the driver's eye by an infrared source and appropriate optical system. Solutions based on LEDs and laser diodes were tested. Coupling the light emitter to a narrowband interference filter minimized environmental light influences, making an autoiris unnecessary.

Two objectives were set for the design of the helmet ensemble:

- Make the system as light and comfortable as possible.
- Limit the environmental influences on the quality of the images.

The first objective was very important: the system had to be light in order to minimize driver fatigue; compact, to allow the driver's head to move freely in a restricted area; and comfortable, in order not to affect the driver's general behaviour. Budget restraints forced the use of commercially available CCD cameras.

The second objective was also essential for allowing the interpretation of the pupil images in a reliable way. Ergonomic considerations advised against the use of an automatic iris on the pupil camera. However, the amount of near infrared light present outdoors (sometimes there is even direct sunlight on the driver's face) and the speed of light changes, e.g. going through tree-shaded areas, required a careful design of the pupil camera lighting and filtering systems.

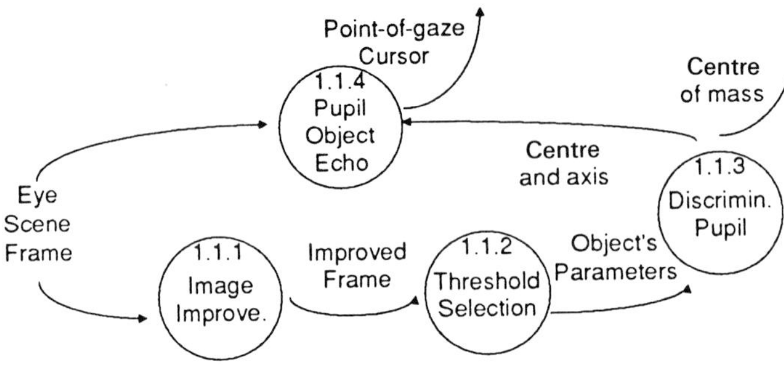

Figure 9.6 Image analysis substages.

9.3.2 SAISMI image-processing algorithms

The process of eye point-of-gaze determination starts with the completion of a data acquisition session, from which two image sequences are obtained with scene and pupil images on two videotapes.

The laboratory data preprocessing stage assigns a number to each of the videotape's frames, determining also the presence of any time lapse between tapes in order to allow for the synchronization of both sequences' analysis processes.

The image analysis stage can be decomposed into three substages as shown in Figure 9.6:

- *Image improvement.* Its goal is to modify the images provided to the analysis module by avoiding images which are under- or overexposed and by maintaining a sufficient contrast level. An algorithm based on histogram modification has been used as an improvement technique, but it is limited to a small window to determine the parameters needed in the enhancement of the pupil image.
- *Threshold selection.* This stage chooses a binarization level appropriate for pupil detection. The threshold must be placed between the iris and pupil grey levels. The histogram information obtained before is used at this stage.
- *Pupil discrimination.* After selecting a threshold and a processing window which are large enough to ensure that the pupil will remain within them, the image is analysed by the Max Video boards. The hardware performs a connectivity analysis and computes a number of geometric features. A number of blob attributes are used to discriminate the pupil from other blobs present on the image. The most reliable features are based on grey level intensity, roundness, non-contiguity to the window limits, and an area which is similar to that which was previously measured.

When the analysis stage is finished (Figure 9.7), the next step is the transformation of the pupil's centre of area to the coordinates of the cursor

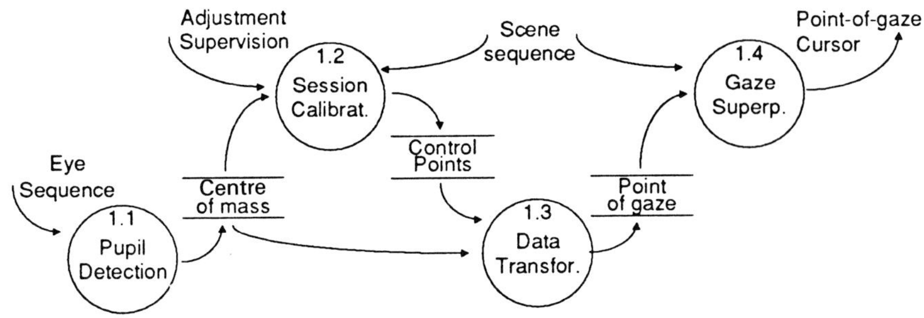

Figure 9.7 Eye point-of-gaze measurement process.

specifying eye point-of-gaze. This transformation is done using a numerical function computed at calibration time.

The calibration requires the recording of special images at the beginning of the image acquisition stage, during which the driver looks at a number of known points. Later in the laboratory, these known points are used to adjust the numeric function from the correspondences between pupil positions and known points by an interactive process.

9.4 SAIDDL MOBILE SUBSYSTEM

Lateral vehicle position is determined by measuring the vehicle's position with respect to the painted line on the shoulder of the road. Panoramic road images are taken by a camera attached to the car and are taped on a portable VCR. The camera uses an autoiris lens to compensate for variable lighting conditions.

The process of vehicle lateral position measurement starts with the completion of data acquisition and preprocessing stages. Coding the video frames is still necessary, because the measurements computed have to be related to the actual time when the image was taken.

The distance measurement is made by determining the position of the painted line on the shoulder of the road, although other references could have been selected. Again, a calibration stage is used to translate image pixels to real-world distances.

The following sections describe the stages in which the lateral position measurement is decomposed.

9.4.1 Contrast enhancement

The images to be analysed might present sudden illumination changes. Sometimes, the contrast between the painted line and the road is very small, due to sunlight reflections, presence of dirt, or poor paint condition.

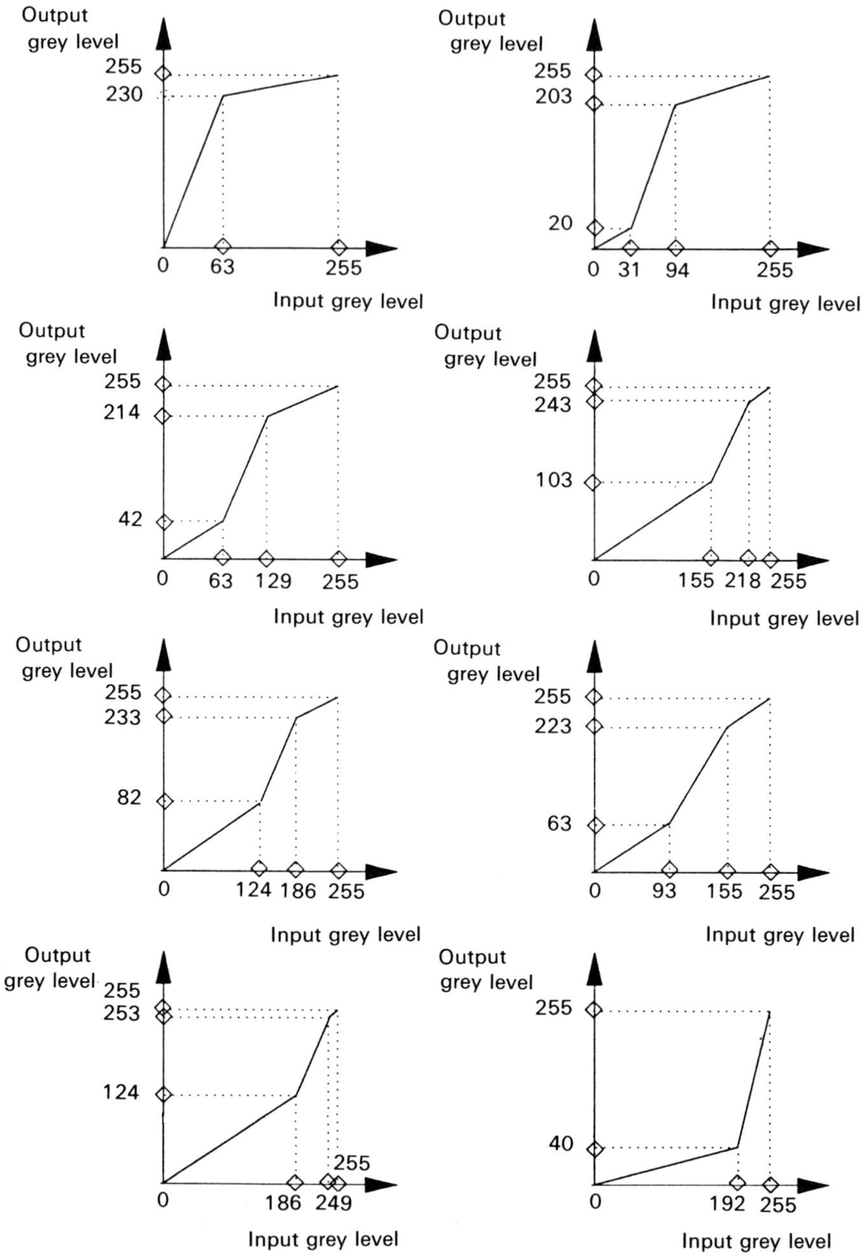

Figure 9.8 Look-up tables.

Figure 9.9 Original road image.

Figure 9.10 Image with enhanced road-to-line contrast.

In order to enhance the difference between the road and the painted line intensity levels, a contrast-stretching histogram modification technique is used. To be able to attain real-time response, eight predefined look-up tables are loaded using the image acquisition and processing hardware (Figure 9.8). The selection is made by sampling and analysing a small number of points around the expected painted line position.

Figures 9.9 and 9.10 show an image of a road with an appreciable amount of dirt on and around the painted line, before and after undergoing a contrast stretching operation.

9.4.2 Painted line detection

The process of the detection is complicated by problems such as the presence of shadows or dirt, variable illumination and atmospheric conditions, and variable paint condition. Most frequently, the line is located around a small image region, and this *a priori* knowledge can be used at line detection time. However, there are periods of time when sharp changes occur, as on sharp turns, or when passing other vehicles.

Three main techniques have been attempted for this stage, based on edge detection, template correlation, and texture analysis. Although other similar systems are based on thresholding the image [6], we found it difficult to perform reliable line detection under unrestricted conditions.

- *Edge detection.* Since the line produces an abrupt grey level change, edge detection techniques seem applicable. The method can be refined, introducing aspects such as the width, road and line grey levels, gradient magnitude on the road–line boundaries, and gradient direction.

  In general, edge detection methods work well on well-contrasted images. Shadows projected on the road (e.g. from trees), low-contrast images or presence of dirt on the shoulder of the road tend to produce erroneous results. In particular, dirt patches or other areas, such as those shown in Figures 9.9 to 9.13, might present grey levels and gradient magnitude and direction similar to the expected values. In fact, they can even be closer to the ideal values than the actual line's values, producing incorrect detections.
- *Correlation.* A line-like template was applied for locating the line position on the shoulder of the road. The results produced by this method were similar to those obtained using a gradient operator. The method worked well on well-contrasted images but performed worse for low-contrast images. In particular, there were cases where dirt on the roadside would give stronger responses than the actual line.
- *Texture analysis.* The techniqes described above are based on searching the image for regions showing features which are similar to the painted line. The problem with this approach is that some regions can be more 'line-like' than the actual painted line, as explained before.

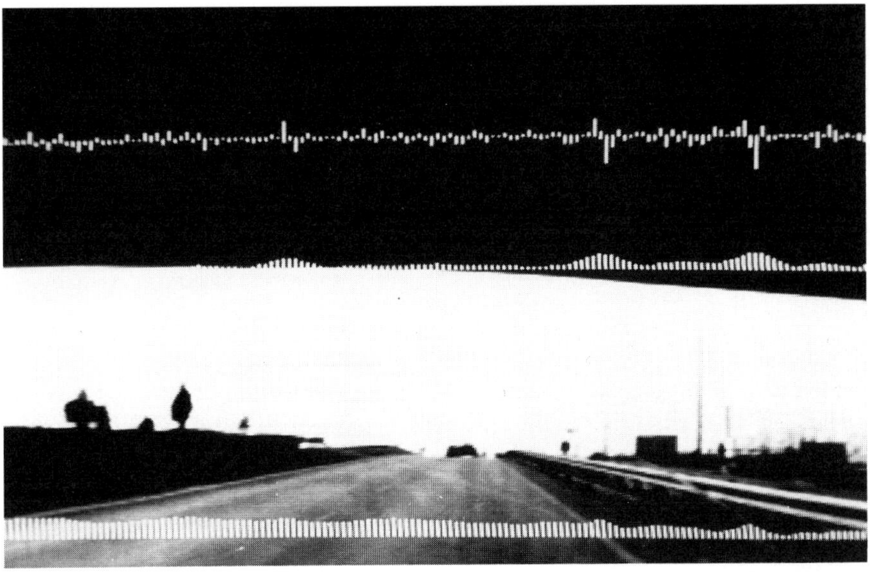

Figure 9.11

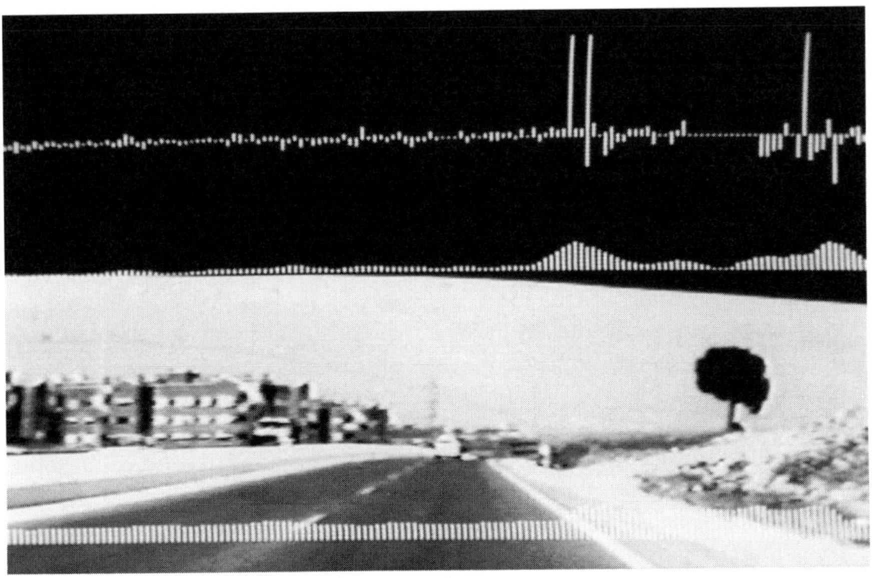

Figure 9.12

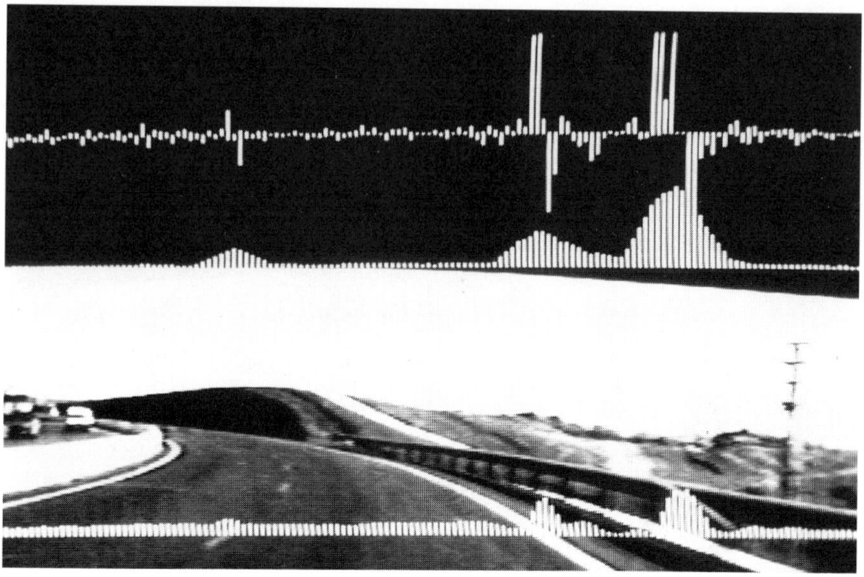

Figure 9.13

An alternative approach is to look for features that can help discriminate the shoulder of the road from the rest of the road or the ground next to the shoulder. Texture can be used for this purpose, because there is a strong texture difference between the road and other kinds of surfaces that might appear in the image. There is even a texture difference between the road and the shoulder, since the latter tends to accumulate dirt (i.e. in the areas where dirt patches make line detection more difficult).

A variety of approaches is described in the literature for analysing texture [7]. A major problem in our case was the requirement of working in near real time with the available hardware. This restricted the methods that could be tested. In particular, simplified versions of the GLRL (grey level run length) method [7] and edge density methods [8] were selected.

A method that has given good results is based on computing the gradient magnitude standard deviation on a small window centred on each point. The window can be rectangular or trapezoidal to adapt to the painted line perspective. Road areas give small standard deviations, while mixed road/line areas, shoulder and terrain give larger standard deviations. A typical pattern for the painted line is an area with low standard deviations followed by a local maximum, and a narrower area with relatively high standard deviations, possibly with other local maxima.

Figures 9.11, 9.12 and 9.13 show three examples, together with the results obtaining using the best three methods. There are three graphics superimposed on each image showing (from top to bottom) the computed gradient magnitude, gradient magnitude standard deviation, and correlation value. The line analysed is the image line over which the bottom graphic is displayed. Gradient standard deviation values are smoothed to decrease the number of local maxima found.

These images show some typical problems: low contrast, due to sunlight reflection on the road surface or to dirt accumulation, or false detections due to dirt or other areas on the shoulder.

9.4.3 Line discrimination

After detecting candidates for the painted line, a discrimination process is necessary. Three approaches are possible:

- *Line features.* Data such as line width, position, grey level, gradient magnitude, direction and texture have already been mentioned.
- *Feature distribution on the scanned line.* The shoulder of the road is a source of errors. The line can also be characterized as having a wide area on its left, where the features searched for give small responses, and a narrower area on its right giving higher responses (Figures 9.11, 9.12 and 9.13).
- *Expected features.* Predictions can be made about where the painted line is located or about the expected values of the line features used for discrimination, such as line width, grey level, etc. These predictions can be used for discriminating among possible candidates.

Other algorithms use simpler detection methods over larger image areas [6].

9.4.4 Image-to-road distance translation

A calibration stage is used to determine a calibration chart, from which actual distances are estimated. More sophisticated techniques are feasible if more accuracy is desired.

After determining the vehicle lateral position, a file is created containing the measured values. Since those will be used at a later stage, a filtering process can be applied to discard values which are incompatible with the rest of the measurements.

9.5 INFORMATION INTEGRATION

As stated in previous sections, Argos is divided into two subsystems which are not physically connected. The SARS subsystem would not be useful if it could not extract conclusions about the data gathered by the vehicle-embedded computer. Two aspects will be explained in the following paragraphs:

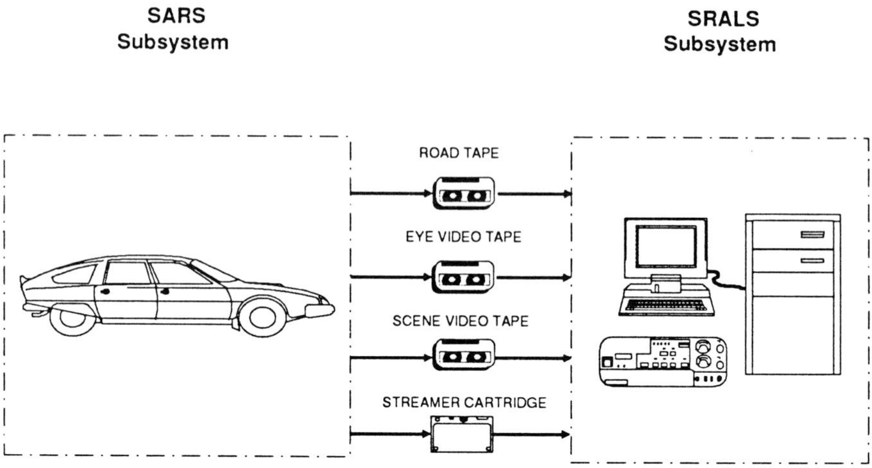

Figure 9.14 Information flow from the mobile to the laboratory subsystem.

- the information exchange between the laboratory and the mobile subsystems;
- the exploitation of the gathered information.

The integration of the information gathered by SARS is performed using two kinds of magnetic supports (Figure 9.14):

- a computer cartridge;
- three video tapes.

The cartridge is a streaming magnetic tape where the vehicle-embedded system stores the following information:

- *Data sampled by the sensors.* These data are sampled at two rates: 20 and 200 Hz. The first rate is used for the signals that do not change their value rapidly. That signal set is composed of the environmental and vehicle signals listed in Table 9.1. The signals sampled at 200 Hz are basically the driver's biophysical signals, as shown in Table 9.2. Both sampling rates allow the acquisition of 65 000 data items per minute.

  Every record of both signal types has a timing information field. This variable is incremented ten times in a second for time measurement, and it is used for synchronizing in the laboratory the information taped in the cartridge with the images recorded by the videorecorders at SARS.
- *Identification record.* The driver's companion must introduce a set of information items for identifying the driving session, such as session date, starting time, and, optionally, some comments for identifying the purpose and the conditions of the session.
- *Comments and operator codes.* In a driving session some incidents may

Table 9.1 Environmental and vehicle signals sampled at 20 Hz

Signal	*Units*
Speed	km/h
Engine turning rate	rev/min
Steering wheel	Degrees
Accelerator position	%
Clutch position	Digital
Brake	%
High beams	Digital
Low beams	Digital
Parking lights	Digital
Right blinkers	Digital
Left blinkers	Digital
Horn	Digital
Environmental temperature	°C
Environmental humidity	%
Environmental luminosity	Lumen

Table 9.2 Driver's signals sampled at 200 Hz

Signal	*Units*
Electrocardiogram	mV
Electroyogram	mV
Electroencephalogram	mV
Skin resistivity	mV

occur that are especially interesting at session reconstruction time. These events are noted in the streamer cartridge as predefined operator codes such as sudden braking or driver's heedlessness. Since there might be situations which differ from those previously defined and coded, the system provides the user with the possibility of introducing text comments that will be stored in the streamer tape.

In addition to the signals sampled and recorded on the computer cartridge, three videotapes are recorded by SARS. The first two are used for determining driver's eye point-of-gaze, while the third one is used for determining the vehicle's relative position with regard to the painted line on the shoulder.

The main problem arising from the different information sources was the need to synchronize them all. This was solved by recording the same timing information both in the streamer (as described above) and the videotapes through a modulated signal introduced on one of the audio channels. Essentially, the synchronization process consists of recording at 1200 bits per second two bytes containing the timing information with the format

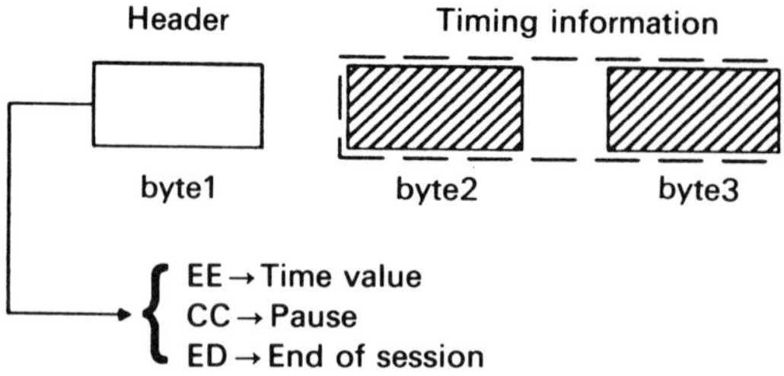

Figure 9.15 Timing information format.

shown in Figure 9.15. The size of the synchronization information (2 bytes) allows continuous recording during 110 minutes in a driving session.

Since the laboratory subsystem is based on a general-purpose computer running UNIX, it is difficult to directly perform real-time studies from the data registered in the streamer and videotapes. Therefore, a data transformation process is executed on the recorded data in the following steps:

1. A set of UNIX files is created from the cartridge tape.
2. From the videotapes, information such as the driver's point-of-gaze and lateral position is computed.
3. A file is created containing information for the synchronization between data from the sensors and images stored in the videotapes.

The first stage requires the sequential reading of the streaming tape, creating simultaneously two files: the first contains all of the signals sampled at 20 Hz, and the second file contains the signals sampled at 200 Hz. In this step the system also computes some signals which have not been sampled but will be treated as if they had been sampled, rather than computing them during session reconstruction time. The computed signals are listed below:

- *Linear acceleration.* This is computed from the measured vehicle speed variations.
- *Gear position.* Determining which gear is selected directly is feasible, but complex and expensive. The approach chosen was to deduce the selected gear from the car speed, clutch position, engine turning rate, and gear reduction ratio (as provided by the vehicle manufacturer).

The second stage is to obtain a set of files that allows the synchronization between sensor data and images, which is done by creating a file with a synchronism field. For every video frame of the scene image the system creates a record in the file with its corresponding timing data. Moreover, the signals listed below, computed from the images recorded on the videotapes, are also stored in this file:

- *Point-of-gaze coordinates*, defining where a cross-hair will be displayed on the driver panoramic camera images.
- *Lateral vehicle position*, as measured from the painted line position.

9.6 SYSTEM EXPLOITATION

The previous sections described the system and how both subsystems intercommunicate. This section will explain how data can be used, and the possibilities provided by the system for their analysis.

Three main functions are provided by the SRALS subsystem:

- real-time session reproduction;
- automatic study of gathered data;
- other possibilities.

9.6.1 Real-time session reproduction

In the laboratory, an initial study of the driving session can be done by having the laboratory subsystem repeat every action performed by the driver and each event which occurred during the driving process. Usually, there is no need to reproduce the entire session. Therefore, the system provides the possibility of selecting a part of the session, called a segment, in order to be analysed by the user. The segment is identified by two time values (the starting and the ending times). Each of them can be selected independently using one of the following methods:

- *Time.* The user can select the segment's time boundaries by introducing in the computer an absolute time which indicates how long after the beginning of the session the segment will begin and end. For example, the segment starting time will be one minute and 30 seconds after the beginning of the session if '1:30' is the time value introduced in the computer. The ending time can also be set by indicating how long the segment will last. When the session is stopped in order to change some session reproduction parameters, the system saves the current time, and the user can change the starting and/or the ending time in relation to this. For example, when the user introduces '+0:30' for the ending time of the segment, 30 seconds will be added to the original value for the ending time.
- *Frames.* The segment's time boundaries may also be selected by introducing a video frame value which may either be absolute or relative to the current frame value. In this way, the system will compute the appropriate time value using the synchronism information obtained previously.
- *Marks.* A mark is associated with a specific instant in the session (section 9.6.2). The user can establish one of the time limits of a session by specifying the type of mark and when it occurred. As in the two previous

methods, mark occurrences can be specified relative to the current mark. For example, if the user enters '1' and the type of mark selected is 'sudden brake', the system will define the time boundary by associating it with the first occurrence of a 'sudden brake' mark. If '+ 1' is introduced, the system will set the time boundary as the next 'sudden brake' mark occurrence after the current instant.

- *Threshold.* Sometimes the three methods described above are not as accurate as the user wishes. A more accurate way is to select an instant, defined as the moment when a signal overpasses a user-defined value. This instant searching can be done from the current session time to the beginning or the end of the session (forward or backward). For example, the user may want to select the instant when the speed reaches 120 km/h before or after the current reproduction instant.

Two kinds of information have to be displayed at session reconstruction time: the biomedical, environmental and vehicle signals sampled, and the video images acquired during the session. The first are displayed on a graphics monitor with the layout shown in Figure 9.16. Some of the sampled variables are represented as on the dashboard: speed and engine turning rate are represented by car indicators, gear is represented in a similar way as in the car, etc. An extensive place has been reserved for four windows in order to allow the analog signal representation. The primary use for these windows was the representation of biomedical signals. The user has the possibility of representing any signal, such as speed, acceleration, etc., for a detailed study.

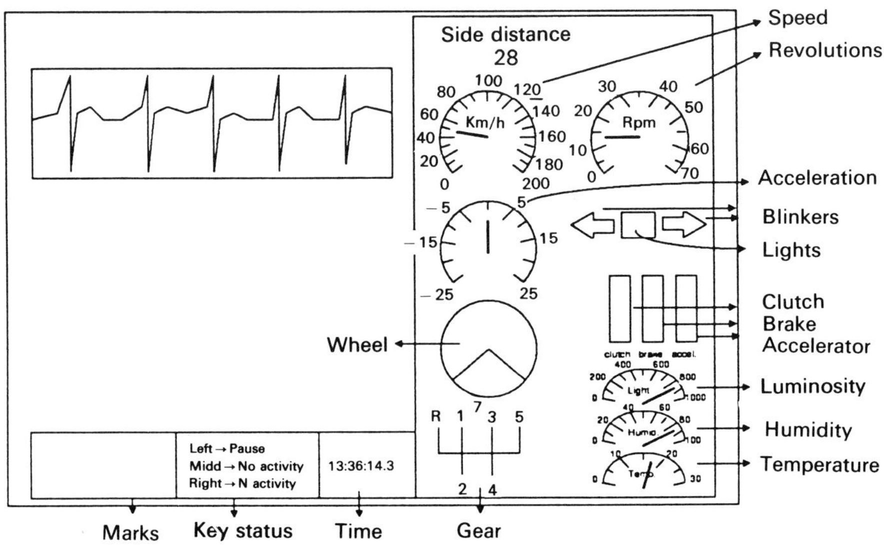

Figure 9.16 Signal representation at the computer display.

These windows are fully configurable by the user: for example, there are three configuration parameters that can be used to obtain a zoom effect. The signal ranges can also be defined by the user as being the maximum and minimum values that can be represented by the system. The time range is the window width available for signal representation.

A computer display portion is reserved for showing the actual time when the represented data were acquired. Another portion is used for displaying mark occurrences while reproducing the driving session (section 9.6.2).

There are some session segments that might be particularly interesting for the user and therefore must be carefully studied. The system also offers the possibility to reproduce those segments at different speeds, allowing the user to choose a slowing factor ranging between 1 and 30.

The video sequences recorded during the driving session can also be displayed at reconstruction time, in synchronization with the signals shown on the computer graphic display. This way, the images shown on the video monitor are those recorded by SARS at the same time that the signals represented on the computer graphic display were acquired. For the analysis of driver's point-of-gaze, a cross, indicating the point-of-gaze, can be superimposed on the sequence of images taken by the helmet scene camera.

9.6.2 Automatic study of data

The automatic study of segments of the driving session allows the detection of deviations in signal behaviour. This study is performed in two ways: marks generation and statistical studies of signals.

Some parts of the session can be specified using marks. A mark is an identifier associated with a specific session time. These marks are introduced to emphasize some events that can be particularly interesting to the user, making the study of the session easier. The marks can be used to establish session segments for reproduction, specific actions, occurrences, etc. There are three kinds of marks:

- *Operator codes.* As explained in section 9.2.1, the passenger can introduce notes during the driving session.
- *Manual marks.* These are introduced at reproduction time. The user can stop the session reproduction and introduce a manual event mark by pressing a mouse button. Mouse button assignments can be dynamically changed during the session reproduction. There are no conceptual differences between manual marks and operator codes. However, manual marks are introduced in the laboratory and can be deleted later on, while operator codes are introduced in the vehicle and cannot be erased.
- *Automatic marks.* These are introduced automatically by the laboratory system as requested by the user. For example, the user may request marking a part of a session when a sudden brake occurs. Two kinds of marks are allowed:

(a) *Threshold marks.* A mark is introduced when a signal reaches a predefined value.
(b) *Derivative marks.* A mark is introduced when a signal changes its value above a specified threshold.

9.6.3 Other possibilities

The laboratory subsystem has been provided with an ASCII dumping option to allow data to be analysed with other tools. This means that specific software does not need to be written when commercial packages are available. For example, the approach taken for the statistical data analysis has been to guarantee data portability for its analysis with standard statistical packages.

9.7 NEW IMPROVEMENTS

A second phase of the Argos project is under development. The Spanish Traffic Agency, while testing the system, noted that although the system is very functional, a higher flexibility level would be desirable. Some possible improvements are being made in the information acquisition and analysis process.

9.7.1 Making a more flexible and portable system

In Argos' first version, it was not possible to include new signals for sampling. The system always gathered a fixed set of signals and each of them was always sampled at the same rate.

Extra flexibility is obtained by allowing the user the possibility of configuring the application. This option is provided by means of a set of files stored both in the SRALS and SARS subsystems.

The SARS configuration file is a binary file that includes one signal characterization per record, created with a menu-based tool. Every record of the file contains a signal name, a signal identifier (for compatibility with the SRALS subsystem), the type of the signal involved in the record (binary, digital or analog), the sampling frequency for that signal, and some hardware information topics for sampling (board and board port and channel in the board for signal acquisition).

The second configuration file is stored in the laboratory computer, which in this case is an ASCII file containing one signal definition per line. Each line contains a set of fields, separated by blanks, which allow the information format to be in agreement with the SARS format, and which allow the user to define signal parameters for signal analysis convenience.

SRALS, the laboratory subsystem, was developed using non-standard software packages. For example, the graphic and window environments were provided by the computer manufacturer. The next version will be based on X window and OSF/Motif for the following reasons:

- increased portability;
- increased user interface configurability according to his preferences or needs;
- possibility of running Argos over a local area network (Ethernet);
- availability of user interface development tools.

9.7.2 New sensor incorporation

New sensors will be introduced to increase the range of parameters acquired by the system:

- An accelerometer will be used to measure the vectorial acceleration (three components), in order to evaluate its influence on the driver.
- A ranging system will be added to measure the distance to the preceding vehicle. Options based on ultrasound, laser and image-processing techniques are being evaluated.

9.7.3 New helmet

The availability of better-suited CCD cameras makes possible the redesign of the helmet ensemble. One option is the use of a commercial helmet for image acquisition within SAISMI. This option permits the digital acquisition of point-of-gaze coordinates at a very high rate (600 Hz) and the determination of the pupil diameter.

9.7.4 Automatic marking

In Argos' present version, automatic marking is very rigid, and permits only a limited set of predefined marks. In the next version, the user will be allowed to create new mark types in two different ways. The first consists of giving a new name to a mark type and selecting the kind of mark (threshold or derivative) and the signal or signal derivative threshold. The second consists of defining a mark relative to another already defined, with the possibility of grouping marks together. The condition for the mark to be activated is that all the associated marks are activated.

9.8 CONCLUSIONS

Intelligent vehicle and highway systems offer the possibilities of increasing driver safety and helping alleviate the severe traffic problems present in heavily populated areas around the world. Before these systems can be widely used, detailed studies on their effects have to be done.

The development of such studies requires the availability of sophisticated systems for driver behaviour modelling and study. The system developed for the Argos project has proven to be an extremely valuable tool for analysing driver reactions to different stimuli under real road conditions.

Regarding the development of image-processing sensors within Argos, the problems found stem mainly from the large degree of variability inherent in outdoor, unrestricted images and illumination conditions. Additional difficulties are caused by the speed at which changes occur (e.g. when driving in a mixed sunny/tree-shaded area). Robust algorithms are needed to work under every possible weather condition.

During the development of Argos, major emphasis has been placed on system exploitation and user interface aspects. Increasing the system flexibility is important for the user to take advantage of the recorded data. Also, a friendly user interface is essential, taking into consideration that the system will be used by specialists without extensive computer expertise.

Argos' first version was finished at the end of 1990. Since then, the Spanish Traffic Department has been using the system. Based on their experiences, improvements are being made for the next system upgrade.

The fundamental and the advanced aspects of the techniques discussed in this chapter can be found in the literature [9–22].

REFERENCES

1. Hill, C. and Davies, P. (1991) IVHS in America, the road ahead in *Proceedings of the 24th ISATA Dedicated Conference and Exhibition on RTI/IVHS*, Florence, Italy, May 1991, pp. 19–26.
2. Pickup, L. (1991) Implementing an integrated road transport environment. The 'Blueprint initiative in Birmingham', in *Proceedings of the 24th ISATA Dedicated Conference and Exhibition on RTI/IVHS*, Florence, Italy, May 1991, pp. 143–50.
3. De Miguel, P., Perez, A. and Rodriguez, S. (1991) Argos: a driver behaviour analysis system in *Proceedings of the 24th ISATA Dedicated Conference and Exhibition on RTI/IVHS*, Florence, Italy, May 1991, pp. 711–18.
4. De Miguel, P., Pastor, L. and Rosales, F. (1991) Image processing sensors for driver behavior characterization, in *Proceedings of the 24th ISATA Dedicated Conference and Exhibition on RTI/IVHS*, Florence, Italy, May 1991, pp. 719–23.
5. VMEmodule Data Manual (1985) Motorola Inc.
6. Asano, K. and Aoki, M. (1991) Lateral vehicular position detection using lane delimiting lines, in *Proceedings of the 24th ISATA Dedicated Conference and Exhibition on RTI/IVHS*, Florence, Italy, May 1991, pp. 193–200.
7. Van Gool, L., Dewaele, P. and Oosterlinck, A. (1985) Texture analysis. *Computer Vision Graphics and Image Processing*, 336–57.
8. Jain, A.K. (1989) *Fundamentals of Digital Image Processing*, Prentice-Hall International Editions.
9. Allworth, S.T. and Zobel, R.N. (1987) *Introduction to Real-Time Software Design*, 2nd edn, MacMillan, Hong Kong.
10. Anderson, C.H., Prout, P.J. and Van der Wal, G.S. (1985) Change detection and tracking using pyramid tranform technics. *SPIE International Society of Optical Engineering*, **579**, 72–8.

11. Bennet, S. (1988) *Real-Time Computer Control: An Introduction*, Prentice-Hall International, London.
12. Cromwell, L. (1972) *Biomedical Instrumentation and Measurements*, Prentice-Hall International.
13. Dally, J. (1984) *Instrumentation for Engineering Measurements*, John Wiley, New York.
14. De Miguel, P., Pastor, L. and Gascón, C. (1988) *Informe Técnico. Proyecto Argos*, Facultad de Informática de la Universidad Politécnica de Madrid.
15. De Miguel, P., Pastor, L. and Perez, A. (1990) *Informe Técnico. Sistema de Reproducción y Análisis de Señales en Laboratorio*, Facultad de Informática de la Universidad Politécnica de Madrid.
16. Deravi, F. and Pal, S.K. (1983) Grey level thresholding using second-order statistics. *Pattern Recognition Letters* **1**, nos. 5,6, 417–22.
17. Doebelin, E.O. (1983) *Measurements Systems: Applications and Design*, McGraw Hill, New York.
18. Kittler, J. and Illingworth, J. (1985) Threshold selection based on a simple image statistic. *Computer Vision. Graphics and Image Processing*, **30**, 125–47.
19. Nakagawa, Y. and Rosenfield, A. (1979) Some experiments on variable thresholding. *Pattern Recognition* **11**, 191–204.
20. Otsu, N. (1979) A theshold selection technique from gray-level histograms. *IEEE Transactions on Systems, Man and Cybernetics*, **SMC-9**, n. 1, 62–6.
21. Van der Van, S., Cannon, R.L., Bezdek, J.C. and Cameron, V.C. (1986) Intermediate level heuristics for road guiding algorithms. *IEEE Transactions on Systems, Man and Cybernetics*, **1**, 292–7.
22. Young, L. and Sheena, D. (1975) Survey of eye movement recording methods. *Behaviour Research Methods and Instrumentation*, **7**(5), 397–429.

CHAPTER 10

Development of dashboard information display system for automotive vehicles

10.1 INTRODUCTION

The driveability and reliability of the automotive vehicle (AV) and its usage costs directly depend on the information display system (IDS) applied, and diagnostics is one of the elements of this system. During use the AV systems are subject to constant wear. The individual control of the AV systems prevents total failure of the entire system (independent control advantages).

In the field of systems intended for information acquisition by the driver, new concepts of dashboard information display systems for automotive vehicles are under development.

In modern AVs sophisticated measuring systems, converting stored data and displaying them for the driver, have been used. One of the basic problems is the mode of data transmission to the driver. A general usable mode for driver information is optical signalling, which can be realized with the aid of dashboard IDSs.

The dashboard IDSs are based on the colour multi-filament vacuum tube (FVT), multi-electrode glow discharge tube (GDT), cathode ray tube (CRT), liquid crystal (LC), plasma, vacuum fluorescence (VF) and electroluminescence (EL), providing the driver with the total amount of information available from the system [1].

These IDSs are controlled by an on-board single-chip microcomputer-based controller fitted under the dashboard. It transmits and receives as well as processes signals and data from the automotive sensory systems on the AV (e.g. propulsion, braking, suspension and guidance controls and so on) as:

- driving information (e.g. water temperature, engine rotational speed, vehicle travel speed, etc.);
- maintenance and diagnostic information;
- a mobile traffic information (MTI) road map on which the current location of the AV is displayed at the large screen centre – driving route, traffic jams, congestion, parking places, slip roads, traffic regulations, etc. may be indicated.

This information is transmitted to the dashboard IDS and an audible tone alerts the driver each time new problems are monitored. The speed, length and direction of the traffic build-up can be seen clearly as the IDS zooms in for a detailed close-up.

The driver may access all the additional information available with the aid of a mobile keyboard.

The main interface between the driver and the on-board microcontroller and the choice of the display affect the success of dashboard IDSs. A dashboard IDS for AVs must be characterized by:

- high brightness;
- high contrast;
- large pixel numbers;
- low cost.

It is currently reported that the CRT dashboard IDS is undoubtedly the best choice while in the not-too-distant future the EL type will be more often applied.

Generally, dashboard IDSs can be divided into active and passive ones, i.e. emitting and/or modulating light. The electroluminescent IDSs belong to the former class.

The electroluminescent dashboard IDS (shown in Figure 10.1) accepts basic information and failure signals can also be transmitted to the driver, appearing to the right-hand side of the large screen. In the case of correct operation of all AV components, the message 'NORMAL' appears, signalling the acceptable performance of AV supervisory control systems.

One of the best-known compounds applied on a large scale which has electroluminescent properties is zinc sulphide, which, depending on the doping used can emit green, yellow or red light [2, 3, 4].

The electroluminescent dashboard IDSs used up to now are fed by a power supply voltage from 100 to 300 V DC or AC. In this chapter a new concept

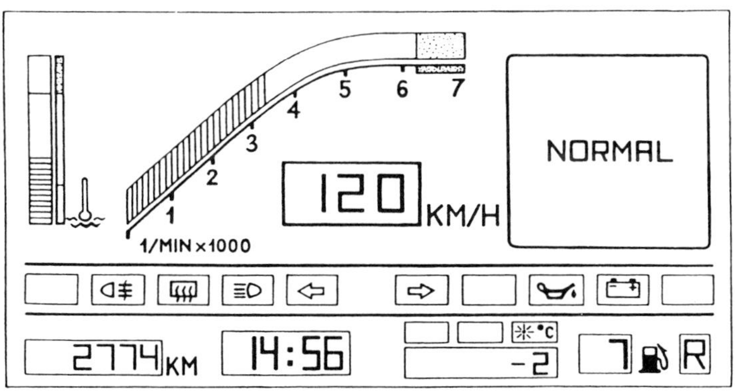

Figure 10.1 An exemplary electroluminescent dashboard information display.

of electroluminescent dashboard IDSs is proposed. In this case, DC and AC power supply voltages are simultaneously applied. As a result, the DC is essentially reduced to about 25 V.

10.2 MANUFACTURING TECHNIQUE

The dashboard electroluminescent IDS was prepared by vacuum methods on a glass substrate in the form of the tri-electrode system [5] shown in Figure 10.2.

The lower electrode was a thin film of In_2O_3(Sn), deposited directly on the substrate, while the remaining two aluminium electrodes were deposited on the upper surface of ZnS(Cu, Cl, Mn) film.

The In_2O_3(Sn) transparent conducting layer (degenerate semiconductor) was obtained by reactive cathode sputtering of alloy containing 90% In and 10% Sn on the Corning 7059 glass substrate. Its resistance was about 100 Ω/sqr. The ZnS(Cu, Cl, Mn) film was prepared by vacuum evaporation of ZnS powder with copper, chlorine and manganese. The mixture in the powder is as follows: Cu, 3.2×10^{-2} g/g ZnS; Cl, 3.6×10^{-2} g/g ZnS; Mn, 2.5×10^{-2} g/g ZnS [6].

The evaporation was carried out under a pressure of 5×10^{-6} torr from an alundum crucible, heated with a molybdenum coil to 950–1100 °C. The distance of the substrate from the source was 8 cm and its temperature during the evaporation was 200 °C. The processing time was varied from 50 to 120 min, which resulted in layer thicknesses varying from 0.7 to 1.7 μm. The ZnS(Cu, Cl, Mn) films obtained were recrystallized by heating in a vacuum (5×10^{-6} torr) for 30 min at 350 °C.

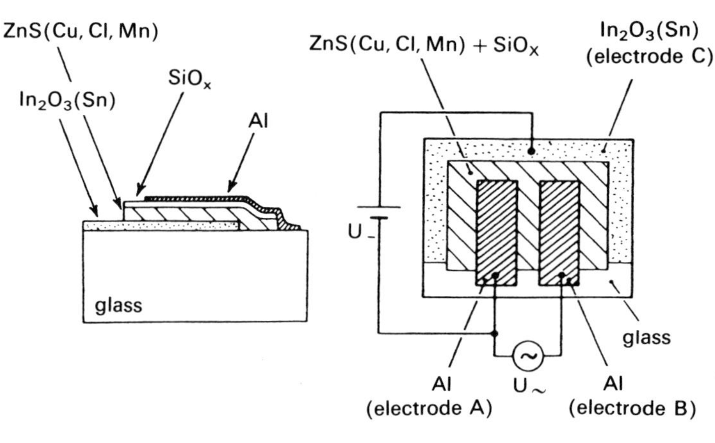

Figure 10.2 Tri-electrode electroluminescent system.

Thereafter, silicon oxide SiO_x film 0.05 μm thick was evaporated on to the ZnS film. Finally, the aluminium electrodes were evaporated onto the SiO_x film (Figure 10.2).

10.3 EXPERIMENTAL RESULTS

A direct voltage between the upper aluminium electrode and the lower In_2O_3(Sn) one, as well as an additional sinusoidal voltage between the two upper aluminium electrodes, was applied to the obtained system (Figure 10.2). For a given value of alternating voltage the dependence of the surface current (between upper aluminium electrodes) on the direct voltage, varied from 0 to 30 V, was measured. The alternating voltage applied to the electrodes placed on the surface of the film had an RMS value of 3 V. The used frequencies of the alternating voltage were 500, 1000, 2000, 3000, 4000 and 5000 Hz.

Figure 10.3 shows the dependence of the alternating current $I_{\sim}$, flowing through the ZnS(Cu, Cl, Mn) film at fixed value of alternating voltage (3 V), on the direct voltage U, applied transversely to the film between A and C electrodes for several values of the frequency of alternating voltage.

For the system under examination, the light emission was observed only at the upper electrode A.

In Figure 10.4 the dependence of the luminance B of light, emitted from the electroluminescent layer, on the direct voltage applied to the A and C electrodes is presented. The alternating voltage applied to the electrodes

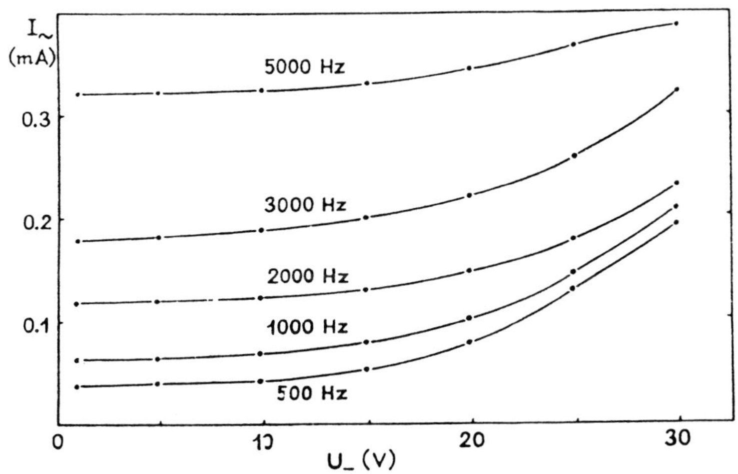

Figure 10.3 The dependence of the AC current on the DC voltage: $U_{\sim} = 3$ V, and $f = 500$, 1000, 2000, 3000 and 5000 Hz.

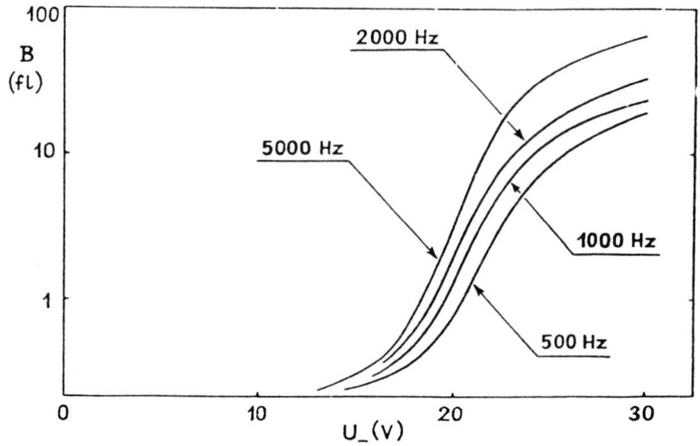

Figure 10.4 The luminance against DC voltage applied to the film for $U_{\sim} = 3$ V, and $f = 500$, 1000, 2000 and 5000 Hz.

placed on the surface of the film had an RMS values of 3 V and frequencies were 500, 1000, 2000 and 5000 Hz.

A photomultiplier with a spectrum range of sensibility from 450 to 750 nm was used for the measurement of light emission. The maximum electroluminescence was observed at about 570 nm (yellow light).

10.4 DISCUSSION OF RESULTS

The obtained experimental results have been in good agreement with the theoretical model of electroluminescent phenomena suggested by Tyagi and Garlick [7]. According to this model, the luminance B of an electroluminescent layer can be described by the formula

$$B = aJ_{\sim} \exp(-b/U^{1/2}) \tag{10.1}$$

where a and b are constant parameters for a given layer, U is the DC voltage applied transversely to this film and $I_{\sim}$ is the RMS value of current flowing through the electroluminescent film.

In the analysis of the dependence of the alternating current $I_{\sim}$ on the constant transverse field and on the frequency of the sinusoidal voltage applied to the electrodes on the surface of the film, the equivalent electric circuit given in Figure 10.5 can be employed. For the equivalent circuit under examination the admittance experienced by the electrodes A and B on the surface of the film can be expressed by the formula

$$Y_{AB} = G_p + j\omega C_p + \frac{(G_{w1} + j\omega C_w)(G_{w2} + jC_w)(G_{w1} + G_{w2} - 2j\omega G_w)}{(G_{w1} + G_{w2})^2 + 4\omega^2 C_w^2} \tag{10.2}$$

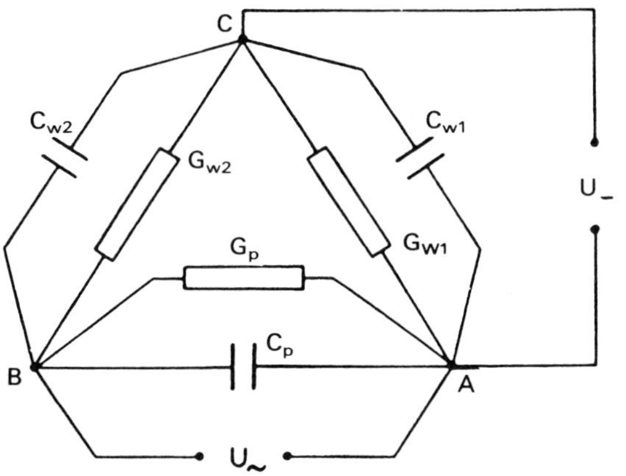

Figure 10.5 Equivalent circuit of a tri-electrode thin film system.

where G_p is the surface conductivity of the ZnS(Cu, Cl, Mn) film for conduction between metallic electrodes, C_p is the capacitance between the electrodes on the surface of the ZnS(Cu, Cl, Mn) film, C_w is the capacitance of the film between the lower electrode and one of the upper electrodes (in practice the relation $C_w = C_{w1} = C_{w2}$ results from the symmetry of the system prepared) and G_{w1} and G_{w2} are the volume conductivities of the film between the lower electrode and the first and second upper electrodes respectively.

The RMS value of the current $I_\sim$ between electrodes A and B placed on the surface of the film will therefore be

$$I_\sim = |Y_{AB}| U_\sim \tag{10.3}$$

where $U_\sim$ is the RMS value of the sinusoidal voltage:

In the absence of the transverse constant field in the film ($U_- = 0$), $G_{w1} \approx G_{w2}$, $G_{w1} \gg G_p$ and $C_p \ll G_{w1}$. Then the admittance Y_{AB} is given by the expression

$$Y_{AB} \approx \tfrac{1}{2}(G_{w2} + j\omega C_{w2}) \tag{10.4}$$

As predicted from Equation 10.3, the surface current $I_\sim$ will increase linearly with the frequency, which is consistent with the characteristic (for $U_- = 0$) given in Figure 10.4.

When the constant voltage U_- has a value of several volts, then in the ZnS(Cu, Cl, Mn) film the Poole–Frenkel effect can occur and in this case the conductivities G_{w1} and G_{w2} will be proportional to

$$\exp(\alpha E^{1/2} + \Phi)$$

where α is a constant, Φ is the potential barrier height and E is the applied field strength [8].

As a consequence of the Poole–Frankel effect, the conductivity G_{w1} will increase such that the relations $G_{w1} \gg G_{w2}$ and $G_{w1} \gg \omega C_{w1}$ will hold; the admittance Y_{AB} will then be expressed by the formula

$$Y_{AB} \approx G_{w2} + G_p + j\omega C_{w2} \tag{10.5}$$

The admittance given by Equation 10.5 is at least twice as high as that given by Equation 10.4; thus the current $I_{\sim}$ should increase at least two-fold, which is consistent with experimental results [5].

The performed measurements show that the luminance B of light emitted by the electroluminescent ZnS(Cu, Cl, Mn) layer depends on the DC voltage applied transversely to this film (electrodes A and C) and on the frequency of the alternating voltage, at a constant amplitude of the AC voltage. As a result, the luminance B of light emitted from the electroluminescent layer can be controlled by adjusting the DC voltage (U) and/or by changing the frequency of the alternating voltage. The tri-electrode electroluminescent system described above can be advantageously used to construct the active display systems of the dashboard driver information and communication system for automotive vehicles.

Any lighting surface (of useful area and shape) can be obtained by the applied vacuum evaporation technology. Luminances up to 200 foot-lambert (fl) [9] can be approached.

Different colours of light can be obtained by suitable doping: green for TbF_3 [4], yellow for Mn [4] and red for ErF_3 [10].

10.5 OPTOELECTRONIC MEMORY CELL CONTROLLED DASHBOARD ELECTROLUMINESCENT INFORMATION DISPLAY SYSTEM

The automotive electroluminescent dashboard IDSs applied for data transmission to the driver may be controlled by optoelectronic logical circuits which are practically insensitive to electromagnetic interference (EMI).

Optoelectronic memory cells are most interesting. They can be used for the electroluminescent dashboard IDS control, especially for the large screen which signals the performances of AV supervisory control systems.

The optoelectronic memory cell can be constructed with two photoconducting (PC) and two electroluminescent (EL) devices. The PC device can be prepared as a sandwich-type system on a glass substrate. The first layer is a transparent electrode of tin-doped In_2O_3 obtained by reactive cathode sputtering. The second layer is a photoconductive CdS film doped with copper and chlorine which is evaporated under vacuum [11]. The third layer, which acts as an electrode, may be produced by the vacuum evaporation of indium.

The EL device can be a thin film capacitor produced by the vacuum evaporation of ZnS doped with copper, chlorine and manganese [6]. The lower transparent conducting electrode is an $In_2O_3(Sn)$ layer and the upper electrode is a vacuum thin aluminium film.

Applied PC and EL devices within the optoelectronic memory element should be able to perform within the determined appearing conditions (PC and EL interdependent functionality), namely optical feedback between them must exist. The value of the feedback coefficient ought to be such as to perform the bi-stable circuit operation [12].

This optoelectronic memory cell should be supplied with a sinusoidal voltage. The input signal is in the shape of rectangular light pulses illuminating the PC device. The output signal is in the form of the light emitted from the electroluminescent cell.

If the PC_1 device shown in Figure 10.6 is illuminated with an input signal L_1, then at the output of the bi-stable PC_1–EL_1 system a signal B_1 appears as the light emitted from the electroluminescent device EL_1 which additionally illuminates the PC_1 device. When the output signal B_1 does not fall to zero, in spite of 'switching off' the input signal L_1, then the PC_1–EL_1 system is in the on state, that is to say it remembers that previously it had been illuminated with the input signal L_1.

Assuming that the PC_1–EL_1 system will be in the on state, the application of the second input signal L_2, illuminating the PC_2 device (Figure 10.6), causes the appearance of the output signal B_2 in the form of light emitted from the electroluminescent device EL_2.

As the PC_2–EL_2 system is the only one without optical feedback, the removal of the input signal L_2 results every time in the decay of output signal B_2 to a value not far from zero.

Results of experiments [13] indicate that the optoelectronic memory cell

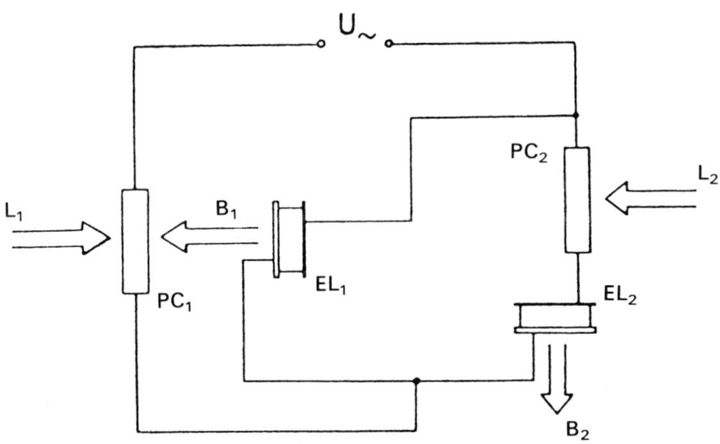

Figure 10.6 Schematic diagram of an optoelectronic memory cell.

presented in this chapter can act as a memory cell under the following conditions:

- The PC_1–EL_1 system should have an optical feebback with a feedback coefficient allowing it to be transposed into an on state.
- The PC_2–EL_2 system should be the single one without feedback.

In an optoelectronic memory cell the PC_1–EL_1 system in the on state can be switched off only by cutting off the supply voltage.

10.6 CONCLUSIONS

In the in-situ tests, it was confirmed that the above-mentioned dashboard IDSs are very useful. However, when evaluation data and well-skilled and/or experienced drivers' opinions are put together, the following questions should be solved in the future:

- promotion of research and development (R&D) to improve visibility of all dashboard IDSs and operationability of the devices;
- promotion of R&D of multi-colour (red, yellow and green as well as blue) dashboard IDSs;
- promotion of R&D of such other functions as setting of straight distance to the destination and display of a good/poor sense of direction and road map retrieval, complex messages, etc.;
- promotion of R&D of additional instruction functions such as display of the shortest course to the destination and route guidance.

REFERENCES

1. Incadorna, M. and Pratico, M. (1989) New concept on multiplexed systems and electronic instrumentation for trucks and buses, in *Proceedings of the 20th International Symposium on Automotive Technology and Automation*, Florence. Italy, May–June 1989, Pap. No. 89 130.
2. Konopiński, M. (1987) *Elektronika w technice motoryzacyjnej*, WKL, Warszawa (in Polish).
3. Okamoto, K. and Hamakawa, Y. (1979) Bright green electroluminescence in thin film $ZnS:TbF_3$. *Applied Physics Letters*, **35**, 408–11.
4. Theis, D. (1981) Application of thin film electroluminescent devices. *Journal of Luminescence*, **23**, 191.
5. Porada, Z. (1991) Electroluminescent information display systems for automotive vehicles, in *Proceedings of the 24th International Symposium on Automotive Technology and Automation*, Florence, Italy, May 1991, Pap. No. 911 222.
6. Porada, Z. and Schabowska, E. (1980) AC electroluminescence of ZnS:Cu, Cl, Mn thin films in the structure In_2O_3(Sn)–ZnS:Cu, Cl, Mn–SiO_x–Al. *Journal of Luminescence*, **21**, 129.
7. Tyagi, R. and Garlick, G. (1966) The enhancement of visible and infrared electroluminescence in zinc sulphide crystals by infrared irradiation. *British Journal of Applied Physics*, **17**, 747.

8. Lamb, D.R. (1967) *Electrical Conduction Mechanisms in Thin Insulating Films*, Methuen, London.
9. Smith, D. (1981) Modeling AC thin-film electroluminescent devices. *Journal of Luminescence*, **23**, 209.
10. Szczurek, I., Łożykowski, H. and Szczurek, H. (1981) Electroluminescence of ZnSe:ErF_3 thin films. *Journal of Luminescence*, **23**, 315.
11. Porada, Z. and Schabowska, E. (1983) Electrical and photoconducting properties of CdS(Cu.Cl) thin films. *Vacuum*, **33**, 179–81.
12. Porada, Z. (1983) Thin film light amplifier with optical feedback. *Thin Solid Films*, **109**, 213–16.
13. Porada, Z. and Schabowska-Osiowska, E. (1991) Application of thin film electroluminescent diodes and photoconducting elements in an optoelectronic memory system. *Materials Science and Engineering*, **B9**, 383–5.

PART SIX

Vehicle Navigation and Communication

CHAPTER 11

Autonomous vehicle guidance using laser range imagery

11.1 INTRODUCTION

The Environmental Research Institute of Michigan (ERIM) was a participant in the DARPA Autonomous Land Vehicle (ALV) project and related programs, developing and refining three-dimensional imaging systems which use a laser radar principle to measure range [1]. Coupling the three-dimensional vision capability to a Cytocomputer pipelined image-processing system allowed us to extend our work to applications in industrial machine vision, such as robotic bin-picking [2], and has facilitated expansion of our efforts in autonomous systems to encompass vehicle guidance, routing and system integration [3].

This chapter describes a road-finding and -following algorithm which uses range and reflectance imagery from a three-dimensional sensor that is processed using a Cyto-HSS high-speed recirculating pipeline image processor [4, 5]. This algorithm runs in near real time (two-second frame-processing rates) by using the morphological image-processing capabilities of the Cyto-HSS system [6], and has been extensively tested on data gathered from roads on an around ERIM's site in Ann Arbor, Michigan. This algorithm was developed as part of a program with TACOM* to integrate autonomous capability into a conventional internal combustion vehicle. Other concurrent developments include a new generation of three-dimensional imaging sensors for robotics and vehicle guidance, automatic route planners, multiple autonomous vehicle modelling, and symbolic map development, planning, and use, and further development of Cyto-HSS systems.

11.2 PAST WORK

Autonomous vehicle navigation can be broken into two quasi-independent functions, vehicle path planning and vehicle path control. Path planning, as

*Work supported by US Army Tank and Automotive Command (TACOM) under contract #DAAE07-84-R138, primary sponsor; also supported by DARPA, US Postal Service and Ford Motor Company.

used here, is the process by which the vehicle system chooses specific 'routes' to move from the starting location to the desired goal location. This process typically requires symbolic and geographic map data, and exploits some combination of AI planning techniques and optimization methods (dynamic programming, relaxation, etc.) [7]. Path control is the process of staying on the desired route, independent of minor unpredicted variation due to unexpected obstacles, changes in surrounding textures (foliage, landscapes, etc.), or variability of lighting conditions (weather changes, time of day, etc.). Path control typically involves use of sensors, specifically variants of computer vision, to monitor vehicle location relative to the route and feedback tracking errors to the vehicle controller. In this chapter we will touch on both components, focusing on the path control component.

Much of the past work on autonomous vehicle computer vision has been derived from earlier image-understanding work, and focuses on direct analysis of reflectance imagery or the derivation of three-dimensional information from reflectance imagery. Davis and Rosenfeld [8] describe early vision algorithms using reflectance-based segmentation, nearest-neighbour edge-preserving smoothing, and Hough transform-based feature identification for roadside objects. They has also pursued derivation of three-dimensional data from stereopsis and computation of optical flow. Quad-tree methods have been applied to represent symbolic image information. Kanade *et al.* [9] have investigated using trinocular stereo to compute three-dimensional information, have worked extensively with reflectance-based methods, and have applied the CMU-developed WARP to speeding up vision algorithms. Riseman and Hanson [10] have studied motion, flow fields, and rule-based interpretation systems (VISIONS). Poggio at MIT [11] has extended the earlier work of Marr by using regularization methods and multi-levelled strategies for early vision.

In the previously described work, three-dimensional information required for vehicle control is derived using stereopsis (kinesis can be thought of as moving camera or objects so that view A is transformed to view B serially), or stadimetry (comparison against standard reference objects or models). It is our contention that direct three-dimensional data are simpler and more reliable for road detection, tracking and obstacle avoidance. The simple rule, 'the road is a flatter and smoother three-dimensional surface than the NOT road' is more powerful than, 'the road is a more uniform reflector than the usually more nonuniform, textured NOT road'.

In particular, active laser range imaging is advantageous because:

- There are no occlusions due to parallax between the illuminator and the detector.
- Direct measurement of range reduces image-processing time required prior to interpretation by eliminating any need to compute or infer depth.
- Standardized illumination allows for true reflectance measurement and ambient light-independent performance.

From our work we have learned that processing range imagery is somewhat different from processing reflectance imagery, and is not simply an application of previously developed methods. Different preprocessing and smoothing are required. We use edge-preserving morphological smoothers which can be tailored to remove typical ranging errors and smooth out noise. We have also found that three-dimensional erosion and dilation by three-dimensional structuring element probes has greater applicability to range data (these operations measure 'fit' between the structuring element and the mating area in the three-dimensional image). For example, structuring element fit has been used to build generic bin-picking systems for industrial use [2]. Range image interpretation has also created a requirement to detect broader classes of discontinuities (edges) than are required in many applications using reflectance imagery. Range imagery routinely contains low-amplitude impulses (from texture), steps (from object occlusions), and ramps (from connected faces). These requirements drive the need for edge finding based on surface detection, rather than local gradient techniques [12].

It should be noted that other researchers working on autonomous navigation problems, notably those at CMU [13], OSU (Ohio State University), and Martin Marietta [14], have integrated laser range imagery into their navigation algorithms. CMU in particular, has worked extensively on correlating video and range imagery from different sensors. However, to the authors' knowledge, only OSU has investigated plan view transformation, and none of them have investigated the additional channels of information available from range imagers. It is in these areas that ERIM's work diverges from these research leaders.

For real-time road-following applications, speed of image processing is a key performance index. For a sensor which 'sees' 120ft (61 m), a $1\frac{1}{3}$s frame rate is required to support a 30 mile/h (47 km/h) vehicle speed. This corresponds to two looks before travelling over each location along a path. After one look, the vehicle has moved 60 ft (18.4 m), allowing another frame time to avoid an obstacle after it has been detected. Vehicle control without dependency on internal maps requires at least this much path data, and perhaps more for stability. This rate of processing requires algorithms designed to take advantage of parallel hardware processing systems. For this reason we have focused our algorithms on methods which exploit cellular (local) transforms and morphology, making the Cyto-HSS pipelined image processor useful [6].

11.3 IMAGE-PROCESSING SYSTEM

The image processing required to find and track roadways described subsequently uses the Cyto-HSS image-processing workstations developed by ERIM to support morphological image processing. The functionally complete image algebra executed in the Cytocomputer can be formulated from a simple set of postulates and laws. The algebra is based on the notion that the images (even three-dimensional or grey tone images) are binary in an n-dimensional Euclidean space. These images are represented with discrete values, corresponding to the binary or grey tone pixels of the images.

Inclusion and equality of these images and subimages are defined in the usual sense with relations satisfying the reflexive, transitive, symmetric and substitution laws. Definitions of simple set-theoretical logical and geometric operations which satisfy closure complete the algebraic foundation [15].

In particular, the complement of a binary image A is the binary image which interchanges the black and white pixels in A. This is expressed as

$$\neg A = \{p \mid p \in E \text{ and } p \notin A\}$$

where E represents the universal binary image and p is a point in the image.

The intersection of any two binary images A and B is the binary image which is black at all points p, which are black in both A and B, or

$$A \cap B = \{p \mid p \in A \text{ and } p \in B\}$$

The union of A and B is the binary image black at all points p which are black in A or black in B or black in both. This can be written as

$$A \cup B = \{p \mid p \in A \text{ or } p \in B\}$$

The geometric operations include vector sum, difference, and product of point images, and scaling of images by a numeric constant.

Vector sum:

$$A + B = \{a + b \mid a \in A \text{ and } b \in B\}$$

Vector difference:

$$A - B = \{a - b \mid a \in A \text{ and } b \in B\}$$

Vector product:

$$A \times B = \{a \times b \mid a \in A \text{ and } b \in B\}$$

Scaling by a point p

$$A \times p = \{a \times p \mid a \in A\}$$

Summing (differencing) each point in an image A to a point p produces the translation of an image, or

$$A + p = \{a + p \mid a \in A\}$$

and

$$A - p = \{a - p \mid a \in A\}$$

Scaling each point in an image by -1 produces the reflection of the image through the origin, or

$$-A = \{-a \mid a \in A\}$$

Additional non-linear operations are available for finding minima, maxima, thresholding, and pixel state labelling.

Using these logical and geometric operations it is possible to define two additional and extremely useful operations, dilation and erosion, sometimes called Minkowski addition and subtraction, respectively.

The dilation ($\oplus$) of image A by image B is defined as the union of the translation of A by the points of B, or

$$A \oplus B = \bigcup_{b_i \in B} A + b_i$$

Erosion ($\ominus$) of image A by image B produces an image that contains all the points which when summed with B produces an image contained in A:

$$A \ominus B = \{p \mid (B + p) \subseteq A\}$$

Erosion and dilation allow the definition of a variety of useful feature detectors and smoothers, including opening an image A by B (which causes small features in A to disappear):

$$A \text{ opened by } B = (A \ominus B) \oplus B$$

and closing and image A and B (which causes small gaps between features in A to be filled):

$$A \text{ closed by } B = (A \oplus B) \ominus B$$

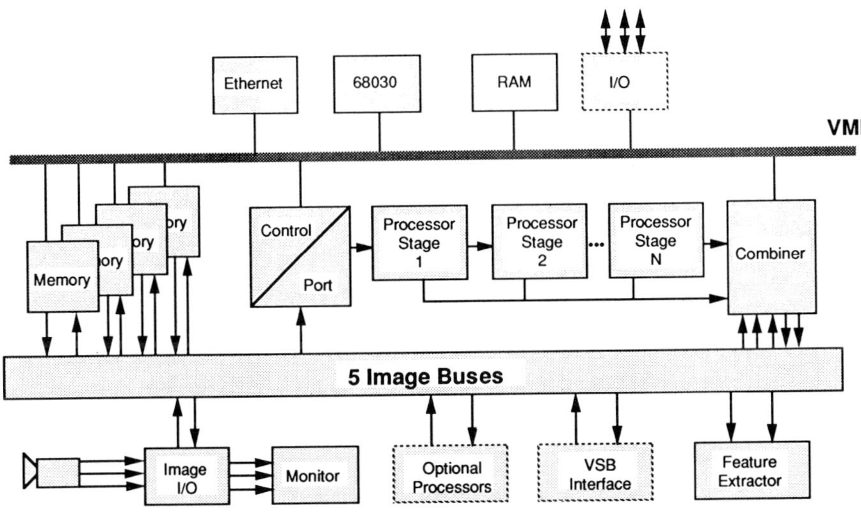

Figure 11.1 Cyto-HSS system architecture. Features: 10 M pixel/s transfer rate on all data paths; five independent 8-bit image data paths; parity checking on all data paths; modular, open architecture; host-processor independent.

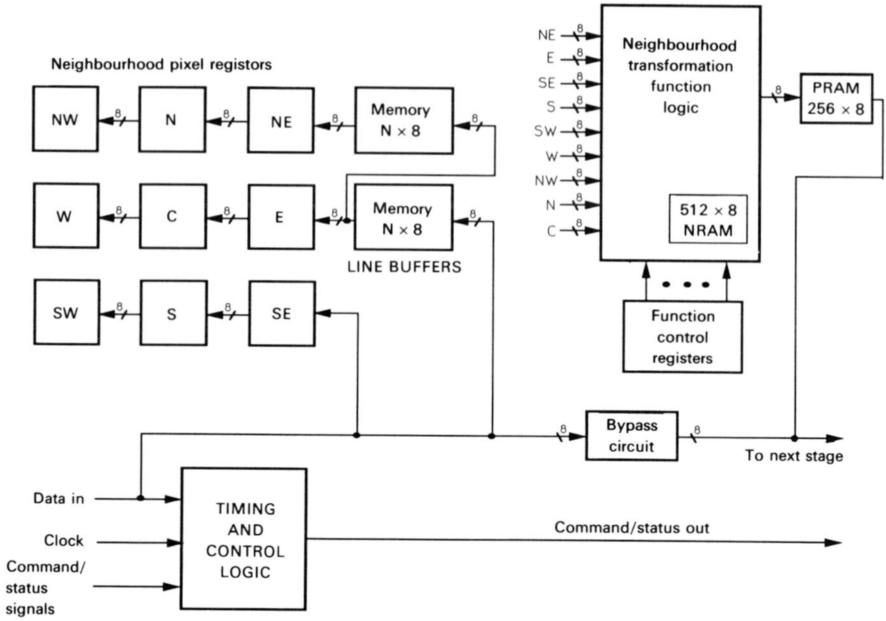

Figure 11.2 Cyto-HSS neighbourhood processing stage.

To support these operations efficiently ERIM has developed a parallel pipelined architecture built around a variable number of primitive 3 × 3 neighbourhood processing stages. The system also contains a three source and two destination image arithmetic unit (Figure 11.1). The stage operations can perform simple dilations and erosions at a 10 MHz rate, and pipelines over 100 stages deep have been built to provide over 10^{11} DEC VAX equivalent operations per second (Figure 11.2). The image arithmetic unit can combine up to three images at the 10 mHz rate.

11.4 THREE-DIMENSIONAL IMAGING SYSTEM

11.4.1 Phenomenology of the laser sensor

The ERIM laser range imager is an optical-wavelength radar system (Figure 11.3). It uses a laser diode source operating in the near infrared region (820 nm), and is nominally eye safe. The laser diode is amplitude modulated and scanned across the field of view using a nodding mirror and a rotating polygon mirror (Figure 11.4). The infrared light is reflected off the desired target, gathered by the receiver optics and focused onto the detector. The output of the photodetector is at the same frequency as the laser source, but

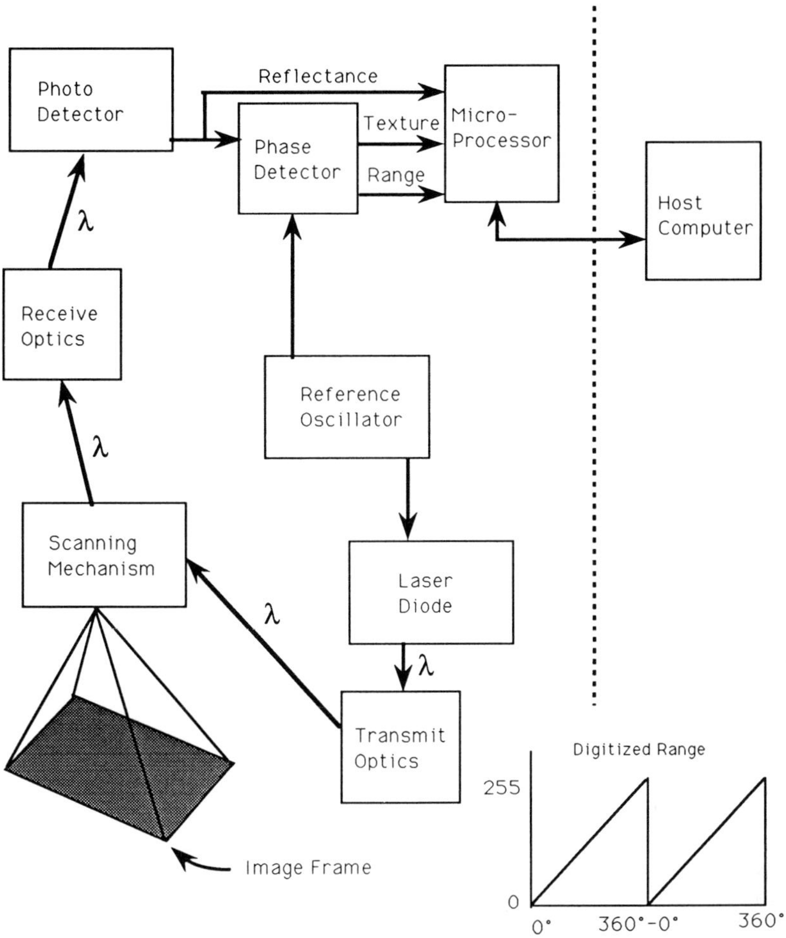

Figure 11.3 Three-dimensional laser range scanner block diagram.

is shifted in phase proportional to the distance travelled by the beam. Using a digital phase detector the range is assigned for each pixel along the scan line. The amplitude of the carrier wave is attenuated by the range, the cosine angle between the normal to the surface and the beam, and the reflectance of the region scanned.

Since the ambient illumination is not modulated at the carrier frequency, the signal is from the laser radiation only. The relative amplitude is recorded as 'reflectance' data. This chapter will concentrate on the range data but will show examples of the reflectance data, and how they can be used.

An additional bore-sighted channel is obtained using coherent cancellation to measure range dispersion within the instrument field of view (IFOV). This

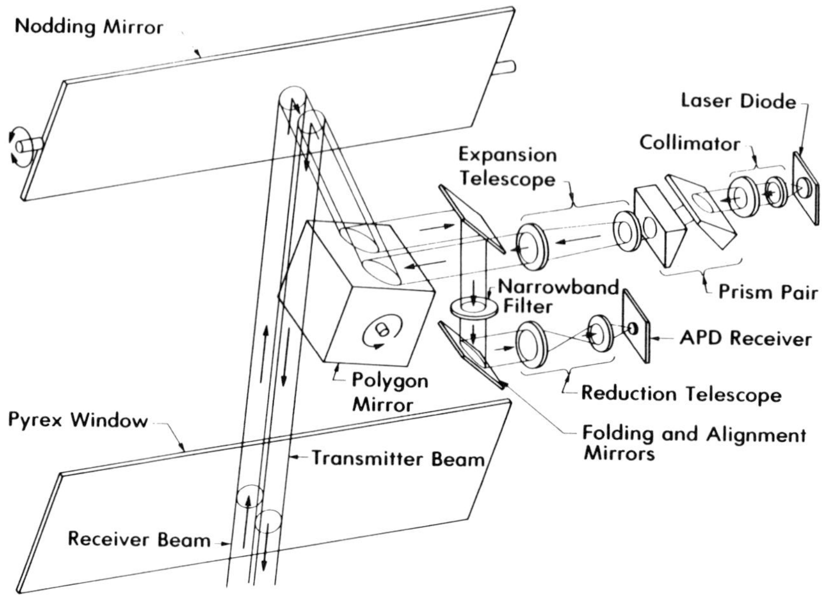

Figure 11.4 Scanner optical arrangement.

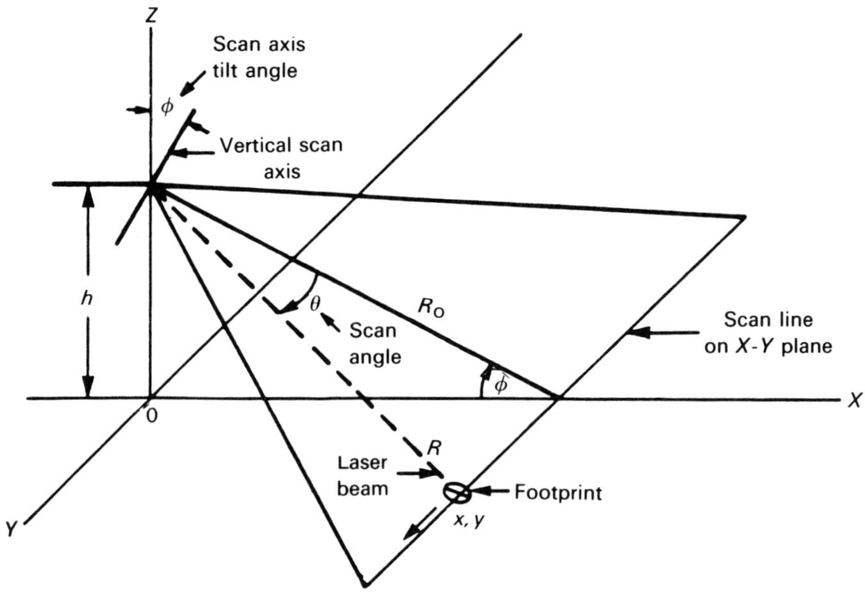

Figure 11.5 Scanning geometry.

has been named, informally, 'texture' data, as regions of uniform range (within the IFOV) are low-valued, whereas regions of mixed range (such as edge points) are high-valued. ERIM has done some experimental work with texture data, but has not incorporated these results into its vehicle navigation work.

Range data is acquired in 256 × 64 pixel images and digitized to 8 bits, at a rate of two frames per second. Figure 11.5 shows the relationship between sensor mounting and the ground. Range measurements cover 80° in the horizontal direction (azimuth, θ) and 30° in the vertical direction (depression, ϕ). The location of each point in real coordinates relative to the image array can be determined in a spherical coordinate system specified by the angles shown in Figure 11.5.

11.4.2 Data characteristics

There are several sensor-specific characteristics which an algorithm designer must consider before full use can be made of the sensor. The most obvious characteristic is the 64 foot ambiguity interval. If the ranges to two objects in the images are, for example, 74 and 10 feet, then they both will be measured as 10 feet because of the phase ambiguity. Some laser imagers use several bore-sighted lasers at different carrier frequencies to greatly increase the ambiguity interval (up to 12.43 miles) without loss in range resolution. For less complicated systems, ways must be found to remove some or all of the range ambiguity, in a process of range recovery. In addition to range recovery, this section will also address the process of coordinate conversion which takes into account corrections for platform motion and changes in vehicle attitude.

Range recovery

The output of the sensor at time t is a pair of images, range and reflectance. The content of the range is the modular range. That is, if the real range is

$$R = \left(\sum_{i=-\infty}^{\infty} b_i 2^i \right) \times 3.0 \text{ inches}$$

then the range image contains

$$r = \left(\sum_{i=0}^{7} b_i 2^i \right) \times 3.0 \text{ inches}$$

The low-order bits, $\{b_i\}_{i=-\infty,-1}$, which are lost in the scanning process, can be modelled as a uniform 1.50 inch error with a standard deviation of $3.0/\sqrt{12} = 0.866$ inches. This loss is typical with any analog to digital conversion. The high-order bits $\{b_i\}_{i=8,\infty}$ are more serious. The difference

$R-r$ is dominated by

$$\left(\sum_{i=8}^{\infty} b_i 2^i\right) \times 3.0\,\text{inches} = \left(\sum_{i=0}^{\infty} b_i 2^i\right) \times 64\,\text{feet}$$

The effect is that r is wrong by some multiple of 64 feet. The approach is to produce an estimate, R', of the true range, R, whose error is less than half an ambiguity interval in magnitude. Then, one uses R' and the sensor's r to obtain a good estimate of range. By good, we mean one whose standard deviation is close to that caused by the loss of the low-order bits. The procedure is to calculate the integer k which minimizes $|R'-(r+64k)|$. The range estimate is $R''=r+64k$ feet.

We have used three models for R' which give reasonable results. The first is called the 'flat earth' model where R' was the range from the sensor to a flat plane h feet below the sensor, where h is the height of the mounted sensor above the ground. The model is represented as a fixed image of range points along a vector δ_i in depression angle and α_j in azimuth angle for pixel (i,j).

The second method is called the continuous model where R' for pixel (i,j) is the range computed for pixel $(i-1,j)$ if i was above the bottom row of the image and $r_{i,j}>r_{i-1,j}$, else $R'=R'+64$ feet. Note that we make the assumption that the scan line closest to the vehicle is giving the correct range. This model works as long as the distance between consecutive scan lines remains less than 4 bits, or 32 feet.

The third method will be describe in detail in the next section. It is basically an enhanced version of the continuous method, with a corresponding increase in computational complexity. In practice, a combination of these approaches may be appropriate, depending on the application's need.

Coordinate conversion

The laser sensor outputs data in a polar coordinate system (range, azimuth angle, depression angle). The two levels of decision, path control and path planning, have different natural coordinate systems. Path control would work best if the information was resampled into the vehicle's coordinate system, say up, down range, cross range. These points would be referenced to time, for example, at the start of frame acquisition. This time reference is important because of the slow frame time. A fast-moving vehicle may travel several feet or turn several degrees within the time of a single frame acquisition. In this reference frame, it is convenient to describe an obstacle as being 40 feet down range and 3 feet left of centre at time t.

The path planning level has objectives most reasonably given in terms of a map system. An image-extracted landmark such as a turn or an intersection should be converted to the same units as the stored map information. This would be done by enhancing the path control data with absolute, fixed coordinate information.

The azimuth and depression angles for pixel (i, j) are:

$$\alpha_j = (j - 128.5)\frac{40}{128}\text{ degrees}$$

$$\delta_i = (i + 0.5)\frac{30}{64}\text{ degrees}$$

We define the rotation matrices:

$$\boldsymbol{R}_x(\theta) = \begin{pmatrix} 1 & 0 & 0 \\ 0 & \cos\theta & -\sin\theta \\ 0 & \sin\theta & \cos\theta \end{pmatrix}$$

$$\boldsymbol{R}_y(\theta) = \begin{pmatrix} \cos\theta & 0 & -\sin\theta \\ 0 & 1 & 0 \\ \sin\theta & 0 & \cos\theta \end{pmatrix}$$

$$\boldsymbol{R}_z(\theta) = \begin{pmatrix} \cos\theta & -\sin\theta & 0 \\ \sin\theta & \cos\theta & 0 \\ 0 & 0 & 1 \end{pmatrix}$$

If $\boldsymbol{w}$ is a three-dimensional vector then $\boldsymbol{R}_x(\theta)\boldsymbol{w}$, $\boldsymbol{R}_y(\theta)\boldsymbol{w}$ and $\boldsymbol{R}_z(\theta)\boldsymbol{w}$ are rotations by θ about the x, y and z axes respectively. The following steps can now be defined which will transform the polar coordinates to a more useful coordinate system. The coordinates of the laser return area with respect to the sensor are

$$\begin{pmatrix} x_{\mathrm{s}} \\ y_{\mathrm{s}} \\ z_{\mathrm{s}} \end{pmatrix}_{ij} = \boldsymbol{R}_x(\delta_i)\boldsymbol{R}_z(\alpha_j)\begin{pmatrix} 0 \\ \boldsymbol{R}_{ij} \\ 0 \end{pmatrix}$$

There are five degrees of freedom associated with mounting the sensor on the vehicle. The first three are the location of the sensor in the vehicle coordinate system. We simplify this by defining the sensor's position as the origin of the vehicle. The other two degrees of freedom are the orientation of the sensor with respect to the vehicle. The sensor used to collect the data has a fixed orientation with a zero degree pan angle (i.e. facing directly forward) and a 1° tilt below horizontal. (Note: these parameters could be subject to computer control.) The coordinates of the laser return with respect to the vehicle are

$$\begin{pmatrix} x_{\mathrm{v}} \\ y_{\mathrm{v}} \\ z_{\mathrm{v}} \end{pmatrix}_{ij} = \boldsymbol{R}_x(\text{tilt})\boldsymbol{R}_z(\text{pan})\begin{pmatrix} x_{\mathrm{s}} \\ y_{\mathrm{s}} \\ z_{\mathrm{s}} \end{pmatrix}_{ij}$$

In our case, where the pan angle is zero, we have

$$\begin{pmatrix} x_{\mathrm{v}} \\ y_{\mathrm{v}} \\ z_{\mathrm{v}} \end{pmatrix}_{ij} = \boldsymbol{R}_x(\text{tilt} + \delta_i)\boldsymbol{R}_z(\alpha_j)\begin{pmatrix} 0 \\ \boldsymbol{R}_{ij} \\ 0 \end{pmatrix}$$

The above transform is only correct if the vehicle is not moving during the image acquisition time. This is not a realistic assumption. Vehicle movement can be obtained from some form of inertial navigation system (INS) on the vehicle. The requirements for the two control levels are different. The vehicle control level requires only roll, pitch, heading (yaw) and vehicle speed. At this level the requirement is to construct a map in the vehicle's coordinate system, referenced to time. The planning level requires in addition to those terms an absolute heading value and odometer information or some other accurate measure of forward motion to construct maps in absolute co-ordinates. The incorporation of GPS (global positioning system) data into INS systems will greatly improve the accuracy of such maps. In general, GPS and landmark recognition can serve to limit absolute measurement error which accumulates with INS systems over time. The effect of such enhancements will be the reduction of the error of the term p_i, defined below.

If one assumes that the vehicle does not move significantly during a single scan line ($\frac{1}{160}$th of a second) then we can model the vehicle's motion as a jump of $(\Delta y)_i$ during scan line i. The change in the vehicle's position during scan line i is

$$\boldsymbol{R}_z(\text{heading})\boldsymbol{R}_x(\text{pitch})\boldsymbol{R}_y(\text{roll})\begin{pmatrix}0\\ \Delta y_i\\ 0\end{pmatrix}$$

in a fixed XYZ coordinate system, where Δy_i is the change in the odometer during the acquisition of scan line i. The change in the vehicle's position, from the start of frame time to the time ending in the acquisition of scan line i, with respect to the vehicle's position at the start of frame time, is

$$\sum_{j=64}^{1} \boldsymbol{R}_z(\Delta heading_j)\boldsymbol{R}_x(pitch_j)\boldsymbol{R}_y(roll_j)\begin{pmatrix}0\\ \Delta y_j\\ 0\end{pmatrix}$$

where $\Delta heading_j$ is the change of heading from the start of frame time to the time of acquisition of scan line j. The values $pitch_j$ and $roll_j$ are the pitch and roll angles measured during the time of acquisition of scan line j. The above relations allow one to update the vehicle's position both in a fixed XYZ format and with respect to the vehicle's position at the start of frame time. Denote these position estimates at scan line i as p_i' and p_i. The fixed XYZ coordinates of the pixels (i, j) are

$$\begin{pmatrix}x\\ y\\ z\end{pmatrix}_{ij} = \boldsymbol{R}_z(heading_i)\boldsymbol{R}_x(pitch_i)\boldsymbol{R}_y(roll_i)\begin{pmatrix}x_v\\ y_v\\ z_v\end{pmatrix}_{ij} + p_i$$

The vehicle's coordinates of the pixels (i, j) are

$$\begin{pmatrix}x\\ y\\ z\end{pmatrix}_{ij} = \boldsymbol{R}_z(\Delta heading_i)\boldsymbol{R}_x(pitch_i)\boldsymbol{R}_y(roll_i)\begin{pmatrix}x_v\\ y_v\\ z_v\end{pmatrix}_{ij} + p_i'$$

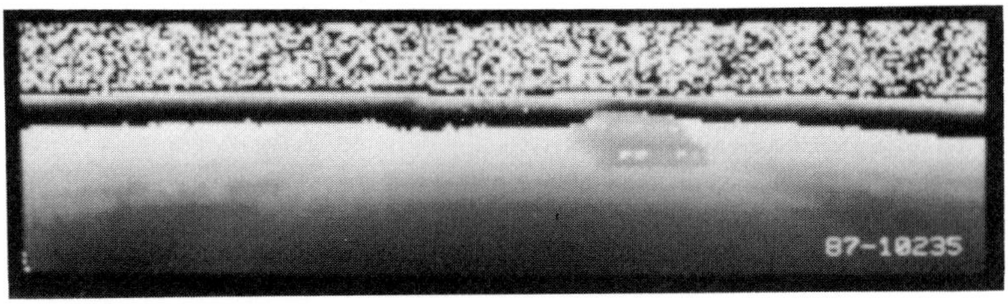

Figure 11.6 Original range image.

Figure 11.6 shows an example of a range image. Note the automobile on the right side of the image. The range recovery techniques are used to convert the ambiguous range information into unambiguous range. The last equation above is applied to the range image and the INS heading to produce (x, y, z) for every pixel for which range recovery was successful. We form Figure 11.11 by placing z in map cell (x, y) where the cells are fixed-resolution squares. It is this image which is processed to find road edges.

It should be noted that other transformations are potentially useful. In particular, a transformation to a corrected forward-looking viewpoint would be useful for identifying sides of buildings (and vertical surfaces in general) and perhaps the outline shape of traffic signs.

11.5 ROAD-FINDING ALGORITHMS

The road-finding/following algorithms accept raw range and reflectance data from the three-dimensional scanner. The first operation performed is to remove (tag) unreliable range data. The next operation is finding and labelling the second ambiguity interval (thus adding one more bit of range information to the 8-bit raw data). Then the forward-looking (scanner) view is transformed to a plan (map) view. The map view is processed to find rough regions, and then the largest smooth region beginning at the known vehicle origin is found, and thinned down to a proposed path. This path can then be passed to the vehicle control systems.

In areas of a scene where little or no reflected light returns to the scanner, range accuracy significantly deteriorates. This typically happens when scanning the horizon, very distant range points, or points which are on surfaces illuminated at small grazing angles. These noisy range points can be easily eliminated through thresholding the intensity image at a low reflectance value, and using the resulting binary image to mask the corresponding range image:

$$\text{Masked } A = A \text{ minimum } (B \text{ threshold by } t)$$

where A is the raw range image, B is the reflectance image, and t is a threshold.

The images in Figures 11.6 and 11.7 show the original range data and reflectance data, respectively. Figure 11.8 shows the masked range data. These data were taken from a vehicle moving at approximately 5 mile/h (8 km/h).

In some applications of three-dimensional scanner, edge-preserving morphological filtering by opening and then closing by spherical or cylindrical structuring elements of large size has been useful to smooth out small noisy areas:

$$\text{filtered } A = (A \text{ opened by } B) \text{ closed by } B$$

where A is a range image and B is a spherical structuring element.

After clean range images have been formed, ambiguity intervals can be identified. This algorithm determines the proper ambiguity interval for each pixel by first locally comparing its value with its eight neighbours to determine and construct interval boundaries, and then globally bleeding an interval designation value over the pixels, which cannot penetrate these impermeable boundaries. This process can be applied recursively to find other ambiguity intervals in the image, though in practice, there are rarely any useful data beyond the second ambiguity interval.

The location of ambiguity intervals is approximated by assigning end-of-interval states (END) to high-valued pixels (200 to 254), and beginning-of-

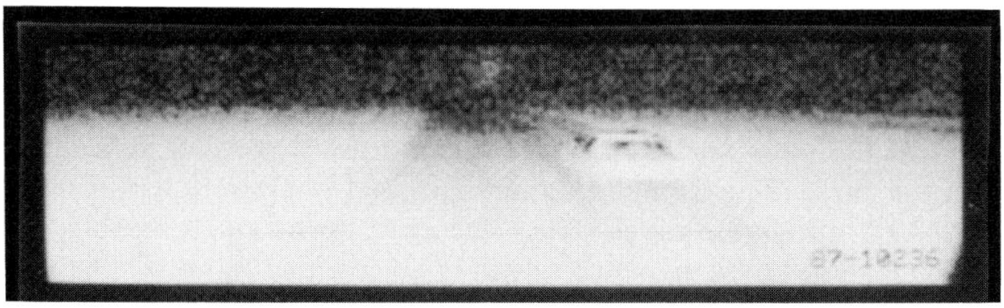

Figure 11.7 Original reflectance image.

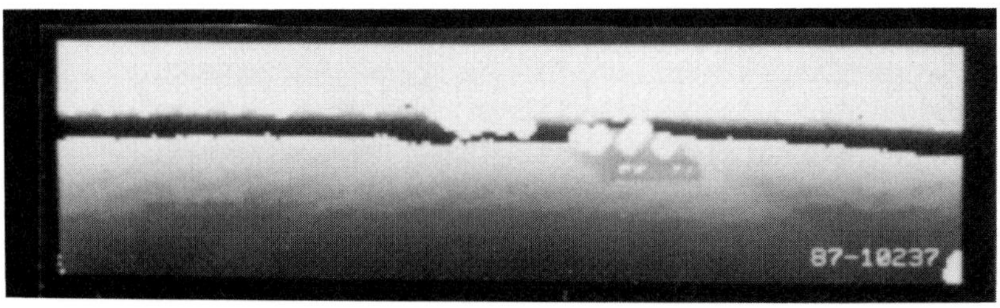

Figure 11.8 Masked range image.

interval states (BEGIN) to low-valued pixels (0 to 127). Next, a boundary is built where any BEGIN state is next an END state. These boundary states are marked as BOUND and are conditionally dilated (grown outward) to fill any gaps.

The first ambiguity interval is found by labelling the bottom line in the image to a unique value (0), and conditionally dilating (growing outward) over any pixel not set to BOUND. The second ambiguity interval is found by setting BOUND pixels to 1 and growing over any pixel not set to 0. The resulting image can be thought of as the high-order bit of a 9-bit range image (when combined with the original 8-bit range image prior to processing for intervals). Figure 11.9 shows the first ambiguity interval marked in light grey, and the second marked in dark grey.

After preprocessing, the range imagery is transformed from a scanner view to a map view by applying the transform discussed earlier. This transformation is performed to bridge the gap between symbolic map information and acquired range data. It also facilitates frame-to-frame comparison for map creation, because vehicle motion causes no perspective changes after the transformation. The transformation is suitably up-scaled to minimize loss of range precision.

After the forward-looking to plan view transformation, road boundaries are identified by finding rough or non-flat areas. Several methods could be used for this operation, including difference of gaussians, difference of steps, average edge density, etc.; however, for convenience of computation we used the difference between the maximum range value over a large area and the maximum value over a smaller area. In previous experiments, we have used the difference between a large spherically filtered image and a non-filtered image and have obtained similar results:

$$\text{Rough } A = (\text{filtered } A) - A$$

or alternatively

$$\text{Rough } A = (\text{max. } A \text{ over large area}) - (\text{max. } A \text{ over small area})$$

After finding rough patches, the image is closed using a cylinder which is

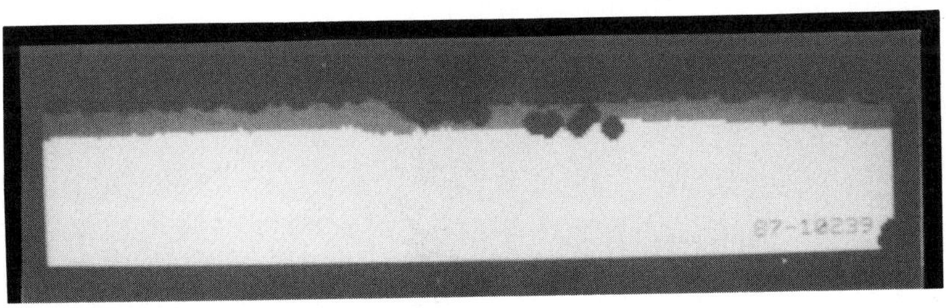

Figure 11.9 Ambiguity intervals marked.

nominally the size of the vehicle to eliminate passages too narrow for the vehicle to traverse. Figure 11.10 shows the top-down view of the range data.

To find the road centre, a unique pixel value is set at the vehicle origin, and conditionally dilated over traversable values. The result is an image of traversable areas accessible to the vehicle from its present location. These regions are thinned from the sides to narrow, single-pixel lines, which designate path centres for the vehicle to follow. In Figure 11.11 low areas are darker than high areas, and the road centre is marked with a black line.

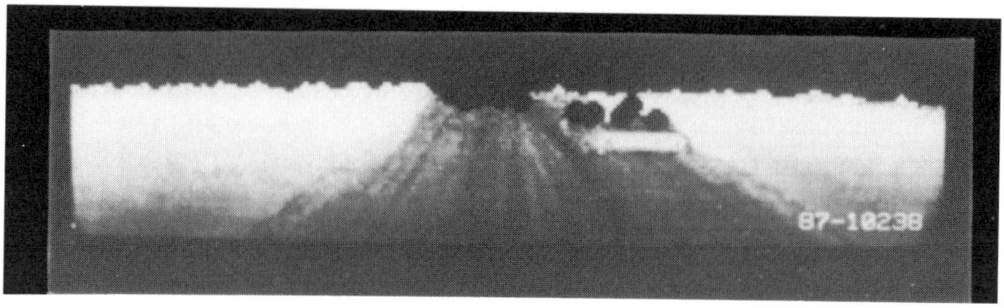

Figure 11.10 Plan view of range data.

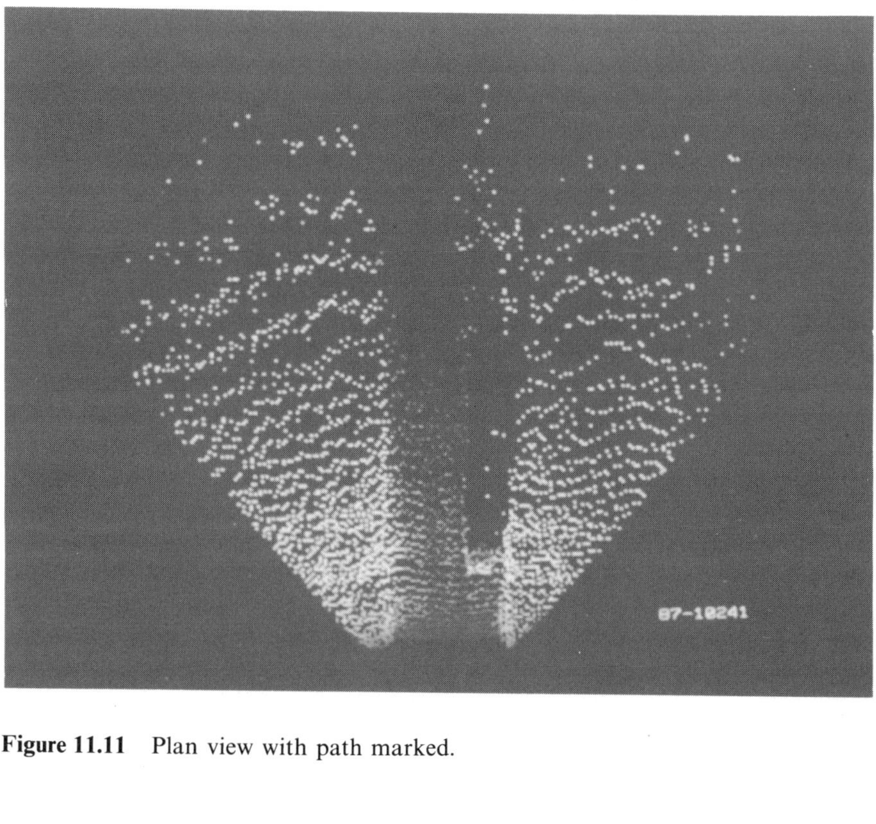

Figure 11.11 Plan view with path marked.

11.6 FUTURE WORK IN ROAD-FINDING ALGORITHMS

Although the road-finding algorithm described above is somewhat costly, particularly for the perspective transform operation, these costs are believed to be justified. By transforming the data from an egocentric to an exocentric coordinate frame, successive framed of data can be easily integrated into a common 'world map', given that the vehicle contains some kind of INS. The resulting 'temporal fusion' can be used by both sensing modalities (range and reflectance), as both can be plotted in plan view. In addition, the reflectance data can be range normalized to produce a more robust data set (Figure 11.2). Although ERIM has some experience in map generation and maintenance based on regular sensor updates and limited memory capacity [7] this model must be expanded and modified to account for multiple sensing modalities and real-time processing requirements. This strategy will allow for the decoupling of data acquisition of the sensor from data processing by the path-planning mechanism of the vehicle.

11.7 PATH PLANNING

As a higher level process than road or path finding, path planning can be organized with the use of maps, either generated dynamically by the sensor

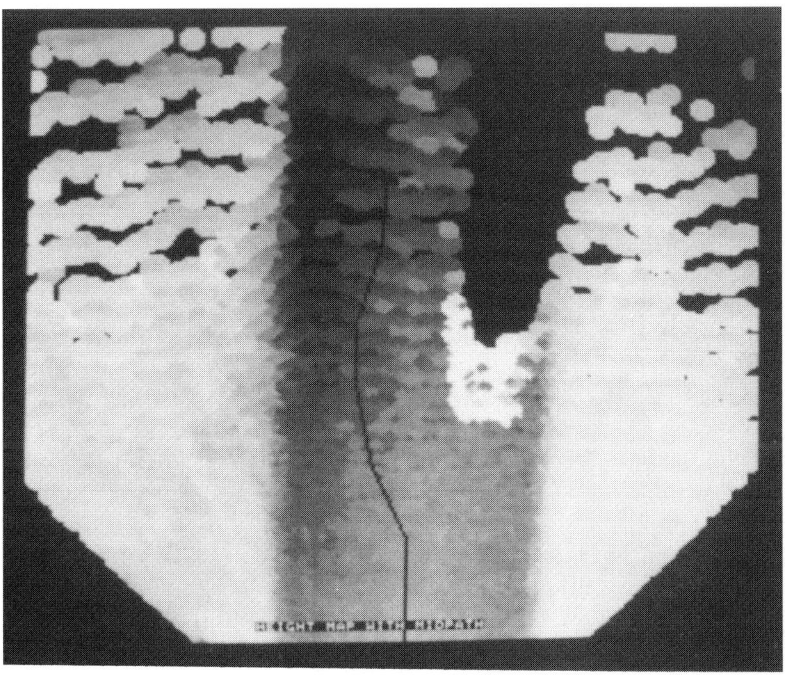

Figure 11.12 Range-normalized reflectance image.

or produced from a pre-existing database. In the former, new information is acquired to update or augment the map. In the latter, old information is used to make reasonable decisions on as-yet unseen terrain. New information (from the sensors) is only used to verify the map's state of the world. Should an inconsistency be found, then the path must be replanned. Both types of information can be integrated into the same structure, the map.

A map, in its basic form, is merely a two-dimensional array of elevation points of a fixed resolution, referenced to an absolute coordinate system. Typical uses of the map structure are to indicate potential paths of unseen terrain, and to adjust for errors of absolute coordinate measurement (such as GPS) through the use of landmark recognition. A landmark is anything which is reasonably 'novel', can be readily detected by the sensor suite, and for which a mapping or matching operation exists to compare the sensor information with some sort of internal model. Examples of reasonable landmarks for this system are road curves (the vehicle's path can be referenced), buildings (the vertical surfaces, orientation and size are all readily recognized with the range sensor, though this takes additional processing) and intersections. Less reasonable landmarks could include trees (hard to model accurately) and fences (potentially too small to get accurate return by the sensor.)

An immediate apparent limitation to the map structure is the fixed resolution. Map databases may not have the high-resolution data contained within them, and if they do, the system may not have the capacity to store or manipulate this information. On the other hand, for the purposes of road following, the highest available resolution is generally desired.

This problem is dealt with by maintaining a hierarchy of maps. A set of maps of varying resolutions are opened and maintained as the vehicle traverses some terrain. High-resolution information is gathered from sensor views and combined with either previous views, previously stored low-resolution information, or both. Eventually, the system's capacity may be exceeded, whereby some of the high-resolution data are downsampled and shifted to a lower resolution map. This process is known as 'scrolling' and provides for the controlled exchange of information between the different data domains.

In the extreme, the lowest resolution map could be largely symbolic in nature. In this case, special 'scrolling demons' are needed to convert symbolic information into expected physical manifestations. The reverse is potentially possible, and would allow for the vehicle to perform its own landmark creation.

While this last concept is admittedly speculative, the need for autonomous vehicles to understand, build and maintain maps is not. For the problem of path planning, the task of truth maintenance of a multi-resolution map database is not a trivial one, with each major component of the system a plausible source of error. It remains to be seen precisely how an autonomous system capable of high-level path planning incorporates map maintenance into its system operation. It is our contention that a reliable source of three-dimensional data is required for any such system to be realized.

11.8 CONCLUSIONS

This chapter has presented a system of autonomous vehicle guidance with a laser range imager as the primary sensing component. From this perspective, it has been shown that such systems have far more potential than simply as a method of obstacle detection. In particular, the use of reflectance data provides a robust shadow-free reflectivity image suitable for more conventional data processing.

The algorithms described have run on the Cyto-HSS image-processing system with an execution time of less than two seconds per frame, and have proven to be quite robust. Additional work is currently underway to incorporate more intensity imagery corroboration of road location to serve as a back-up to the range algorithms. Also, work is underway to connect the image format map information to the map information processed by a symbolic computer. As part of the systems integration program to demonstrate a working vehicle system, enhancements are being made to the Cyto-HSS support software system (which currently resides partly on a timeshared DEC VAX) to allow the full road-finding operation to reside locally in the workstation. When this is completed the image-processing rate supported will be commensurate with the rate of range data aquisition (two frames per second).

REFERENCES

1. Bair, M. (1983) *Workshop on Ground Vehicles*, Final Report, DoD, Delivery Order 0725, University of Maryland, Appendix C.
2. Coleman, E. and Sampson, R. (1985) Acquisition of randomly oriented workpieces through structure mating, in *IEEE Proceedings on Computer Vision and Pattern Recognition*, San Francisco, CA, June 1985.
3. Zuk, D., Pont, W.F., Franklin, R. and Dell'Eva, M. (1985) *A System for Autonomous Land Navigation*. Rep. no. 179 100-10-F, ERIM, Ann Arbor, MI.
4. Beyer, J. (1985) Geometric perspective transformation and techniques in range recovery, in *Workshop on Processing 3-Dimensional Range and Reflectance Information* (ed. K. Eberle), ERIM, Ann Arbor, MI.
5. Pont, W.F. (1985) Road feature extraction from range data, in *Workshop on Processing 3-Dimensional Range and Reflectance Information* (ed. K. Eberle), ERIM, Ann Arbor, MI.
6. McCubbrey, D. and Lougheed, R. (1985) Morphological image analysis using a raster pipeline processor, in *IEEE Computer Society Workshop on Computer Architecture for Pattern Analysis and Image Database Management*, Miami Beach, Florida, November 1985, Institution of Electronic and Electrical Engineers.
7. Quek, F., Franklin, R. and Pont F. (1985) *A Decision System for Autonomous Robot Navigation over Rough Terrain*, ERIM, Ann Arbor, MI.

8. Davis, L. and Rosenfeld, A. (1985) Image understanding techniques for autonomous vehicle navigation, in *Proceedings of the Image Understanding Workshop*, DARPA, Dec. 1985.
9. Kanade, T. and Shafer, S. (1985) Image understanding at CMU, in *Proceedings of the Image Understanding Workshop*, DARPA, Dec. 1985.
10. Riseman, E. and Hanson, A. (1985) Summary of progress in image understanding at the University of Massachusetts, in *Proceedings of the Image Understanding Workshop*, DARPA, Dec. 1985.
11. Poggio, T. (1985) MIT progress in understanding images, in *Proceedings of the Image Understanding Workshop*, DARPA, Dec. 1985.
12. Besl, P. and Jain R. (1985) Invariant surface characteristics for three dimensional object recognition in range images, in *Computer Vision, Graphics, and Image Processing*.
13. Crisman, J. and Webb, J. (1991) The warp machine on Navlab. *IEEE Transactions on Pattern Analysis and Machine Intelligence*, **13**, No. 5.
14. Turk, M., Gremban, K, and Marra, M. (1988) VITS – a vision system for autonomous land navigation. *IEEE Transactions on Pattern Analysis and Machine Intelligence*, **10**, No. 3.
15. Sternberg, S. (1980) Language and architecture for parallel image processing, in *Proceedings of the Conference on Pattern Recognition*, Amsterdam, The Netherlands (eds E.S. Gelsma and L.S. Kanal), North-Holland.

CHAPTER 12

Vision systems for vehicle guidance

12.1 INTRODUCTION

This chapter describes not merely the function of several systems for real-time vision in high-speed vehicle applications; it also tries to explain the design process itself, to show how the specific solution to the problem might be found in a goal-oriented framework.

Therefore the terminology of goal-oriented systems and the motivation for this approach in robotics are introduced to serve as a basis for the explanation. Within this conceptual framework the components of a software/hardware sensing system for high-speed autonomous vehicle guidance are discussed. The chapter emphasizes the use of vision and will give some hints on modelling. Dynamic vision constitutes the main theme. The basic paradigm will be explained and several modules fitting into this framework will be described. To show the tangled interconnections between software and hardware, both algorithms and the hardware architecture will be presented. Also some examples of the performance of realized systems will be shown.

12.2 GOAL-ORIENTED VISION FOR ROBOT VEHICLES

Designing a computer vision system for mobile robot applications not only gives rise to a number of detail problems, but also casts some light on the nature of vision itself. Thus, we start from the point of view of design philosophy and architecture and proceed to the particulars of several implementations.

The first question of someone who is confronted with the above problem might be: 'How do the mechanisms of organic vision function?' Answering this question should provide the chance to guide the design by just copying the mechanisms underlying the vision processes in already existing systems. Unfortunately, this question has proven in the past to be very hard to answer. This is why the design of the vision system family to be explained in this chapter focuses on the question: 'What are the reasons *why* organisms see?'

The difference between both questions might reflect a real change in paradigm [1] and the success in several applications justifies the emphasis put onto this point.

Asking the second question means focusing on intentions, goals and tasks and not on descriptions. Regardless of the differences in the point of view, goals and tasks will be treated as nearly synonymous in the following because tasks can be assumed to be the goals of an external task-setting system, whereas a distinction between description and encoding will be made to emphasize the communication with respect to the representative aspects of information.

Ethological research may give us some first hints to the answer. Speaking of goals means to abridge the fact that the tendency to prefer specific situations and to try to provoke them by appropriate behaviour results in an evolutionary advantage [2].

Ethologists [3, 4] have shown that most animals process very often only rather simple key stimuli to achieve their goals. Especially in co-operative behaviours, it is much more effective to use a codified signalling system than to try to produce complete internal descriptions of the surroundings. The colouring of a large number of animals is only one example. Also, in non-co-operative situations, only small sections of the potentially possible description suffice to allow appropriate reactions.

In linguistics there is the same distinction between the pragmatic and the semantic (descriptive) dimensions of communication. The pragmatic dimension simply includes the possible reactions towards a certain environmental configuration.

If, a system for example, for which red lights indicate a dangerous situation, like a car, saw a red light, it would not need to be able to describe the event as a circular segment of a plane emitting a specific wavelength of light. Stopping in front of the red light would be completely satisfactory. Thus, a filter tuned to the event for initiating an appropriate behaviour or a collection of appropriate behaviours might be sufficient. Possibly, but not necessarily, knowledge about circles in space and colour optics could support the *design process* of the filter. But maybe, the infrared emission of the red light could initiate the same behaviour. Generally, a number of different possible sensorial modalities usable as behaviour triggers exist and what may be an appropriate combination depends heavily on the task, with respect to the goal. Furthermore, the example shows that description can take place at several levels of detail. Also, different purposes can require different descriptions. One very specific intention causing explicit descriptions of situations is the goal when transmitting information to another subject.

In evolution the descriptive dimension of communication emerged relatively recently. Of course, some other descriptive systems existed much earlier, e.g. in insects, but explicit flexible descriptions were only produced when humans, having an extremely flexible variety of possible filters they can choose from, started to react not only to environmental situations but also to other humans dealing with these situations, i.e. when language was invented, allowing discursive behaviours.

Furthermore, most problems of complete descriptive reconstruction of environmental features need a very substantial amount of computation, due to the inherent parallelism and non-linearity of the real world. But real computing systems can afford only restricted resources in time, hardware and complexity. Thus, systems, regardless of whether they are technical or organic, will have an advantage whenever they are able to execute the same *task* with less resources.

Human introspection at first seems to oppose the above argument against completeness. Intuitively, humans seem to have a consistent and uniform descriptive model of the surrounding world. But even this very flexible and effective model can be proven to be goal oriented.

An everyday example will illustrate this. Imagine a heavily crowded downtown shopping area. If someone who went through this area were to be asked about his memories afterwards, he would only be able to reproduce a small amount of the potential variety of impressions, mainly determined by his interest. This illustrates, at least for memory (and the same may hold for sensing itself), that recognition is mainly an act of selection. Specific cues trigger recognition processes.

In contrast, a descriptive attempt would try to synthesize a model which was as complete as possible. Applied to the above example, this means at least reconstructing the surfaces of all shops, including their albedo and reflectance coefficients, to obtain the basis for reasoning on the letters painted on the shop signs. Of course, this type of knowledge is important and helpful for the *design process*, but at run time may have been 'compiled down' to other strategies by exploiting, for example, the known system goals. This compilation process would not mainly consist of a tricky implementation of existing methods; rather it means to control the complete design process from the viewpoint of goal orientation.

Psychophysics also presents a number of examples of goal-oriented recognition in humans negating the existence of one omnipotent recognition process. For example, colour and grey level edge detection have different resolutions in the human sensory system (a fact utilized in TV signal coding) but humans normally do not recognize a misplacement of colour edges. Rather, we are mostly completely unaware of the fading of colours in the twilight if we can attain our goals. A flashing light source in the peripheral region of the field of view is another example. Regardless of the fact that the spatial resolution is very low in the fovea area, nobody will receive a blurred impression of the spot, but the event will initiate an unintentional eye movement because it might be of some importance to the system goals.

Because goal-oriented systems have to respond to their operating environment, the main goal is to filter out those invariants depicting specific reactions. A filter in this sense is a system module recognizing and signalling the existence of the situation it is designed for. Filters may be as highly complex as situations. Thus, the complete goal-oriented recognition system may be interpreted as a complex and often extremely non-linear filter. This terminology is much more general than the common usage of the term 'filter' in signal processing.

Organic systems evolved by developing the necessary filters for their goals. During evolution, common parts of the numerous filters might have been merged, giving rise to the next steps of complexity; for example, the vast amount of pictorial memories stemming from imaging sensors might have collapsed into a representation in space and time. It may be hypothesized that shared resources like algorithms or hardware structures were untied from the specific context to form increasingly general systems. These general systems finally showed the *descriptive* power that humans possess.

The idea of filtering out special events triggering reactions already existed in the 1950s. Similarly, neural networks are now undergoing a revival. Neural network approaches try to implement distributed fine-grained and possibly adaptive filters for specific tasks. A network tuned to specific input patterns, representing relevant situations, is intended to generate appropriate reaction behaviours; Pomerleau [5] gives as example for road driving. Road border recognition as another example of a filter in the above sense is treated in Kuhnert [6].

Brooks [7] was among the first to investigate a specific robotics *architecture* for coupling fixed behaviours to sensorial patterns. There is now an increasing tendency towards goal-oriented approaches, which Aloimonos and Rosenfeld [8] have termed 'purposive' in a recent survey on vision. Obviously, a number of hints exist that intentionally designing goals or tasks into robotics systems is going to prove fruitful by delivering a new design paradigm.

Furthermore, the design process for a system can be clearly distinguished from the real execution of the task. For the design, complex and extensive descriptive formulations of the laws governing the task environment are advantageous. In this framework physics or, specifically for vision-guided robots, inverse optics, are examples of descriptive methods delivering valuable tools for the design.

For the real execution in the target environment economical usage of such restricted resources as time or information transportation and processing capacity becomes the crucial point, and emphasizes again the necessity of understanding goal and task structures.

In this sense, work on understanding and building vision-guided systems is ruled by two principal approaches:

- Specific aspects of the physical world can be seen in their physical context and the laws governing them might be described.
- Not only are the physical laws evaluated, but also how to imbed goal structures under conditions of restricted resources.

During active operation an autonomous robotics system searching for situations conforming to the goal has to deal mainly with the tradeoff between completeness and selecting important information rapidly. How this tradeoff can be handled for mobile vision-guided systems will the basis of the following discussion on dynamic vision.

12.3 CONCEPT OF DYNAMIC VISION

In the course of organic evolution, vision has become a main information source. Vision as the ability to obtain information about distant objects is inevitably entangled with motion. Only a mobile observer can take real advantage of knowledge it has gathered through its sensors. Thus, vision and motion constitute a problem area advantageously treated as a whole.

With this definition of the field, already the specific problems are defined. In principle, complete information is advantageous. Therefore, the sensor potentially should be able to deliver as much information as possible. But due to the restricted resources in time, information transportation and computing capacity, a selection and/or filtering process has to be applied. Performing the allocation of resources on line and parallel with the data acquisition would be the best possible case.

In contrast to this, early attempts at building mobile robots equipped with vision sensors tried to connect a pattern recognition system to a control scheme by completely neglecting any considerations about allocation of resources or the systems task. The basic idea was to obtain an almost unified and thus – formally – simple description of the environment as a basis for reasoning on possible actions. Such a universal system would be able to perform any arbitrary task. For this design approach, very naturally the derivation of the descripition and the planning in the representation space becomes the main concern.

Obviously, unified descriptions are unlikely to be obtained from a complex and diversified world observed by imperfect sensors and also the effort in terms of resources might grow alarmingly, even if only one type of description were to be applied.

Also, dynamic computer vision is concerned with the evaluation and the analysis of image sequences with regard to the control of real systems, but the image sequences taken with video cameras are evaluated with respect to the systems goal or task. Thus, it is nearly impossible to refer to the method without stating the area of application. This statement might provoke the impression that dynamic vision constitutes precision working on implementations and may suffer a lack of generality. But assessing task structures is a task in is own right, and submodules which are suited to specific subgoals may be reused in different systems having the same subgoals. So, for instance, a module dedicated to the fusion of sensorial modalities into a model in space and time [9] might have very wide applicability.

A comprehensive summary of the method of dynamic computer vision is presented by Dickmanns and Graefe [10, 11]. In the following, the important aspects and the architectural viewpoints will be briefly explained and illustrated by examples of algorithms. Several experimental implementations of vision systems will also be discussed.

Because the task at hand is extremely important in dynamic vision, autonomous road cruising and the important sub-theme of robot road following will be treated as examples of applying the method to illustrate the abstract approach throughout the chapter. Other applications utilizing the same paradigm and some of the corresponding computational modules are a model satellite docking experiment [12] and a simulated aircraft landing manoeuvre [13].

High-speed autonomous road cruising on the German 'autobahn' is the most well-known application area of dynamic vision up to now [14, 15, 16]. The general aim is to build a system able to drive on a road, supported mainly by the information coming in from a video camera and only utilizing the infrastructure normally supplied to the human driver. For robot road following the vehicle must recognize the road path. For the more general cruising task, obstacles and other road conditions must be analysed too. Autonomous road cruising cannot be complete without other behavioural competencies, e.g. obstacle detection or lane changing.

Several attempts, stemming from robotics and aiming to produce experimental systems, which have been made to master the problem of autonomous road cruising [14, 17–22], have led to a number of experimental autonomous outdoor vehicles. Also the more optics-oriented approaches of computer vision have been applied to road cruising [23, 24]. But at present in real-time experiments the second type of algorithm suffers from its complexity.

Within the robotics approaches, mostly classical vision algorithms have been combined with an ordinary control scheme. Specific features were detected in the image and combined to form an environmental description. Edges, lines and uniform-colour and -texture regions have been utilized as cues for the road position. In most approaches these elements were linked, to form geometrical elements like straight lines, parabolas or clothoidal curves, producing a two-dimensional representation. Sometimes this information directly fed the controls, whereas sometimes a three-dimensional model generated by inverse optical projection was employed intermediately. In almost all cases the selection of features is performed heuristically. Very often the ambiguous inverse projection causes some problems with numerical stability and singularities.

Time and hardware capacities are the most restricted resources in mobile, real-time applications of vision. The images must be interpreted fast enough to allow an appropriate reaction to the actual state of the environment. The huge amount of data delivered by the sensor in one video cycle and the measurement noise cause the most severe problems.

Exploiting the knowledge about the task's environment or about operational constraints leads to an escape from the central dilemma of real-time image sequence processing: firstly, one intends to obtain image data of the widest possible bandwidth, because spatial and temporal resolution should be high in order to recognize fast-moving objects at large distances; but on the other hand, it is required to react to external events within a predetermined short reaction period.

In a number of cases, the available reaction time allows delays of between 100 ms and 1 s, if one would like to reach a performance comparable to that of a human being. Only under such conditions is an autonomous system really able to reach its objectives.

In classical image processing the common approach tried to extract all the *possible* information out of an image by having as little knowledge about the world as possible at the same time. This implies a contradiction by itself: if no knowledge is available about the laws underlying the regularities that may be found in the class of images under concern, it might be impossible to obtain the specific information needed for executing the task. To deal with autonomous car driving, it is obviously necessary to have knowledge about roads, tracks and moving objects. So, it is not necessarily important to extract all possible features out of the image, because the task-specific knowledge allows a selective tuning of the system, delivering a reduction of the imposed effort.

Thus, dynamic computer vision may be characterized by the fact that it extracts from the vast amount of incoming image data those parts which allow an appropriate reaction in real time, and it extracts these data by explicitly modelling the regularities of the world (especially the dynamics) and the statistical properties of their variability.

Fortunately, changes in the real world do not occur completely randomly and may be described by the laws of physics. These laws essentially encode redundancy in possible observations. Thus, the amount of information which needs to be processed can be cut down by avoiding redundant measurements. Furthermore, normally only a small part of the events in the sorrounding world will influence the system's goals. Only those events relevant to the task of a robot have to be observed. These two restrictions very often allow a substantial reduction of the input information. Thus, generally, the efforts of information transportation and evaluation have to be concentrated on informative (not redundant) and task-relevant portions of the world.

Of course, the real world (including the sensory system) may by understood as an information-encoding mechanism and, in general, it might be very difficult or even impossible to extract the needed information from the encoded or maybe encrypted messages only, without any knowledge about the encoding process itself. The available and well-founded physics may be utilized to construct the decoder.

Naturally, the physics of real bodies may deliver a reliably founded corpus of knowledge. It is a well-known fact that bodies having inert masses with a given consumption of energy can realize only a limited acceleration. Thus, real movements must be steady. Additionally, the laws of optical projection within the sensor are known. Combining these two facts leads to the conclusion that in the plane where the light-sensitive elements of the sensor are located most projected movements may be expected to be steady. Thus, observable parts of objects will move only slightly between two consecutive frames of an image sequence, if these are taken within an appropriately short time interval. Furthermore, the object may be tracked easily because consecutive images may be expected to be very similar to each other.

As mentioned, physical laws may be used in designing the analysis algorithms and these, in turn, determine the structure of the system, especially the hardware to be utilized. But this does not mean that dynamic vision is solely an exercise in physics.

In real systems *ideal* physical equations do not suffice for a realistic description. Therefore, the ubiquitous noise of the real world is also explicitly built into the model of dynamic vision.

A priori knowledge exists in *various* forms; for example, by modelling the physics and geometry of the sensor itself a highly accurate estimation of the position of individual object images or images of features in the sensor plane may be gained [25].

So, although from the beginning we try to model the incoming data (e.g. by their physics), at the same time, only the relevant information has to be extracted. Relevant in this context means that the task to be realized respective to the goal to be undertaken determines how to select the information from the environment.

While the task of describing the environment is covered by physics, at the moment no general theory about the construction of vision-guided goal-oriented autonomous robots exists. So, regularities underlying this domain should be found experimentally. Empirical knowledge gathered in real experiments with a complex system may guide the assessment of the relevant aspects of the problem field. This was the main reason for developing and realizing a family of image-processing systems suited to dynamic vision applications [26].

12.4 STRUCTURE OF VISION ALGORITHMS

The first step in deriving appropriate behaviours consists of filtering out the task-relevant cues. For road cruising the positions, distances and directions of other objects like the road path, adjacent tracks or other cars should be known. Unfortunately, a vision-guided system first of all relies on the ambiguous two-dimensional projection of the three-dimensional world imposed by the lenses of the camera. If the objects of the three-dimensional world were described by an extensive model incorporating relative position and surfaces interlarded with local curvature, albedo and reflection coefficients, the strong interrelations of the physical effects and their non-linearities would allow no practical evaluation at this time.

Describing the road (the most important object of the road-following task) as a band infinitely extended in space allows a much more phenomenological model. Near to the observer the projection of the band will present a homogeneous area with smooth borders, if, additionally, smoothness of the curvature along the main spatial extension is assumed. Thus two-dimensional methods for either region or border detection are applicable and have been tried by several authors.

From image coding it is known that homogeneous image regions possess more redundancy than edges. Edges of road borders therefore offer the chance to fulfil the requirements for an informative task-relevant cue. They may be broken, disturbed by dirt, ambiguous or even invisible in some specific situations but in many driving situations at least parts of them are observable. The same emphasis on road edges is reflected by the standard road-marking methods.

If safety of recognition were to be of utmost importance, as for instance in a passenger car, the main concern would be the combination of cues to back up the system with sufficient redundancy to cope also with *very* ill-conditioned situations. The system, as it is presented here, was built with the intention of showing that a reliable and robust operation can be reached in a not too restricted environment such as an 'autobahn' or a rural road. For such environments markings are standardized, or, even if no markings are present on a rural road, some edge-type grey level changes remain visible in almost all cases, e.g. between road pavement and adjacent soil.

Another important class of objects is constituted by further traffic participants or obstacles. This class includes a wide variety of three-dimensional structures and two-dimensional images.

Because the class covers an extreme diversity of object images it might be advantageous to find universal filters for which intuitive concepts already exist. End-points of lines and corners present examples. Both feature types were investigated in the framework of dynamic vision. End-points of lines may be found by a non-linear combination of linear spatial filters [27].

A method to describe corners as Dirac peaks of the curvature of contours was developed by Dickmanns [28]. A real-time implementation of curvature approximations for corner detection is described in Kuhnert [27].

The large variety of images may also be handled straightforwardly by designing several filters, directly adapted to specific features of the class members. Dickmanns and Graefe [10, 11] give an example for corners with a definite orientation in the image. Such specific filters have already been utilized, for instance, in classical letter or face recognition, but in dynamic vision the evaluation of speed and robustness is the first consideration in the design process.

Moreover, a search strategy is incorporated into the filter to improve the efficiency of the computation. Also, several filters may be joined in a search network to recognize highly complex cue patterns.

All these local methods exploit the fact that singular points like corners possess a fairly clear correspondence between their position in three-dimensional space and their two-dimensional projection generated by the camera on the sensor plane, and also for different images of an image sequence. In a number of cases the correspondence will be not exact but, in general, features of this type possess a graceful behaviour which allows an approximate treatment as attributes fixed on the surface of the object.

In an image, such object features may be recognized via locally operating processes. The term ‘process’ will denote an abstract entity characterized by processing information. It might be implemented in hardware, software, or both. One way to derive an architecture consists of breaking down the task into processes and mapping this configuration onto a real system afterwards. If the mapping can be kept simple due to an appropriate structure of the target system a high degree of efficiency is to be expected in the implementation. A more detailed explanation and generalization of this idea is given in Graefe [29].

As mentioned already, to drive a vehicle on the road it is essential to gain relevant information from the environment for this task. To illustrate the paradigm of dynamic vision at the direct image-processing level, Figure 12.1 shows an example of extracting the borders of a natural road scene in stylized form.

Six border-detecting processes, which are indicated in Figure 12.1 by diagonal large rectangles, try to recognize single points belonging to the

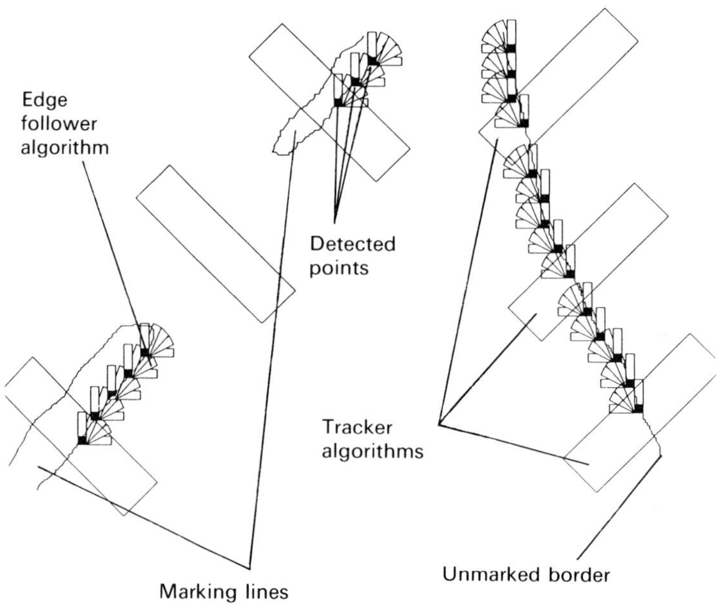

Figure 12.1 Example algorithm for road border detection.

road border. Position and direction of these processes are controlled at start-up time by knowledge about the expected typical appearance of the road and at run time by the information gained from the previous images.

If a road border point is detected in the example, an edge-following algorithm is started to exploit the spatial continuity of road edges and to deliver several adjacent border points. This algorithm is indicated by the small locally rotated rectangles. Detected points are depicted by small black squares and finally combined into a more analytical form of image description (e.g. a straight line or a parabola).

In the example only the analysis of one single road trace is treated. For a real application several other processes must operate on the same image, e.g. for obstacle or traffic sign detection, but the general approach may be the same as is shown in Figure 12.2, a more general situation during highway cruising.

Image regions that may be treated as an entity because they require and allow local processing will be named 'windows'. Through these windows normally a few processes look onto the image. At least one process has access to the image data. In a minimized system only those parts of the image covered by windows have to be transported to become available for the attached processes. Several different processes may work on the same window if different tasks have to be executed in that region of the image.

Normally, windows are rectangular because this contour is easier to realize in a hardware implementation. The restriction to rectangles does not impair generality very much because most compact regions of arbitrary shape may be circumscribed by a rectangle easily.

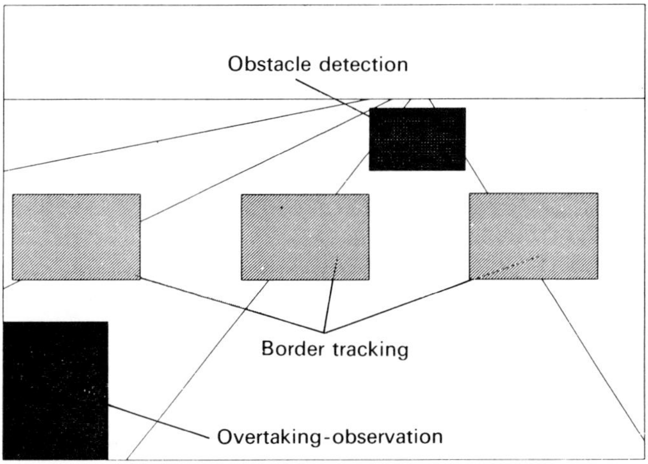

Figure 12.2 Several tasks in highway cruising.

Non-compact regions may be partitioned into a number of compact ones. In Figure 12.2 three windows are applied to track the border of the two tracks of the road. At the same time one window is looking for vehicles that may overtake and another one for those which may be obstacles. It can be seen easily that in typical applications several different window-bound processes should work at the same time, delivering their results to processes which specialize in fusing the information.

Expressed in more conceptual terms, the complete analysis consists of partial time-parallel processes arranged in a hierarchical order. For most applications, a full parallelization is not necessary. Parts of the hierarchy corresponding to sub-hierarchies, e.g. representing specific filters, may be executed sequentially. The decisions that are based on these *local* analyses will control the global progress of the entire analysis process.

12.4.1 Filters for dynamic vision

As discussed in the preceding section, local filters are needed to extract and track task-relevant features. They serve as 'elementary' processes for feeding the visualization of complex features and the modelling in space and time. Within this framework, the method of controlled correlation was developed, a generalized form of the well-known two-dimensional correlation function with respect to the 'squared sum of differences' (SSD) method.

The discrete non-normalized form of the two-dimensional correlation function is defined as:

$$C_{i,j} = \sum_{k} \sum_{l} I_{i+k,j+l} M_{k,l}$$

where

$I_{x,y}$ = intensity function of the image sampled at the point (x, y)
$M_{u,v}$ = template function for comparison sampled at the point (u, v)
$0 < i < I + 1;\ 0 < j < J + 1;\ 0 < k < K + 1;\ 0 < l < L + 1$

The function determines for each translation (i,j) the non-normalized linear correlation between the light intensity function of the image I and a template function M. The template function is normally constructed directly from the intensity function of a specific cue. Often the template function is also referenced as 'mask' function. The intensity function as well as the template function can be defined to be continuous. Thus, if the sampling theorem fits, both may be reconstructed entirely by appropriate interpolation from their discretized form shown above.

By computing the correlation function for all possible translations not only the correlation coefficient depicting the similarity for a specific translation will be found. Also the position in the image with the highest linear similarity may be determined by searching for the global maximum.

Computationally, the process is expensive because it needs $(I - K + 1) \times (J - L + 1)KL \times$ const. The constant factor summarizes the number of multiplications, and additions, and the effort of searching the global maximum of the correlation function.

Thus, the necessary number of operations per pixel ranges between 10^7 and 10^8 for reasonable image and mask sizes. Even for modern high-performance processors or special machines this constitutes a heavy burden.

On the other hand, cross-correlation offers some desirable properties. Communication theory has proven that correlation for a relatively large class of disturbed signals represents the optimal detection technique for linear filtering. Additive ergodic white noise is allowed by the theory.

Noise presents, besides speed, the most severe problem in filtering the data of real images. All physical analog systems are inevitably subject to thermal noise. Real electronic systems, furthermore, are influenced by disturbances like crosstalk or quantization. But not only deficiencies of the recognition system contribute to a non-ideal digitized image. Disturbances in the real environment, like dirt, haze and shadows, distort the features that a filter is designed for. As these few examples show, it is rather unlikely that all disturbances will fit into the white noise model for which optimality can be proven.

Thus, in images generated by a real video camera the assumptions of the theory very often are violated to a certain degree. Furthermore, it seems to be difficult to find a closed statistical description of these images which could serve as a basis for deriving theoretical results of the edge detection behaviour of the (correlation) method.

Hence, a comparison with other methods is of interest. Therefore, an experimental investigation of the recognition behaviour of several methods, including some standard edge-detecting techniques and correlation, was set up. To compare the different methods a quality measure was defined taking into account the spatial deviation from a known edge position and the likeliness of recognition of the edge. Synthetic edge images disturbed by pseudo-random noise, and varying the statistical parameters of the noise, the edge direction, and the signal-to-noise ratio of the edge systematically, were processed by the different edge-detecting techniques.

The same images were also presented to several human observers who were told to describe their impression. From the verbal statements edge images were produced to which the formal comparison procedure was applied afterwards. Some care had to be taken to prepare the experimental set-up in such a way that no biasing of the statistics of the images was introduced from ambient illumination or non-linearities in the image display system.

The general result of this procedure delivered a quality measure value for several image processing methods relative to the human visual system and the objective data because the image generation process of the synthetic images was completely known.

First of all, correlation under a wide variety of common statistical conditions produces a good and very noise-resistant compromise between detection and localization because it can incorporate a wide variety of edge models.

The most interesting result was, however, that the number of quantization levels of the mask function influences, at least for step edges, only very slightly the quality of recognition. Even if only three levels of quantization were allowed to the coefficients of the mask e.g. -1, 0, $+1$ the quality was even better than that of most classical edge detectors.

Correlation, furthermore, offers an additional degree of freedom, because the type of match function may be chosen in accordance with the knowledge that exists about the task. The method may be flexibly and easily adapted to a wide variety of situations. Thus, not only for those specific situations covered by the theory but also for a large variety of other similar situations, the method exhibits a robust and reliable behaviour.

As mentioned, two-dimensional non-normalized correlation has problems in real-time application due to the computational burden. To overcome this obstacle the method of controlled correlation was invented.

Controlled correlation consists of transforming the mask function of the position (0, 0) into a set of mask function M^i. The transformation may be described by the mapping $M(x, y) \Rightarrow M(z_x(x, y), z_y(x, y))$. Arbitrary transformations are possible in general, including the translation, rotation and arbitrary deformation of the sampling grid, but up to now only affine transformations have been implemented. To achieve correct results discrete mask functions should be interpolated and resampled to the original sample grid after the transformation to remain applicable in real time.

Normally, the complete transformation is performed in advance of the real-time run. Thus, at run time only a standard cross-correlation has to be computed. The method is only applicable and competitive with others like the Newton–Raphson search or parameter estimation if knowledge exists about the possible appearances of relevant features in the image and if the mask set or multiple mask set can be arranged in memory in such a way that the search region in the parameter space is kept small and the masks are accessible rapidly at run time. Generally, the method employs an exchange of memory space with computing time.

The transformation is chosen in accordance with the *expectation* about the appearance of a feature in the next image. Due to steadiness of motion, only small changes in position will be expected and the amount of translation or rotation in the image can be limited to a certain region. For implementation these regions have to fit into a window.

If, additionally, information on the probability of specific parameter combinations exists, e.g. if a certain type of visible motion is more likely to occur than others, this knowledge may be exploited to order the set of masks, searching first for the most probable images.

Again, due to the steadiness of motion the ordering is mostly sequential. If one point of the original untransformed mask function is chosen as a reference point (e.g. the origin of the coordinate system of the mask function) in the sequential ordering, the movement of this point constitutes a line, which will be called a search path.

By defining regions of possible motions the amount of computation may be decreased by two orders of magnitude with respect to plain two-dimensional correlation. By introducing sequential search paths for edges, a factor of 10 can be gained.

For the economy of filtering a hierarchy of filters presents a compromise between flexible descriptive power concerning selectivity and computational effort. Very large filters with arbitrary match functions could recognize different cues very selectively, but would force a recomputation of similar or equal parts of the filters at the same time. If these common subfilters are determined and computed at first, to be combined afterwards, computational effort is substantially saved. Flexibility is constrained by the hierarchical approach because the number of elementary filters has to be kept small to gain the real advantage from the decomposition and it becomes impossible to encode arbitrary match functions. Therefore, filters at the smallest resolution level should encode properties likely to be of universal usefulness.

Short-edge elements constitute an example. They may be utilized for cues of object or albedo boundaries and combined into complete contours.

Correlation masks of edge elements may be adapted for different needs. For example, by extending an original mask along the expected contour the spatial selectivity and the noise suppression will be increased. Rotated mask sets may be used to compensate for the rotational motion of the observing camera. Also, properties of the imaging situation may be taken into consideration. Visual motion blur or focusing of the camera could influence the final coefficients of the mask function.

Reducing the amount of computation is essential in reaching real-time performance. Compiling all the available knowledge about parameter ranges and variations into mask sets in advance and executing only those parts which are really needed at run time is the key idea of controlled correlation. Further adaptations of the method are possible to optimize speed.

As already mentioned, very coarsely quantized correlation masks exhibit nearly the same recognition performance as arbitrary finely quantized ones. Certain symmetry conditions must be observed [30] to guarantee that the positions of all local maxima are placed at the same coordinates as for normalized correlation. The number of quantization levels may be as low as three if the desirable features are to be maintained. Therefore, masks with coefficient values of -1, 0 and $+1$ (ternary masks) are the most widely utilized ones in dynamic vision. Most mask functions quantized to a ternary mask will contain a number of zero elements. At these locations the original image function will have no influence on the result of the filtering. Therefore, those positions measured in mask coordinates do not need to be taken into account at all during the evaluation of the correlation.

This class of algorithm may be implemented straightforwardly. At run time only simple arithmetic and memory access are to be executed. The structural components of the algorithms may be implemented via lists of memory addresses.

A special compiler was developed [31] which supports the interactive generation of masks and their conversion into efficient machine code by several optimization techniques. For example, if equal parts of masks have to be computed on the same image region, multiple computations may be avoided by restructuring the mask set utilizing the method of common subexpression elimination. A graphical interface supports the image-oriented input to the compiler.

The sketch of an implemented filter algorithm for dynamic vision road boundary detection is shown in Figure 12.3. Only one window covering the left shoulder of a road is magnified. Thus, the depicted region represents only a small part of the complete image.

The feature that the filter is designed for consists of the edge of a white marking line at the left side of the pavement. The line is assumed to be broken in the example. Five mask sets have been generated by translating the elementary edge element filter, shown at the right side of the figure, perpendicularly to the expected position of the edge, indicated by the dotted line. Only 2016 pixels are involved in the computation. After applying the mask sets to the image and deriving five sets of correlation values, the local maxima within each set are determined.

Even if the shape of the elementary filter is chosen with the intention of increasing the signal-to-noise ratio in the real image it is possible to avoid noise disturbing the correlation function. The fact that a spatial order is defined at the mask sets may be exploited by a non-linear low-pass filtering of the correlation function before searching the local maxima. Afterwards, each mask set delivers the position of the one edge element.

Obviously, one element tracker missed the correct position in the example. Errors of this kind have proven to be very typical due to multiple lines, dirt, shadows and other disturbances. They are also completely unavoidable in

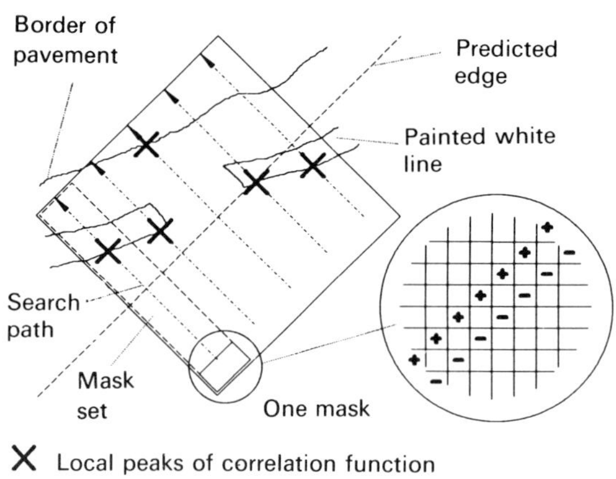

Figure 12.3 Local border tracker.

the long sequences of many thousands of images which must be processed in real experiments. Only a combination and a cross-checking of redundant measurements allows us to reach the needed robustness. Additional knowledge about the appearance in the image of the three-dimensional structure and motion of an object has to guide the fusion process, e.g. constraints of steadiness in space or time may be applied. If white lines are known to be only slightly curved in the projection, outliers as in the border recognition example may be diagnosed and handled separately [30]. Modelling in space and time [9] describing the relations between the measurements by differential equations is an even more powerful dynamic vision method.

Generally, it is necessary to acquire a large number of unreliable measurements and to fuse them using global knowledge to obtain a robust system. The high sampling rate in a video system becomes especially important because the effort for feature tracking increases severely with increasing sampling period.

Several variants of the described algorithm with different numbers of mask sets, mask orientation and mask function were applied in real-time systems. Extensive experiments with videotapes and 8 mm films, recorded from a driving car or a landing aircraft, showed the robustness of the method.

Corners present another type of cue having an even higher informational content than edge elements because their positions may be determined unequivocally. Mostly, they have to be searched in a planar region. Only if sufficient knowledge is available may the movement be predicted and the amount of possible transformations of the mask function be restricted. Therefore, the effort of corner detection normally increases by second order with the amount of uncertainty, whereas the detection of edge elements with sufficiently small curvature may be reduced, as was shown above, to a one-dimensional problem. In this case the curvature has to be small with respect to the uncertainty of the movement measured along the edge.

One algorithm to detect corners computed the curvature along an object contour by an approximation and detected local maxima of this function. Several different approximations were implemented.

Also, controlled correlation can detect and locate corners. A filter for a specific type of corner was implemented to show the flexibility of the method. This type of feature detector is mainly suited to tracking positions of moving objects like cars.

To summarize the above, algorithms and especially controlled correlation, which are well adapted and optimized for use with dynamic vision, have been presented.

12.5 HARDWARE ARCHITECTURE

Figure 12.4 illustrates the general architecture of the hardware: several processes are running in parallel (dotted lines depict the information flow,

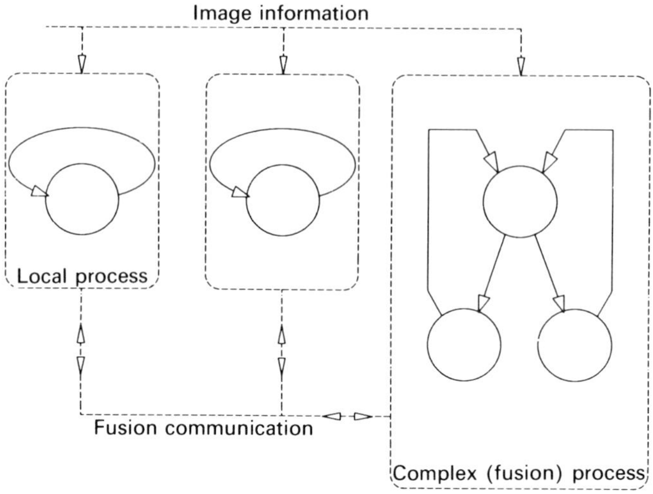

Figure 12.4 General processing scheme for task-oriented real-time vision.

full lines show the processing flow) because their information is needed concurrently whereas several subprocesses within one process may be executed sequentially for efficiency reasons. Thus, they can share the same hardware resources. All this holds also for the higher levels of interpretation. Therefore, they will not be treated separately.

Different mappings of the processes onto the hardware are possible. One window process may be related to one processor. Also it would be possible to attach a processor to each subprocess. Furthermore, computationally intensive parts of the decision tree could be mapped on one processor and the decisions themselves on another. Indeed, all these configurations were realized.

At this point granulation of the system comes into consideration. Principally, it would be possible to break the subprocesses down to smaller pieces, deriving a massively parallel hardware architecture [32, 33, 34], but from the beginning the system was intended to be built out of rather conventional 'easy to handle' hardware. Therefore, a coarsely grained solution was preferred and mostly standard microcomputers were employed.

The demands on the architecture consist of a parallel access to the image data. Also, the processed information needs to be moved around in the system, and therefore the configuration should be flexible enough to allow, for example, several types of hierarchy. Furthermore, *overlapping* regions of the image might be accessed at the same time by different processes.

As mentioned, a natural task distribution could be realized by applying one processor to each local image region. For a natural scene, however, it is not guaranteed that those local image areas are placed at a fixed grid. In fact, just the opposite is the case. The characteristic features are normally spread very unevenly across the image. The hardware should thus offer the possibility of distributing its computing power according to its submitted task.

The various processors of a multiprocessor system therefore should be able to access individual areas with previously unknown locations in the image. Clearly, all image data are originating from one source: the camera. If they are stored in *only one* image memory, then access conflicts of the various processors to this one image memory are unavoidable. These access conflicts may be overcome by dividing the image memory into segments and allocating one processor to each section of the image memory. Such a rigid configuration with fixed segmentation leads to the problem that uniform-feature regions may be divided into several segments, and therefore artificially intersected, causing the different processors to communicate additionally with each other in order to compensate for the rigid partitioning of the data. Besides, different local-feature regions may jut out of one memory segment, increasing even more the burden upon particular processors and causing uneven utilization of the various processors.

If it is assumed that the needed amount of processing power is connected to the area of the attached region roughly via a monotonous non-falling function, it could, furthermore, be necessary to work with several processors on the same window if it is large, or different algorithms should be applied to the same region.

A distributed image memory presents a solution to these problems. A

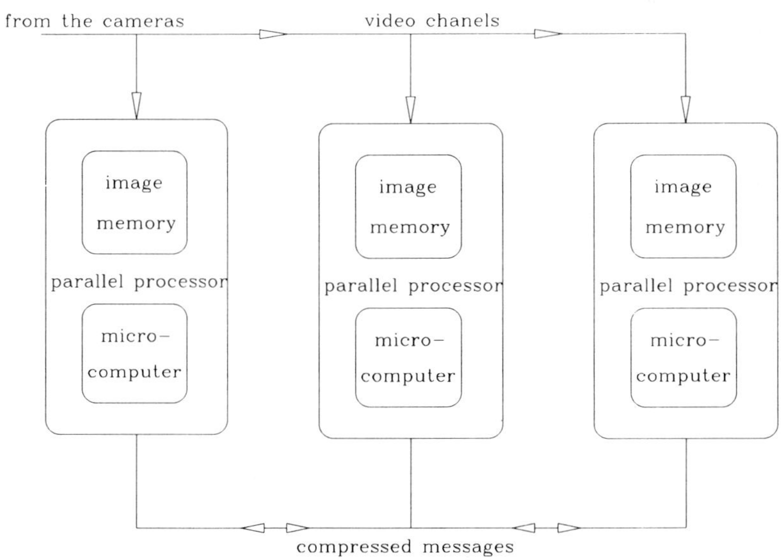

Figure 12.5 General hardware structure of the BVV image processor family.

memory, large enough to contain a complete image, is possessed by each processor, and all these memories are fed, in parallel, from the camera to the analog to digital converter. Flexibility is the main advantage of this concept. The processors communicate with each other via a second, separate communication bus (Figure 12.5). The architectural concept of a coarse-grained multimicroprocessor system, containing two specialized buses and a distributed image memory, is described in Graefe [35] and Graefe and Kuhnert [36].

The utilization of current microcontrollers permits a compact realization of this concept, which is explained below.

12.6 HARDWARE REALIZATION

The BVV family realizes the claimed properties. A videobus system transports the video data from analog–digital converters to the several parallel processors, granting them only read access.

The BVV 1 was the first member of the family. It was replaced by the BVV 2 [35] (Figure 12.6) on the advent of 16-bit microprocessor technology (8086). Both systems basically possess the same structure. Thus, only the BVV 2 will be described here briefly.

Parallel processors made up of single-board computers process the information. Mostly they are equipped with a special hardware circuitry videobus interface for acquiring and storing windows out of the sequential stream of video data on the videobus. But parallel processors may also be connected to other special interfaces if necessary.

The videobus itself is designed to transport up to four image sequences independently and in parallel. In an autonomous vehicle application, for instance, several cameras with different viewing directions or resolutions may be dedicated to different tasks, like obstacle detection, observing the road border, or checking for overtaking vehicles. Also, colour cameras might be utilized for filling up two or three video channels.

For control applications the images on distinct video channels are not allowed to be completely independent because defined temporal relations between them are needed to reach stable and reproducible conditions for the control algorithm. A central clock generator therefore provides the synchronization signals.

The videobus is fed either by analog-to-digital converters connected to video cameras or by pixel processors which are dedicated to receive an image from one channel of the videobus, process it and transmit the result over another channel. Pixel processors normally execute linear or non-linear local filter algorithms schematically on a whole image. For convenience of the human operator of the system, the images on the videobus may be displayed on TV monitors. Additionally, information about image regions involved in processing is overlaid on the display.

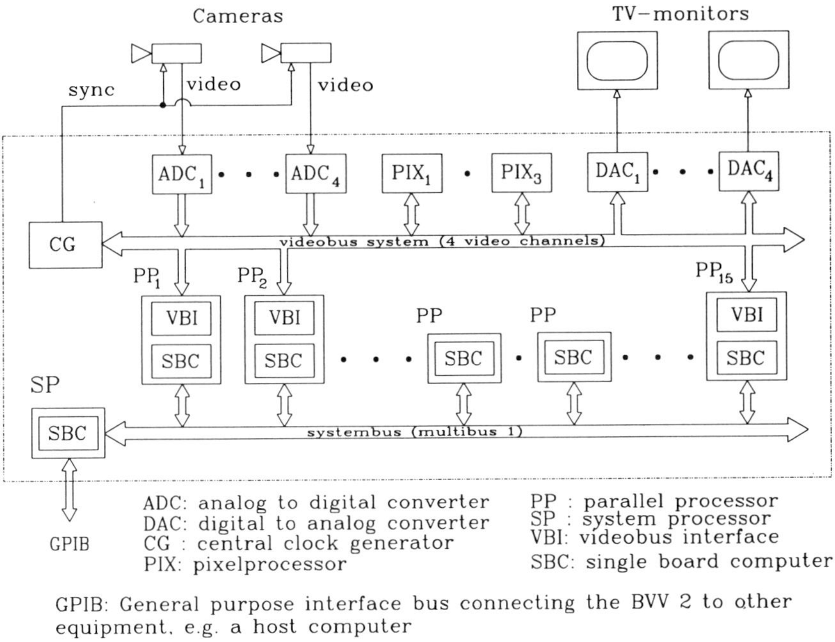

Figure 12.6 The parallel image-processing system BVV 2. ADC = analog-to-digital converter; CG = central clock generator; DAC = digital-to-analog converter; GPIB = general purpose interface bus connecting the BVV 2 to other equipment, e.g. a host computers; PIX = pixel processor; PP = parallel processor; SBC = single-board computer; SP = system processor; VBI = videobus interface.

The parallel processors communicate over the system bus. An asynchronous message-passing scheme is utilized for this purpose. Each processor may send out a message at any time while the acceptance of messages takes place cyclically. The communication bus traffic is organized and managed by a separate processor called a system processor, which also controls the message exchange to external computers via the GPIB.

In a number of real-time experiments the performance of the BVV 2 was proven. Execution of the road-border-tracking algorithm described in the preceding section, for instance, needs less than 17 ms for three to five small edge elements on one parallel processor of the BVV 2.

The BVV 3 has been designed by exploiting the experience gained during the experimental work with former systems. Because the basic structure had proven its effectiveness, the internal structure of the parallel processor was changed to bring it even closer to those type of algorithm which had been found particularly useful for real-time vision.

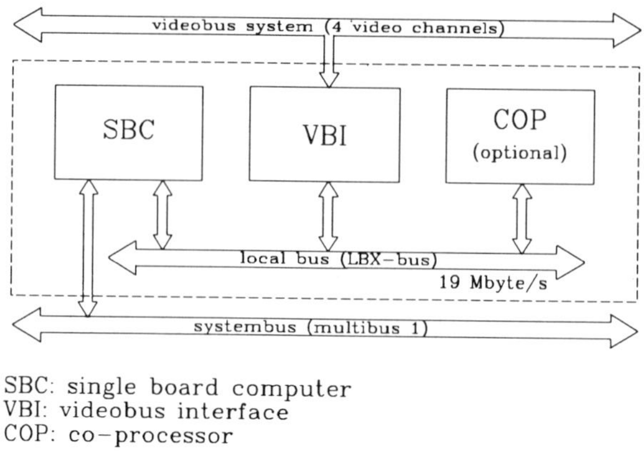

Figure 12.7 Sketch of the parallel processor of the BVV 3. SBC = single-board computer; VBI = videobus interface; COP = co-processor.

Figure 12.7 shows an overview of the new parallel processor. The gradual improvements in microcomputer technology led to a change to another type of microprocessor (80286, 80386) and to the expansion of the capacity of the videobus interface from 2×4 kbytes for the BVV 2 to full images for the BVV 3. Thus, within the new system no restrictions of the window size now exist. In the extreme case one window may cover the whole image.

Also, the processors of the newest version may be optionally equipped with a specific RISC processor (reduced instruction set computer) designated to alleviate especially the burden of sequential arithmetic and other frequently needed operations. The new processor, called a co-processor, constitutes a completely new structural element within the system. It shares the local bus with the microprocessor to have access to the image and the other memories. The co-processor is based on a modified Harvard architecture to support sufficient internal bus bandwidth.

The combination between a general-purpose microprocessor and the co-processor maps very exactly to the process shown in Figure 12.4. Sequential parts of the analysis may be executed on the co-processor whereas overall control of the execution, e.g. recursion and decision making, is performed by the more general microcomputer.

The internal architecture of the co-processor first adapts the hardware instructions closely to the task and allows high-speed execution because memory channel bandwidth is high, due to the existence of separated buses for instruction, data and constants. The machine performs 10^7 complex operations per second. Furthermore, it has a pipeline structure with seven stages where each complex instruction may be composed of seven elementary operations like address generation, addition, multiplication, maximum detection or maximum localization. One of the main applications of the co-processor is in executing controlled correlation. It is, however, also applicable to linear or non-linear filtering, or neural network computation.

At first, a prototype of the co-processor was built as a breadboard model. Recently, an advanced version became operational [37] which is about 15% faster than its predecessor.

Due to the complex structure of the machine, especially the pipelining, programming it in assembler may be tedious and error-prone. A special software package was written to support the implementation of the real-time vision algorithm [31]. The intended computation may be defined interactively by either a graphic or a formula-like description allowing for a compact problem-oriented formulation. The built-in compiler for the co-processor not only takes care of the hardware structure; several optimizations decreasing the execution time, for instance, of controlled correlation are performed too.

12.7 APPLICATION EXAMPLE OF THE CO-PROCESSOR

To show the performance of the co-processor concept, an example of an application will be discussed. The task was to detect the edges of a secondary road. Several different algorithms were combined into one complex filter and implemented on the preliminary version of a parallel processor of the BVV 3 to demonstrate also the flexibility of the system.

An overview of the algorithm has already been given in Figure 12.1. First, road border candidates are detected by controlled correlation utilizing exactly the tracker algorithm explained in Figure 12.3. One mask covers a region of 14 pixels and the mask set for one search path consists of 30 masks. Due to the overlap of the masks only 37×7 pixels have to be evaluated. Because nearly the complete computation can be executed by the co-processor, the result of tracking is obtained after 0.03 ms. This first step delivers not only the image coordinates of one small edge element but also a quality measure which will be utilized later on in the fusion process.

At the border point found by the tracker, a spatial edge-following algorithm is started. The follower matches sets of rotated and translated edge element masks with the image data. To restrict the search space, translations are computed at run time. Thus, the algorithm steps sequentially from one border point being found by rotational variation to the next one, until it is data-dependently terminated.

Each rotational search step evaluates seven masks of 7×3 pixels. Up to eight sequential border points will be found. The microcomputer has to manipulate the coordinates for the translation. Because the microprocessor is much slower than the co-processor, this type of algorithm, allowing only a part of the parameter variation to be computed in advance and deposited in memory, needs 0.25 ms to execute.

Among the whole image, six tracking and six spatial edge-following processes are distributed and run concurrently. Of course, they are mapped onto the sequential hardware by a time-slicing method. Process management needs about 1 ms, causing a total time consumption of 2.8 ms for handling all 12 processes.

The coordinates of 48 border points and their affixed quality measures are fused into borders by two straight line models fixed to image coordinates [36]. By coupling the simple local filters at a higher level, specific knowledge about redundancy contained in the data can be introduced and occasional failures of the only locally operating filters may be tolerated or even compensated.

Scenes of secondary roads were recorded on video-8 tape to be utilized as test data for the complete system. The passenger car carrying the camera was driven manually at a speed between 60 km/h and 80 km/h and the tape finally covered about 15 km of roadway.

Afterwards, the tape was fed to the image-processing system (i.e. an early version of the BVV 3 with only one parallel processor) and processed by the above algorithm. The complete process, including both two-dimensional models, which mainly employ the microprocessor, runs in 4 ms. Thus, it leaves plenty of time for extensions because it uses only 25% of one processor of a multiprocessor system if a measurement cycle rate of 60 Hz is required.

To test the robustness of the system, extensive real-time experiments were conducted. For the right border of the road no problems occurred over the entire data set whereas the left side was lost typically three or four times on each run over the 12 minutes of the tape (corresponding to an approximate 21 000 images). Broken lines and visual ambiguities of features near to the horizon caused the difficulties, but due to the two-dimensional models the algorithm was able to recover correct function after a short delay of typically less than one second. The scenes contained several difficult situations like overexposures caused by reflections of the sun on the moist road surface and cast shadows from trees. Furthermore, no road markings existed on several portions of the way. All these situations were handled correctly.

12.8 FUTURE HARDWARE

The BVV family had reached a certain degree of maturity with its last member. But the new developments in microcontroller technology and the demand for compact hardware in mobile applications, for instance visual indoor navigation [38, 39], led to the design of a new version.

A design goal for the new system was to develop a more compact construction that could be simply integrated into a standard hardware environment. Also, the processing power of the BVV 3 [29] should be improved compared to that of the BVV 2 [35]. Furthermore, simple compatibility was required with the existing software. All these design criteria were fulfilled by the i80376, which is a processor intended for controller application.

The well-proven overall architecture for the BVV was the basis for the design of the new system. For loading programs and testing, all the BVVs need a host computer. One major change in the new system was a PC-AT-compatible machine chosen as the supporting hardware environment. The cost-to-quality relationship of commercially available frame grabbers means that such a board is the ideal choice for feeding a digital video data bus without the need for compromises.

For digitizing of the video input a standard grabber board having a digital videobus output is utilized. The image memory of this board is not necessary for data acquisition during the real-time runs. It may be used only for marking relevant features in real time with a short delay. Each processor board of the new system is a single printed circuit board fitting into a PC-AT-type computer. Thus, the maximum processing power of the system is limited mainly by the number of available slots.

In order to make the data, streaming on the digital videobus, available to the processors, in principle two general design ideas exist. In the first one, as realized in the BVV 2, the image memory (videobus interface) of one of the parallel processors utilizes the double-buffer principle, i.e. it possesses two parallel memory banks. While one of them is connected to the videobus, the other one co-operates with the processor simultaneously.

Which bank is connected to the processor at a certain time is determined by the image frame signal and can be changed at several patterns, e.g. at the end of each full frame both banks may be exchanged. The switching is performed by hardware. Therefore, the input of image data is completely transparent to the program running on the processor.

The double-buffer method has the advantage that, when the image is completely read into memory, this information is available to the processor, virtually at the same moment.

Realization of the double-buffer configuration in hardware is relatively expensive in terms of space and complexity even though special LSI ICs might be used. Therefore, the new design utilizes the built-in high-performance direct memory access facility of the 32-bit RISC microcontroller i80960 to transfer the data from the videobus to the image memories. The circuit board space was thus reduced to 25% of the original size, so that now a complete parallel processor fits onto a standard PC board.

Because the transfer is controlled by software, several acquisition patterns are realizable. Therefore, it is possible to acquire horizontal stripes out of an image. Also, subsampled images consisting only of every second row of the original images may be read in. In this way the part of the processor time necessary for inputting the image data while inhibiting the processor bus can be further reduced.

The different methods of accessing the image data conform clearly with the concept of dynamic vision. Only relevant image regions may be transferred and processed. Further, the selection of the image data can be controlled actively by the analysis process.

Like the BVV 2, processors communicate over a common bus. The PC-

AT works like a telephone exchange, and passes the messages (mainly during real-time runs) between the parallel processors.

The parallel processor can deliver high performance (up to 60 million instructions per second) and has a high internal bus bandwidth for image transfers (up to 59 Mbytes/s). Also, the memory capacity is larger compared to the parallel processor of the BVV 3. A 4 Mbyte dynamic random access memory operated at an average cycle time of 60 ns is utilized. The basic functions of real-time image and message transfer of the new board have been tested.

A disadvantage of the new processor is the fact that a completely different interface between hardware and operating system exists. Thus, the hardware drivers and the compiler had to be exchanged. This work is going on now.

The most interesting next step will be the transfer of an application running on the recent BVV 3 to this new hardware.

12.9 CONCLUSION

Dynamic vision for high-speed autonomous vehicle applications has been discussed within the framework of goal-oriented sensing. The software structure of algorithms for the robust recognition of road borders and specific features on other objects like obstacles has been shown. Also, several examples of implementations of an appropriate hardware architecture have been presented.

Two examples have been discussed to illustrate the performance of the compoundly optimized systems.

One shows the detection of the border lines of a marked road. It was implemented on the BVV 2 and runs less than 17 ms for one complete evaluation. This system configuration was employed in guiding a test vehicle on a German autobahn at a speed of nearly 100 km/h [16].

The other example demonstrates the measurement process of the border of a rural road which is robust enough to cope with a number of difficult visual situations. Due to improved hardware (compared to the BVV 2) the system, even if its complexity is higher by a factor of approximately 20, needs only 4 ms for execution.

Thus, real-time image processing can also be realized with moderate hardware power by utilizing a suitable system architecture.

Finally, we have discussed a newer implementation of the architecture employing an up-to-date microcontroller which additionally offers a very compact and cost-effective realization.

ACKNOWLEDGEMENT

This research was conducted while the author was with the Universität der Bundeswehr München, Fakultät für Luft- und Raumfahrttechnik, Institut für Meßtechnik headed by Professor Graefe. Thanks to all the members of the institute who helped to make the research happen.

REFERENCES

1. Kuhn, T.S. (1962) *The Structure of Scientific Revolutions*, Chicago, University of Chicago Press.
2. Lorenz, K. (1965) *Evolution and Modification of Behaviour*, Chicago, University of Chicago Press.
3. Uexküll, J.V. (1909) *Umwelt und Innenleben der Tiere*, Berlin.
4. Tinbergen, N. (1951) *The Study of Instinct*, Oxford University Press, London.
5. Pomerleau, D.A. (1990) *Neural Network Based Autonomous Navigation. Vision and Navigation: The Carnegie Mellon Navlab*, Kluwer Academic Publishers, 1990, Chapter 5.
6. Kuhnert, K.-D. (1989) Real-time suited road border recognition utilizing a neural network technique, in *Proceedings, IEEE/RSJ International Workshop on Intelligent Robots and Systems* (*IROS '89*), Institution of Electronic and Electrical Engineers.
7. Brooks, R. (1986) A robust layered control system for a mobile robot. *IEEE Journal of Robotics and Automation*, **RA-2**, 14–23.
8. Aloimonos, Y. and Rosenfeld, A. (1991) Computer vision. *Science*, **253**, 1249–54.
9. Dickmanns, E.D. (1988) Object recognition and real-time relative state estimation under egomotion, in *Real-Time Object Measurement and Classification* (ed. A.K. Jain), Springer-Verlag, Berlin, pp. 41–56.
10. Dickmanns, E.D. and Graefe, V. (1988) Dynamic monocular machine vision. *Machine Vision and Applications*, **1**, 223–40.
11. Dickmanns, E.D. and Graefe, V. (1988) Applications of dynamic monocular machine vision. *Machine Vision and Applications*, **1**, 241–61.
12. Wünsche, H.J. (1987) *Erfassung and Steuerung von Bewegungen durch Rechnersehen*. Dissertation, Fakultät für Luft- und Raumfahrttechnik der Universität der Bundeswehr, München.
13. Eberl, G. (1987) *Automatischer Landeanflug durch Rechnersehen*. Dissertation, Fakultät für Luft- und Raumfahrttechnik der Universität der Bundeswehr, München.
14. Dickmanns, E.D. and Zapp, A. (1987) Autonomous high speed road vehicle guidance by computer vision. Preprint, *10th IFAC Congress*, München, Vol. 4, pp. 232–7.
15. Kuhnert, K.-D. and Zapp, A. (1985) Wissensgesteuerte Bildfolgenauswertung zur automatischen Führung von Straßenfahrzeugen in Echtzeit, in *Mustererkennung* (ed. H. Niemann), Springer-Verlag, Berlin, pp. 102–6.
16. Zapp, A. (1988) *Automatische Straßenfahrzeugführung durch Rechnersehen*. Dissertation, Fakultät für Luft- und Raumfahrttechnik der Universität der Bundeswehr, München.
17. Davis, L.S., Le Loigne, J. and Waxman, A.M. (1986) Visual navigation of roadways. Preprint, *Conference on Intelligent Autonomous Systems*, Amsterdam, Dec. 1986, pp. 21–30.
18. Maruya, T., Takahashi, H. and Katemi, A. (1991) Development of an autonomous onroad vehicle, in *Proceedings of the Sixth International Conference on Automotive Engineering*, 1991, Seoul, Korea.
19. Thorpe, C., Herbert, T. Kanade, T. and Schafer, S. (1988) Vision and navigation for the Carnegie-Mellon Navlab. *IEEE Transactions on Pattern Recognition and Machine Intelligence*, **10** (3).
20. Tsugawa, S. (1979) An automobile with artificial intelligence, in *6th IJCAI*, Tokyo, 1979.

21. Turk, M. (1986) Color road segmentation and video obstacle detection, in *SPIE Conference 727 on 'Mobile Robots'*, Cambridge, Ma., USA, 1986 SPIE, Washington.
22. Zimdahl, W., Rackow, I. and Wilm, T. (1986) OPTOPILOT ein Forschungsansatz zur Spurerkennung and Spurfuehrung bei Strassenfahrzeugen. *VDI Berichte Nr.*, **162**, 49–60.
23. Enkelmann, W. (1990) Interpretation of traffic scenes by evaluation of optical flow fields from image sequences, in *Control, Computers, Communications*, Selected Papers from the IFAC/IFIP/IFORS Symposium, Paris, France, 1989 (ed. J.P. Perrin), Pergamon Press, pp. 43–50.
24. Meygret, A. and Thonnat, M. (1990) Segmentation of optical flow and 3D data for the interpretation of mobile objects, in *Proceedings of the International Conference on Computer Vision*, December 1990, Osaka, Japan.
25. Kuhnert, K.-D. (1989) Sensor modelling as basis of subpixel-image processing, in *Image Processing III*, Proceedings of the SPIE, Vol. 1135 (ed. J. Duvernay), SPIE, Washington.
26. Graefe, V. (1990) The BVV-family of robot vision systems, in *Proceedings of the IEEE Workshop on Intelligent Motion Control*, Istanbul, August 1990, IEEE.
27. Kuhnert, K.-D. and Graefe, V. (1988) Vision systems for autonomous mobility, in *Proceedings of the IEEE International Workshop on Intelligent Robots and Systems, IROS '88*, Tokyo, pp. 477–82.
28. Dickmanns, E.D. (1985) 2D-Object recognition and representation using normalized curvature functions, in *Proceedings of the IASTED International Symposium on Robotics and Automation '85* (ed. M.H. Hamza), Acta Press, pp. 9–13.
29. Graefe, V. (1989) Dynamic vision systems for autonomous mobile robots in *Proceedings, IEEE/RSJ International Workshop on Intelligent Robots and Systems* (*IROS '89*), Tsukuba, pp. 12–23.
30. Kuhnert, K.-D. (1986) A modeldriven image analysis system for vehicle guidance in real-time, in *Proceedings of the Second International Electronic Image Week*, CESTA, Nizza, April 1986, IEEE.
31. Graefe, V. (1989) A flexible semi-automatic program generator for dynamic vision systems, in *Proceedings of the Workshop on Machine Intelligence and Vision*, Tokyo, April 1989, IEEE.
32. Hillis, D. (1985) *The Connection Machine*, MIT Press, 1985.
33. Weems, C.C. and Lawton, D.T. (1986) Incorporating content addressable array processors into computer vision systems, in *Proceedings of the SPIE Conference on Applications of Artificial Intelligence*, 1983, SPIE, Washington.
34. Weems, C.C., Levitan, S.P., Hanson, A.R. and Riseman, E.M. (1989) The image understanding architecture. *International Journal of Computer Vision*, **2**, No. 3.
35. Graefe, V. (1984) Two multi-processor systems for low-level real-time vision, in *Robotics and Artificial Intelligence* (eds J.M. Brady, L.A. Gerhardt and H.F. Davidson), Springer-Verlag, pp. 301–8.
36. Graefe, V. and Kuhnert, K.-D. (1988) A high speed image processing system utilized in autonomous vehicle guidance, in *Proceedings of the IAPR Workshop on Computer Vision*, Tokyo, pp. 10–13.
37. Graefe, V. and Fleder, K.H. (1991) A powerful and flexible co-processor for feature extraction in a robot vision system, in *Proceedings of IECON*, October 1991, Kobe, Japan, IEEE.
38. Kuhnert, K.-D. (1990) Dynamic vision guides the autonomous vehicle 'ATHENE', in *Japan–USA Symposium on Flexible Automation.*, Kyoto, July 1990.
39. Kuhnert, K.-D. (1990) Fusing dynamic vision and Landmark Navigation for autonomous driving, in *IEEE/RSJ International Workshop on Intelligent Robots and Systems* (*IROS '90*), Tsuchiura, July 1990, IEEE.

CHAPTER 13

Beacon–vehicle link in the 1–10 GHz frequency range

13.1 INTRODUCTION

Many applications for a road–vehicle link can be imagined, such as automatic toll systems, road pricing, vehicle identification, freight and commercial vehicle management systems, vehicle information systems, road guidance, etc. For these different applications, the link is not the same in terms of:

- range, which can be extended from few metres up to few hundred metres;
- message length (a few bytes in some applications to a few kilobytes);
- access (simple or multiple).

The objective of this chapter is to describe the performance of the physical link in the 1–10 GHz frequency range and not the types of modulation or the coding techniques to improve the quality of the link. We emphasize some intrinsic characteristics such as attenuation in air or due to a conducting layer on the windscreen near the transmitting antenna. The effect of the multipath propagation will than be studied. However, one can expect that the range and the directivity of the antennas will be important parameters. In a preliminary step, an attempt to classify the applications is thus presented.

13.2 CLASSIFICATION OF THE APPLICATIONS

The applications can be roughly divided into three categories, according to the necessary range.

A short-range link, of about 10 m, is needed in tool systems. As represented in Figure 13.1 directive antennas can be put on a canopy above the road, each of them producing a small and well-defined electromagnetic (EM) spot on each lane.

A medium-range link corresponds to communication over a distance of about 70 m and is used, for example, for road guidance applications. Figure 13.2 shows such a configuration. Since the height of the pole supporting the fixed antenna is limited in most cases to few metres, the angle between the beam and the road surface is very small. Reflections and diffractions on the

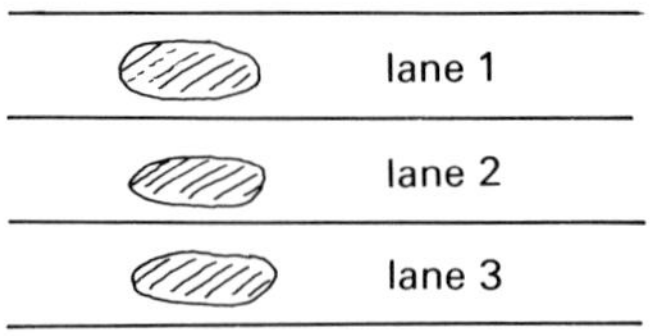

Figure 13.1 Plane view of the EM spots on each lane.

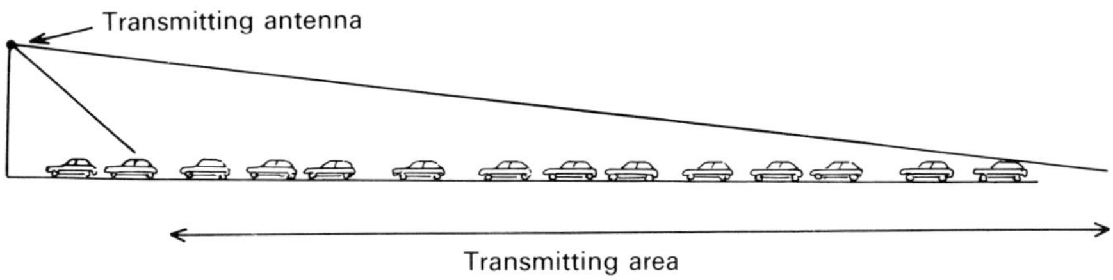

Figure 13.2 Medium-range link: configuration for road guidance application.

vehicles and obstacles will occur and, contrary to the case of the short-range link, multipath propagation will be an important constraint.

A long-range link of a few hundred metres can be used in tunnels for road information such as congestion and accidents. In this configuration, the reflections on the walls of the tunnels are important but we shall see that the linear attenuation is rather small, allowing communication over a large distance.

These three types of communication will thus be successively studied after a presentation of the intrinsic attenuations due to some physical constraints which are the same for these various links.

13.3 INTRINSIC ATTENUATIONS

The first step in characterizing these links is the determination of the intrinsic parameters as attenuation due to propagation in air and through the windscreen and the influence of the weather conditions: snow, rain, etc.

13.3.1 Propagation in air

Figure 13.3 presents the attenuation of the signal due to oxygen and water molecules. We can see that the attenuation in air over a few hundred metres is negligible in the 1–10 GHz band, whatever the weather conditions.

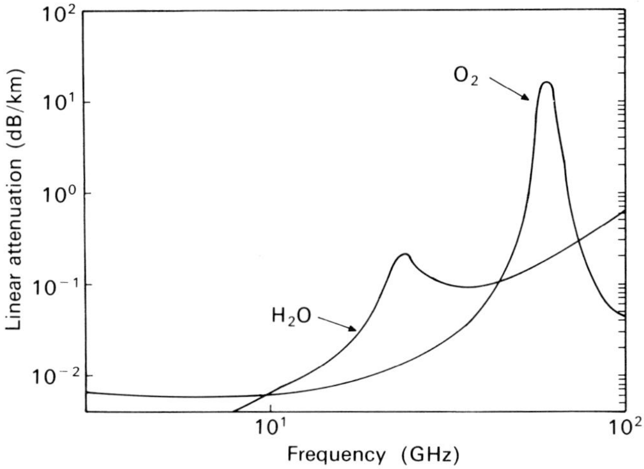

Figure 13.3 Attenuation due to gas atmosphere at sea level. Temperature = 17°C, water vapour = 7.5 g/m³, pressure = 1013 mbar (O_2) or 1020 mbar (H_2O). (After Ulaby *et al.* [1].)

13.3.2 Attenuation due to the windscreen

The windscreen can be made of a few layers of different materials but no intensive measurements have yet been performed on various types of windscreen. Furthermore, the transmission losses depend on the angle of incidence, the polarization and, of course, the thickness and type of windscreen.

However, at 10 GHz, measurements on a 'standard' windscreen (Peugeot 505) have shown that the attenuation is small (about 1–2 dB) for an angle of incidence of about 0° either for vertical (the electrical vector is in the plane of incidence) or horizontal (the electrical vector is perpendicular to the plane of incidence) polarization (Figure 13.4) [2].

More investigations have to be made to evaluate different types of windscreen in various configurations and especially to study attenuation caused by windscreens with a metallic heating structure which could act as an obstacle to microwave propagation.

13.3.3 Influence of the weather conditions [2]

The attenuation due to various thicknesses of water, snow and melting snow on the windscreen has been measured at 10 GHz with an angle of incidence of 0° and a vertical polarization. The electric field on board is measured using a patch antenna put behind the windscreen, the configuration of reference being a clean windscreen.

The attenuation due to 'running' water (rain on the windscreen, with the wipers off) is about 5 dB.

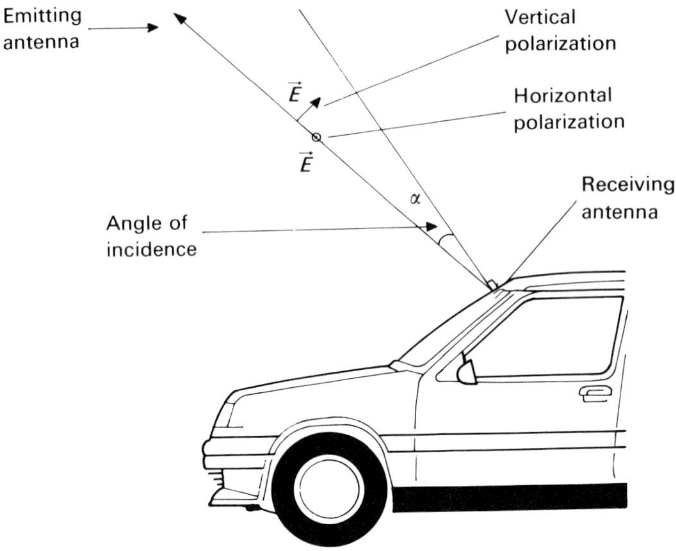

Figure 13.4 Geometrical considerations.

Table 13.1 Attenuation due to powdered snow on windscreen

Depth (mm)	*Attenuation (dB)*
20	9
30	10
40	12

Table 13.2 Attenuation due to melting snow on windscreen

Depth (mm)	*Attenuation (dB)*
1	8
5	24

The results in the case of snow are given in Tables 13.1 and 13.2.

Laboratory tests carried out for various thicknesses of water put on a horizontal plane, and for normal incidence, detected an attenuation of 3.6 dB/mm for a thickness of water less than 5 mm.

The attenuation due to the presence of dirt, a few millimetres thick, is negligible in the frequency range under consideration (1–10 GHz).

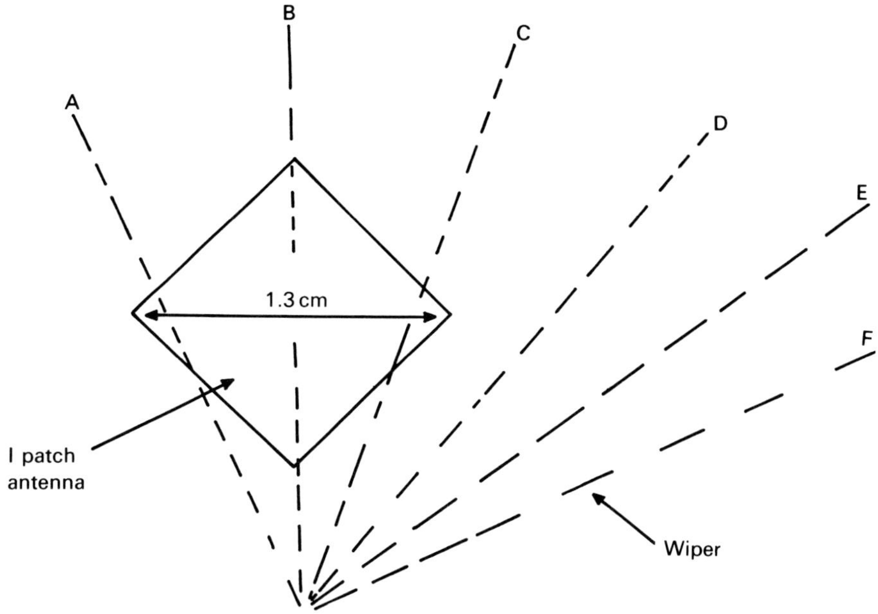

Figure 13.5 Different positions of the wiper.

We can note that the presence of melting snow greatly increases the signal attenuation and it would be interesting to put the receiving antenna behind the cleanest zone of the windscreen.

13.3.4 Influence of the wipers

In order to measure the influence, at 10 GHz, of the wipers on the received signal, six positions of the wipers, labelled A to F, have been successively considered and are represented in Figure 13.5.

The results are given in Table 13.3 (the 0 dB level corresponds to the wipers far from the antenna) [2].

We can remark that when the wipers are situated just in front of the patch antenna we get an attenuation of about 8–9 dB. This variation can be explained by the reflection or diffraction phenomena on the wiper itself but the presence of a metallic structure just near the patch antenna may also produce a further mismatch. So, the total attenuation due to the rain on the windscreen can reach 10–15 dB for some specific positions of the wipers. In the following all these additional phenomena will not be considered, the experiments being done under good weather conditions. We successively consider a short- and a medium-range link, by pointing out the masking and multipath effects.

Table 13.3 Attenuation due to the wiper in various positions

Position	E_{dB}
A	− 6.5
B	− 8.5
C	− 6.5
D	− 6
E	− 3
F	− 1.5

13.4 SHORT-RANGE LINK

As mentioned in the introduction, a short-term application of this link concerns tollway and automatic debiting systems. In order to characterize the microwave link, successive configurations have been studied, starting from the simplest one, called 'ideal conditions', without any obstacles, and ending with measurement in real traffic conditions.

13.4.1 Measurement in ideal conditions

An antenna (a parabolic antenna with 24 dB gain) put on a 5 m high canopy transmits a 10 GHz signal, and the electromagnetic field is measured with a spectrum analyser. Many measurements have been performed for various inclinations and heights of the receiving patch antenna (6 dB gain). The field amplitude is stored every 20 cm when the antenna moves along the x and y axes (Figure 13.6). Different representations of the distribution of the electric field amplitude can be made; one of these uses false colours, as in Figures 13.7 and 13.8.

The central circle corresponds to an important received signal while in the outer dotted zone the electric field amplitude is very small. This allows us to define the main lobe of the transmitting antenna or more exactly the area illuminated by a field amplitude greater than a fixed value.

In this case, the plane of the receiving patch antenna is horizontal and situated at a height of 1.15 m.

In order to observe the masking effect, a car or a van has been put in the illuminated area.

Reflections and diffractions occur on the obstacle and the masking effect will of course depend on the angle of incidence and on the type of obstacle. Figure 13.8, for example, shows the influence of a van.

Figures 13.7 and 13.8 show black and white reproductions of measurements corresponding to a horizontal position of the receiving antenna put at a height of 1.15 m, without car and in the presence of a van respectively. Extensive trials have shown that when the receiving antenna is put parallel to the windscreen, the reflections on the ground and on the hood are very small and the masking effect behind a car remains negligible.

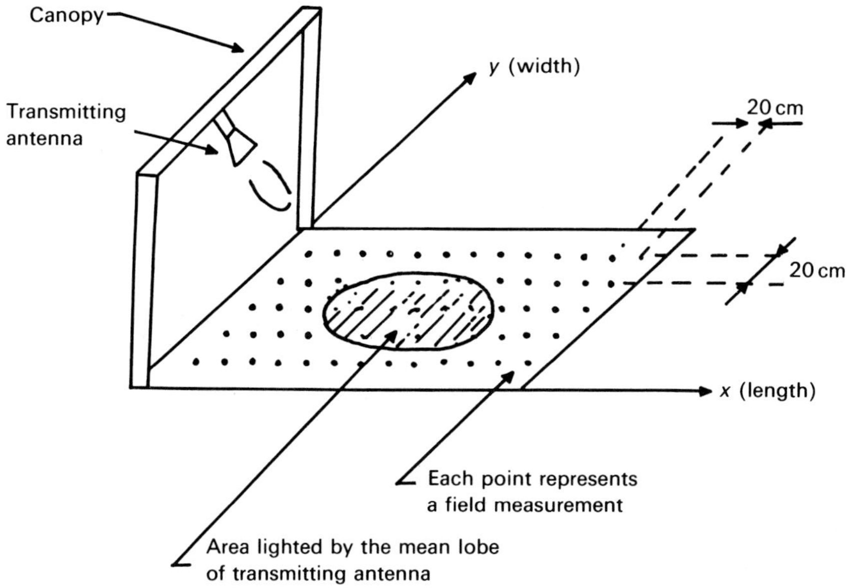

Figure 13.6 Site description.

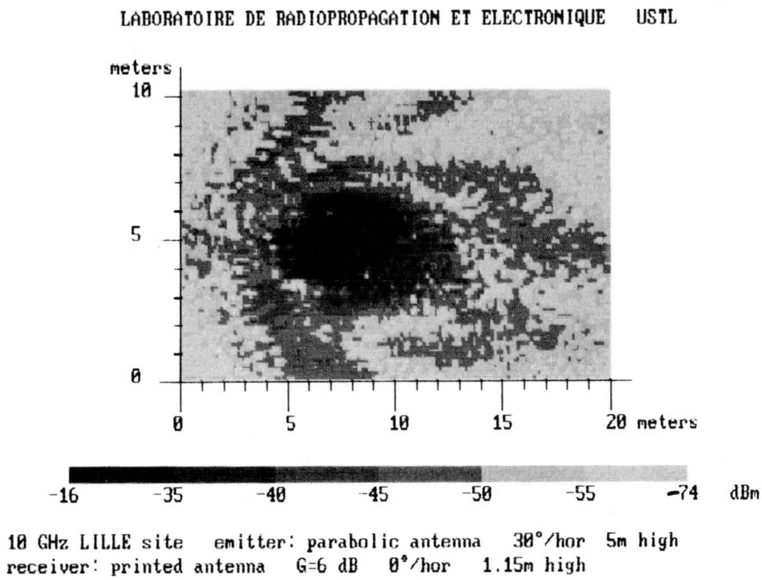

Figure 13.7 Measurements in ideal conditions.

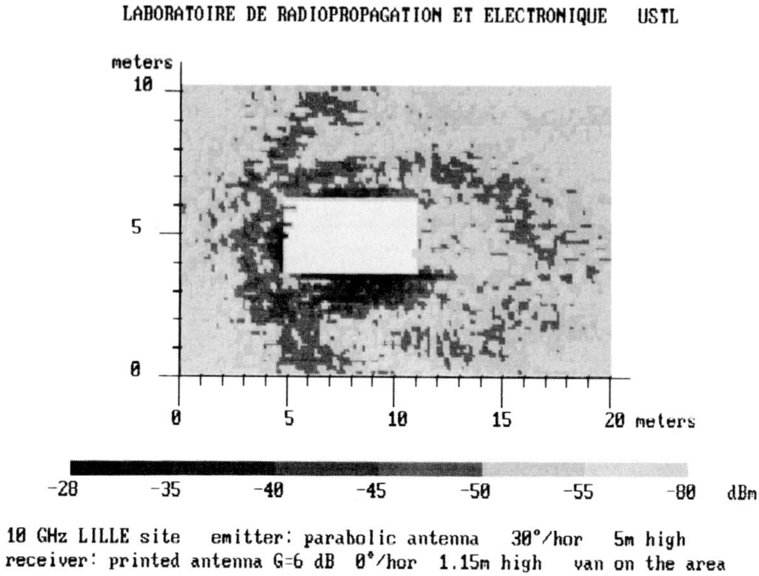

Figure 13.8 Measurements with a van on the receiving area.

13.4.2 Toll simulation [2]

The same site is chosen for simulating a toll, and in this case, the angle between the main beam radiated by the transmitting antenna and the horizontal plane is 30°. The receiving antenna is put in a car behind the windscreen (behind the driving mirror) (Figure 13.9). The electromagnetic field is measured on the main lane 'illuminated' by the transmitting antenna and on an adjacent lane. At first, in ideal conditions, and thus without any

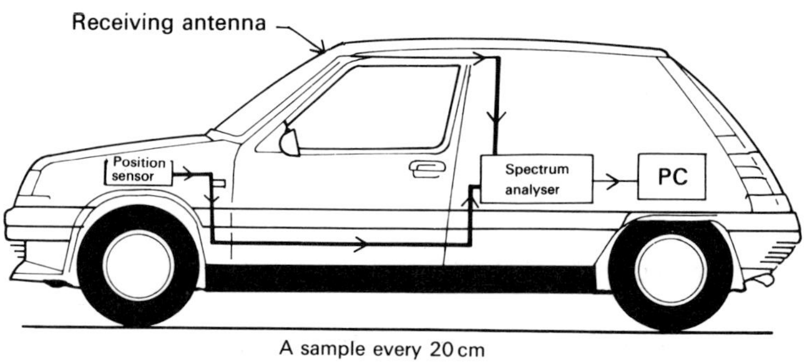

Figure 13.9 Experimental vehicle.

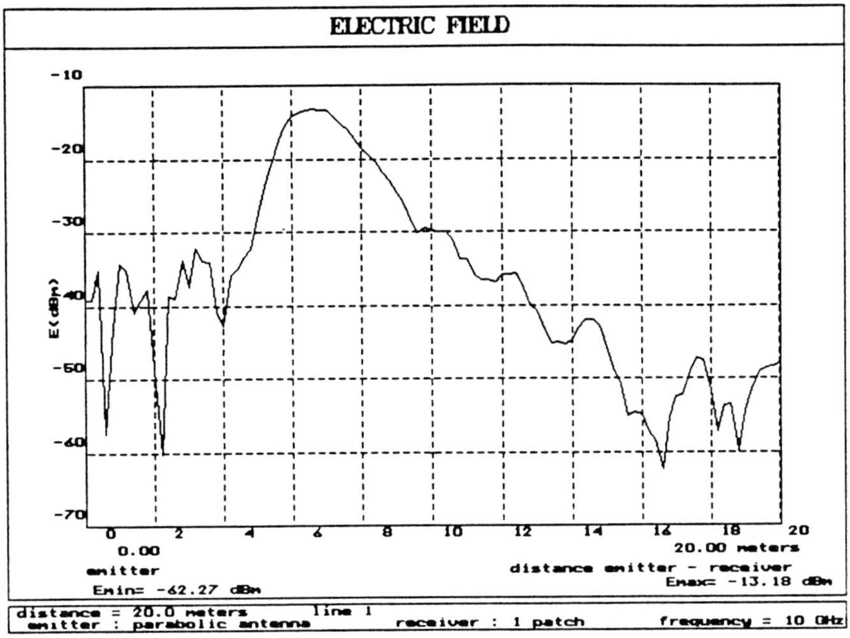

(a)

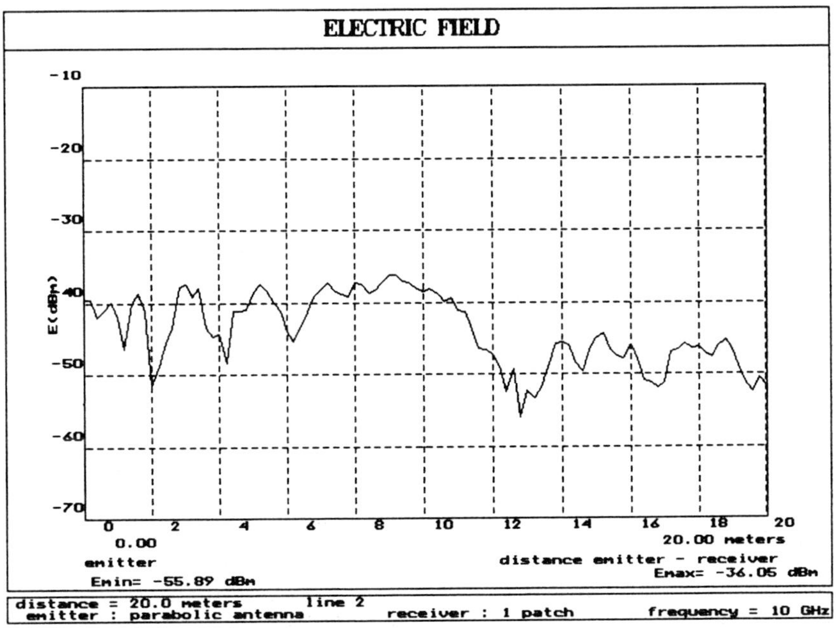

(b)

Figure 13.10 Electric field amplitude versus emitter–receiver distance: (a) measurements on the main lane; (b) measurements on the adjacent lane.

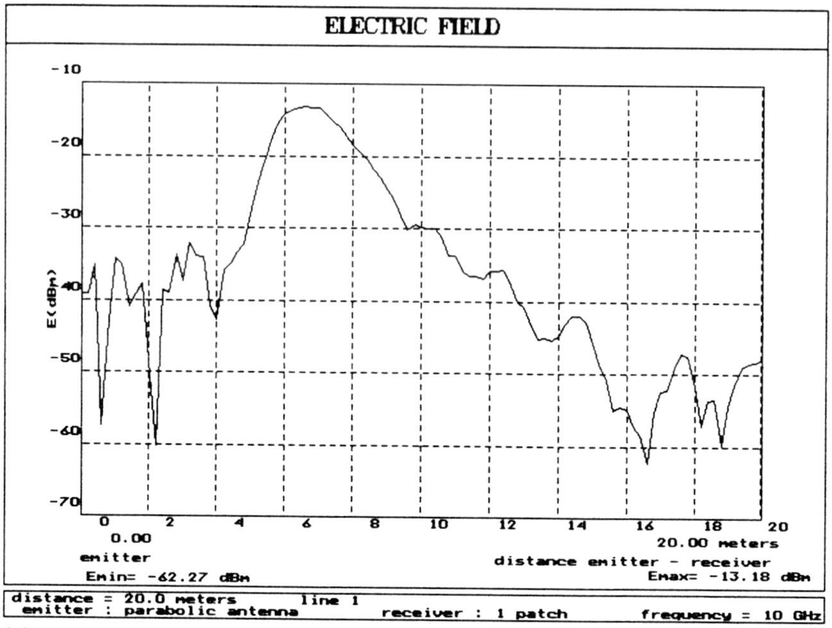

(a)

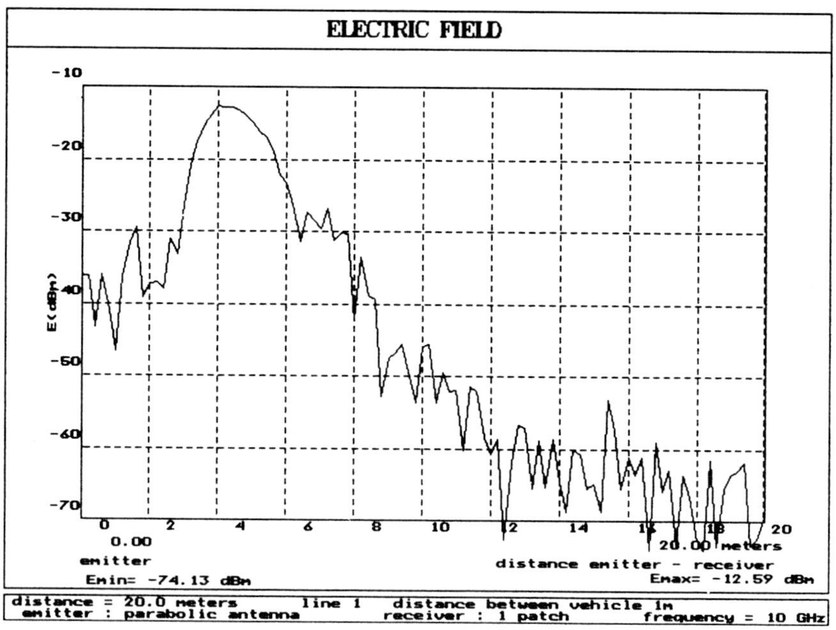

(b)

Figure 13.11 (a) Measurements without car in front of the receiving car. (b) Measurements with a van in front of the receiving car (distance: 1 m).

other car, the results are as shown in Figures 13.10a and 13.10b respectively.

The comparison of these two curves shows that the relative attenuation from the main lane to the nearby one is about 25 dB.

In another configuration, a car or a van has been situated in front of the receiving vehicle, both vehicles moving at the same speed.

In the worst case corresponding to a distance of 1 m between the car and the van (Figure 13.11), we only observe a reduction of the width of the main lobe.

Other experiments have also shown that the reflections and the diffractions produced by a car situated in the main lobe of the beacon do not modify appreciably the field in the other lanes. Additional measurements made in a real traffic environment on an experimental automatic debiting system have led to the same conclusion.

13.4.3 Conclusion

Measurements along the illuminated lane show that along a distance of about 4 m (± 2 m from the maximum receiving point) the signal amplitude varies within a ratio of 10 dB when the transmitting antenna is a parabolic one with a gain of 24 dB and the receiver is a one-patch antenna with a gain of 6 dB. Furthermore, both in ideal conditions (with no car in the illuminated lane) and in real traffic conditions, the received signal in other lanes is at least 20 dB below the maximum amplitude.

Consequently, if the reception level is judiciously chosen, all these disturbing effects can be overcome.

13.5 MEDIUM-RANGE LINK

As mentioned in the introduction, a medium-range link has to be used in a road guidance system. The covered area is larger than for toll application and multipath effects (reflection on the road, reflection and diffraction on the vehicles) are no longer negligible.

13.5.1 Measurement in ideal conditions

In real environments the geometrical configuration can be very complicated, since in many cases multipath propagation occurs. In a first step we can characterize the intrinsic fadings due only to the reflection on the ground.

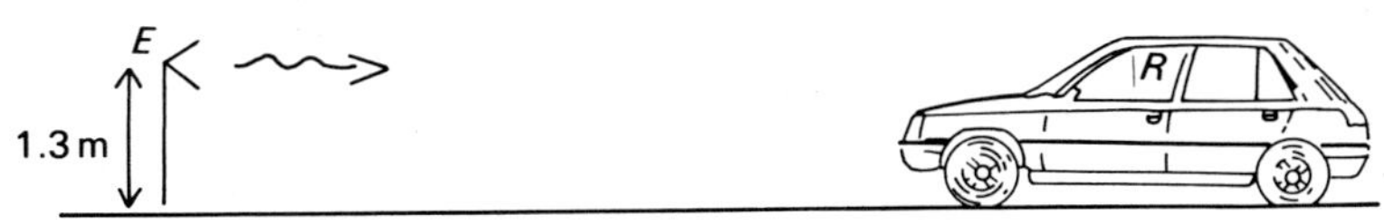

Figure 13.12 Experimental conditions.

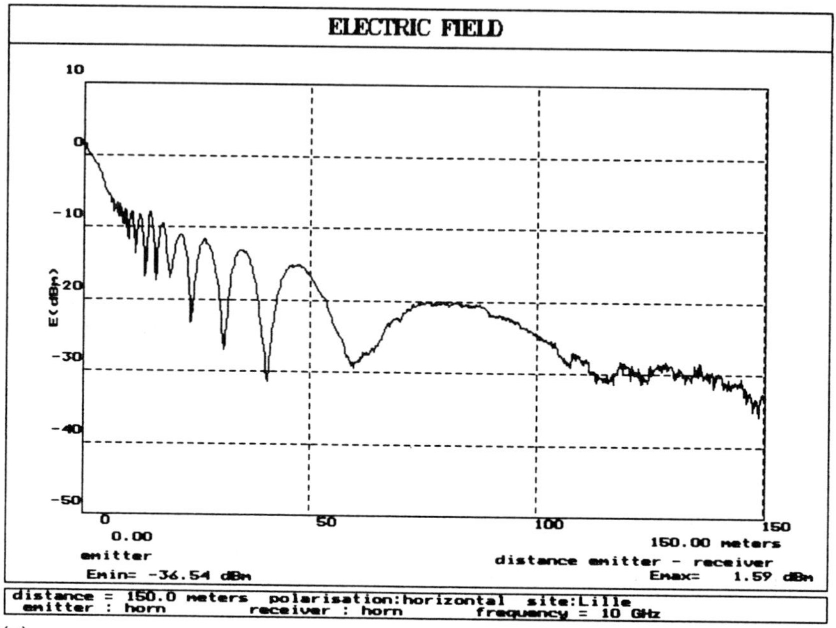

(a)

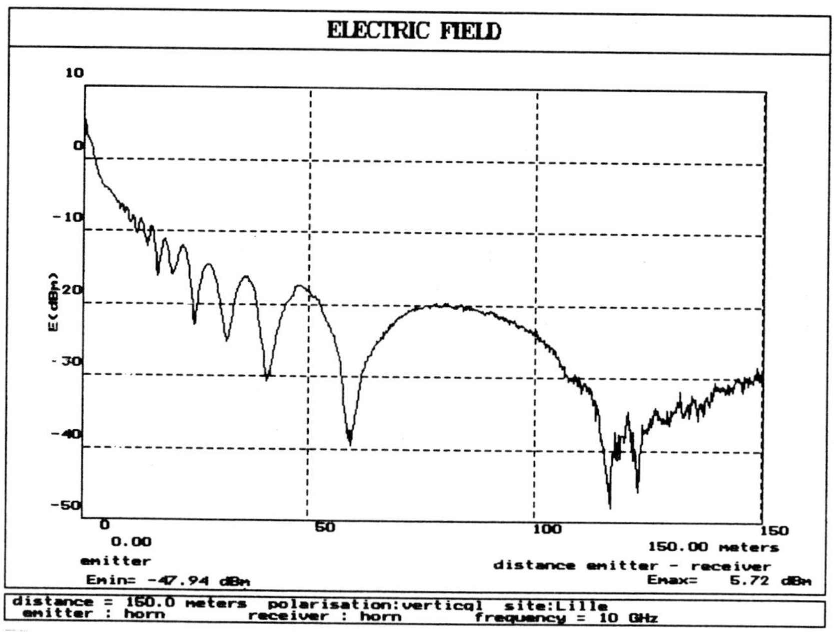

(b)

Figure 13.13 Measurements in ideal conditions with (a) horizontal polarization and (b) vertical polarization.

These experiments (at 10 GHz) have been made in free space without obstacles, the emitter being put at a height of 1.3 m and the receiving antenna being situated inside a car behind the driving mirror (Figure 13.12). The received electric field amplitude was measured with a spectrum analyser triggered by a coded wheel every 20 cm.

The results for the vertical and horizontal polarizations are given in Figures 13.13a and 13.13b.

The experimental results show that there is not an important difference between the two polarizations except for the fact that the depth of the fadings is more important for the vertical polarization. A theoretical approach made with two rays (direct and reflected on the ground) agrees with the experimental results.

These measurements show that the reflection on the ground is an important phenomenon mainly due to the fact that the rays have a grazing angle of incidence on the road surface.

One can also remark that it is very difficult to define a well ' covered' area even in this idealized situation (any car or obstacle near the receiver) and that can be a problem for road guidance systems based on beacon–vehicle communication along a precise zone. In this case, a solution could be the use of a very directive receiving antenna which, unfortunately, cannot be achieved at a reasonable size at 10 GHz. With such constraints on the radiation pattern, millimetric waves seem more appropriate.

13.5.2 Measurements in traffic conditions [3]

Measurements at 10 GHz have been performed downtown, in the centre of Lille (France). The emitter was placed on a pole 8 m high and the receiver was placed inside a car (Renault 5) behind the driving mirror. The car was moving in heavy traffic conditions on a large road (four lanes in the same direction (Figure 13.14).

The vehicle speed was about 60 km/h and the electric field amplitude was stored every 20 cm.

An example of the experimental results is shown in Figure 13.15, where fadings having a depth of 20–30 dB frequently occur.

Such a record shows the well-known characteristics of the signal fluctuations which are the superposition of short-term and long-term fadings. The first ones are mainly caused by the multipath effect due to the reflections on the local obstacles like nearby buildings or vehicles and the last ones by the reflection on the ground, as was seen in measurements in 'ideal conditions' (Figure 13.13).

Figure 13.15 also shows that the range of this kind of communication will not be important, the signal attenuation reaching 30 dB 200 m away from the transmitter. This attenuation (and, of course, the range of the communication) depends on different parameters such as the height of the emitter, its angle of inclination and the environment: small street or large road.

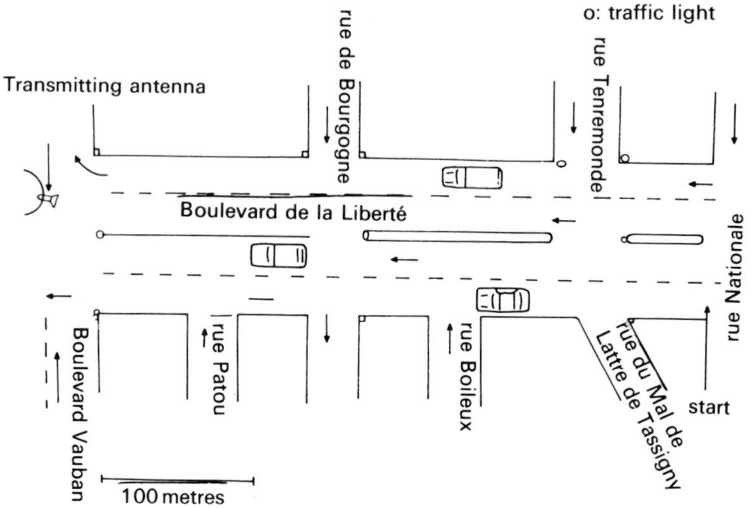

Figure 13.14 Downtown measurements: site description.

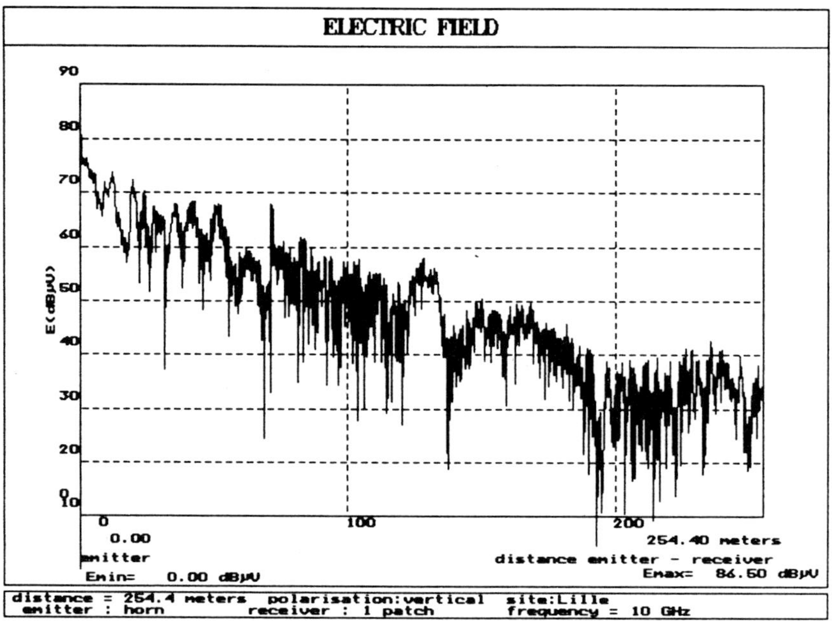

Figure 13.15 Electric field amplitude versus emitter–receiver distance (downtown).

Table 13.4 Characteristics of the tested antennas

	Lobe width	*Gain (dB)*	*Polarization*
Parabolic antenna	7	24	Linear
One-patch antenna	53	6	Linear
Horn antenna	30	17	Linear
Circular antenna	15	15	Circular

The characterization of the link can be made in terms of a statistical analysis.

13.5.3 Statistical analysis [4]

To study the influence of either the radiation pattern or the polarization of the transmitter, various types of antenna have been used (Table 13.4). In order to compare the performance of the link in each case, a statistical analysis of the received signal around its mean value has been performed (Figure 13.16). Some classical parameters such as standard deviation σ and dynamics (peak-to-peak value) are determined. Two other parameters can also be evaluated: the mean time between two successive fadings t_B and the mean fading time t_F (for a given vehicle speed).

The first parameter t_B, is obtained from the determination of the cumulative function of the distance between two successive fadings greater than a given level $|Y_R|$ (Figure 13.16); this function $f(Z_R)$ is given by the probability that the distance Z is smaller than the abscissa Z_R:

$$f_{Y_R}(Z_R) = \{\text{probability } Z < Z_R/\text{fading} > Y_R\}$$

The mean fading time t_F is determined in the same way from the cumulative function of fadings width for fadings greater than a given level $|Y_R|$

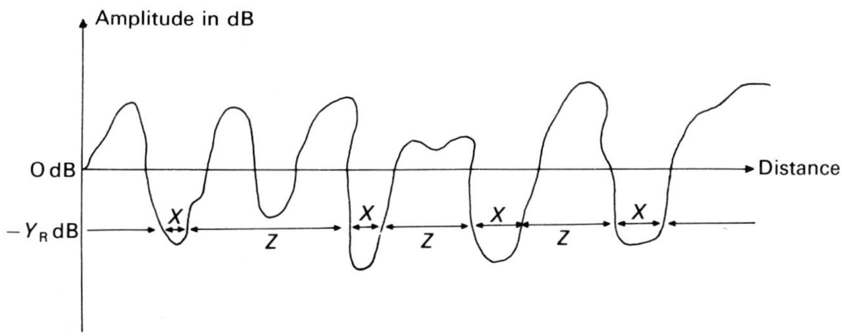

Figure 13.16 Statistical analysis of the received signal around its mean value.

Table 13.5 Statistical analysis

Emitter	*Receiver*	*σdB/μV*	*Dynamics (dB)*
Parabolic antenna	1 patch	6.7	41
Horn	1 patch	6	50
Circular	Circular	6.4	42

Table 13.6 Fading mean time (vehicle speed 60 km/h)

Emitter	*Receiver*	*− 10 dB* t_F*(s)*	*− 20 dB* t_F *(s)*
Parabolic	1 patch	0.027	0.024
Horn	1 patch	0.027	0.026
Circular	Circular	0.028	0.025

Table 13.7 Mean time between two successive fadings (vehicle speed 60 km/h)

Emitter	*Receiver*	*− 10 dB* t_B *(s)*	*− 20 dB* t_B *(s)*
Parabolic	1 patch	0.25	1.67
Horn	1 patch	0.22	1.27
Circular	Circular	0.25	2.40

(Figure 13.16); this function $g(X_R)$ is given by the probability of having the width X smaller than the abscissa X_R:

$$g_{Y_R}(X_R) = \{\text{probability } X < X_R/\text{fading} > |Y_R|\}$$

Some results are summarized in Tables 13.5, 13.6 and 13.7. In Tables 13.6 and 13.7 the given reference level is successively equal to − 10 dB and − 20 dB (statistical analysis of fadings greater than $|Y_R|$) and the vehicle speed is normalized to 60 km/h for all the results.

We remark that the mean fadings time t_F is nearly independent of the type of antenna, and of the polarization (linear or circular). However, the values given in Table 13.6 must be considered as an order of magnitude since the mean value of t_F is only equal to about two sampling periods, which is not enough to determine the statistical distribution of t_F.

If we consider the results given in Table 13.7, we can remark that the mean time between two successive fadings depends on the chosen depth for the fadings. Indeed, the time intervals between two important fadings (20 dB) are 1.3–2.4 s against 0.2 s if 10 dB fadings are considered. Lastly, we can note a slight advantage with the circular polarization since great fadings seem to occur less frequently. However, this result has to be confirmed by additional measurements in other environments.

13.6 LONG-RANGE LINK: MEASUREMENTS IN TUNNELS [4]

Measurements have also been performed in tunnels, at first in an empty tunnel in Liège (Belgium) and then in traffic conditions in the St Cloud Tunnel in Paris (France).

13.6.1 Measurement in an empty tunnel

These measurements have been performed in an ideal rectangular (8 × 4.7 m) empty tunnel. The emitter and receiver were put a height of 1.5 m and we have measured the electric field magnitude with a spectrum analyser versus the emitter–receiver distance (a sample every 20 cm). An example of the experimental results is shown in Figure 13.17.

13.6.2 Measurements in normal traffic conditions

Additional measurements have been performed in a large tunnel (four lanes in one direction, 900 m long) in normal traffic conditions. The emitter was also put at a height of 1.5 m but outside the tunnel, near the entrance, on the emergency lane. The receiving antenna was put, as previously, inside a

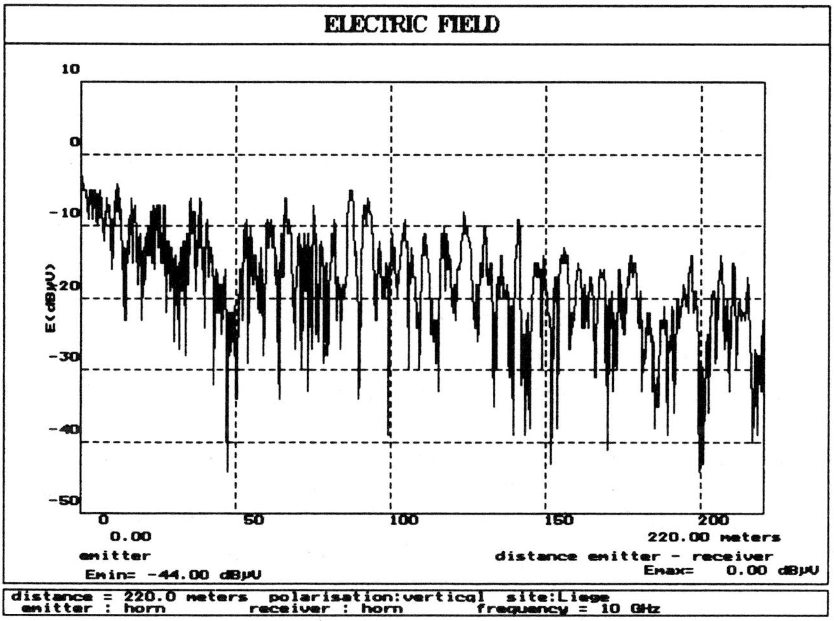

Figure 13.17 Electric field amplitude versus emitter–receiver distance (Liège tunnel).

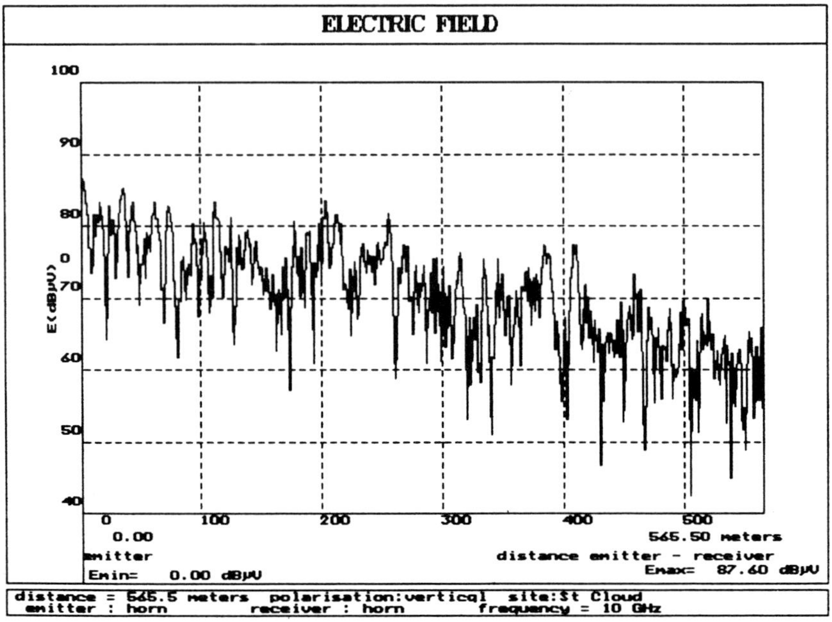

Figure 13.18 Electric field amplitude versus emitter–receiver distance (St Cloud tunnel).

car, behind the driving mirror. The receiving car was moving in the traffic on the first lane. The signal was measured with a spectrum analyser triggered every 50 cm. An example of results is shown in Figure 13.18 and we note that the fluctuations of the signal do not seem more important in traffic conditions than in an ideal empty tunnel which is, in fact, twice as small as the St Cloud tunnel.

13.6.3 Comparison with a theoretical model

Measurements in the empty tunnel are compared with a theoretical simulation using a ray theory. An example is given in Figure 13.19 for a 10 GHz signal.

We can note a good agreement between the theoretical (Figure 13.19) and experimental (Figure 13.17) results. With such a model we can predict characteristics of the link such as linear attenuation and mean time between fadings, and deduce some parameters of the communication.

13.6.4 Statistical analysis

A statistical analysis based on measurements in the empty tunnel is summarized in Tables 13.8 and 13.9, while Tables 13.10 and 13.11 correspond to traffic conditions.

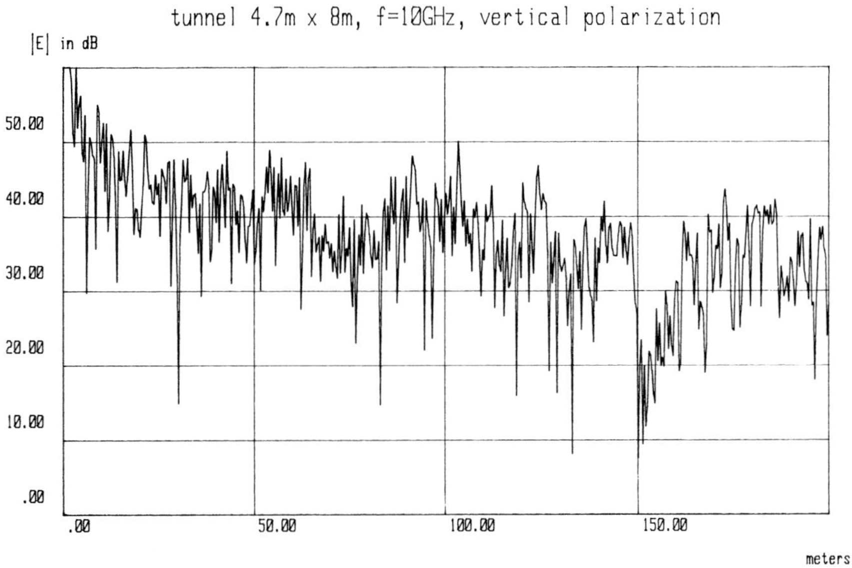

Figure 13.19 Electric field amplitude versus emitter–receiver distance (theoretical results).

Table 13.8 General characteristics of the measurements in the empty tunnel

Emitter	*Receiver*	*Att. dB/100 m*	*σ dBm*	*Dynamics (dB)*
Horn	Horn	6.4	6.16	41.5
Horn	1 patch	6	5.91	46.2
Circular	Circular	5.5	5.93	50.9

Table 13.9 Measurements in the empty tunnel: mean fading time (t_F) and mean time between two successive fadings (t_B) for two reference levels (10 dB and 20 dB) and a vehicle speed of 60 km/h

Emitter	*Receiver*	*– 10 dB*		*– 20 dB*	
		t_B (s)	t_F (s)	t_B (s)	t_F (s)
Horn	Horn	0.25	0.036	2.22	0.028
Horn	1 patch	0.27	0.033	1.48	0.028
Circular	Circular	0.36	0.039	3	0.028

Table 13.10 General characteristics of measurements in tunnel in traffic conditions

Emitter	*Receiver*	*Att./100 m*	*σ dB/μV*	*Dynamics (dB)*
Horn	Horn	3.4	5.8	32
Horn	1 patch	3.9	6.6	41
Horn	4 patches	4.06	6.2	40
Circular	Circular	4.08	6	30

Table 13.11 Measurement in traffic conditions: mean fading time (t_F) and mean time between two successive fadings (t_B) for two reference levels (10 dB and 20 dB) and for a vehicle speed of 60 km/h

Emitter	*Receiver*	*− 10 dB* t_F *(s)*	*− 20 dB* t_F *(s)*	*− 10 dB* t_B *(s)*	*− 20 dB* t_B *(s)*
Horn	Horn	0.11	0.09	1.7	12.6
Horn	1 patch	0.09	0.09	0.95	5.9
Horn	4 patches	0.10	0.09	0.8	8.9
Circular	Circular	0.10	0.09	0.94	8.9

Table 13.12 European normalization for automatic debiting system [5]

Frequency band	5.795–5.805 GHz 5.805–5.815 GHz
Systems	Short-range link Automatic debiting system
Authorized bandwidth	2 × 10 MHz (4 channel of 5 MHz)
Authorized power	3 dBW
Recommended antenna gain	10–15 dBi
Baud rate	1–3 Mbits/s
Modulation	FSK, PSK, AM

The statistical results calculated from data from the empty tunnel and in traffic conditions show that the polarization and the beamwidth have no significant influence on the signal fluctuations.

These results also show that the characteristics of the wave propagation, like fading phenomena (and thus the mean time between successive fadings t_B) and the linear attenuation, are more dependent on the dimensions of the tunnel than on the traffic.

The tunnel behaves as an oversized wave guide and thus the communication between the road and a vehicle can be established on a range longer than in downtown areas. Indeed the linear attenuation is smaller or of the order of 15 dB/100 m for the downtown experiments and 3.8 dB/100 m for the St Cloud tunnel experiments.

13.6.5 Conclusion

Long-range communication in tunnels is possible with a rather low linear attenuation but many types of interference occur between the various modes propagating inside the tunnel and give rise to numerous and important fluctuations of the received level, even in an empty tunnel.

13.7 PERSPECTIVES

Numerous automatic debiting systems have been developed in Europe during recent years, most of them being described in the DRIVE 1 reports, the transmitting frequencies being either 2.5, 5.7 or 9.9 GHz. However, an important consequence of the DRIVE and PROMETHEUS projects has been the definition of a European frequency band for such systems; the recommendations of the CEPT (European Committee of Posts and Telegraphs) are summarized in Table 13.12. This frequency allocation can be considered as a first step in the definition of a European product.

REFERENCES

1. Ulaby, F.T., Moore R.K. and Fung A.K. (1986) *Microwave Remote Sensing: Active and Passive.* Volume III *From Theory to Applications*, Artech House.
2. DRIVE WP3 (1989) *Short range microwave links: present and future.* Evaluation of short range road-vehicle links in the 1 to 10 GHz band. Project DRIVE V 1002 SMILER – WP3 final report.
3. DRIVE WP7 (1990) *Short range microwave links: present and future.* Experimental feasibility evaluation for roadside vehicle communication. Project DRIVE V 1002 SMILER – WP7 trial report.
4. DRIVE WP4 (1990) *Short range microwave links: present and future.* Comparison of centimetric microwave links with other technology. Project DRIVE V 1002 SMILER – WP4 final report.
5. Hess, S. (1991) Presentation of the CEPT conclusions on the frequency bands usable for road application, in *DRIVE Conference* 1991, Brussels, CEC DRIVE Directorate, Brussels.

PART SEVEN

Sensors and Systems in Automotive Manufacture

CHAPTER 14

A perspective on machine vision at Rover Cars Group

14.1 INTRODUCTION

With the growth of industrial automation, it has become increasingly important to validate the quality of every manufactured part during production. Until now, human visual inspection aided by hard tooling or machines have been the primary means to this end, but the speed of today's production lines, the complexity of production equipment and the highest standards of quality to which parts must adhere, frequently make the traditional methods of industrial inspection and control impractical, if not impossible.

Subsequently, new solutions have been developed for the monitoring and control of industrial processes, in real time. One such technology is the area of machine vision. After many years of research and development, computerized vision systems are now leaving the laboratory and are being used successfully in the factory environment. They are both robust and competitively priced as a sensing technique which has now opened up a whole new sector for automation.

Machine vision systems are becoming an important integral part of the automotive manufacturing process, with applications ranging from inspection, classification, robot guidance, assembly verification through to process monitoring and control. Although the number of systems in current use is still relatively small, there can be no doubt that the automotive industry will once again lead the way with the implementation of machine vision just as it has done with robotics technology.

This chapter presents real applications of machine vision within the Rover Group and will describe how the technology is being used to manufacture automobiles. The chapter will present a broad overview of the installations, documenting their concept, construction and operation within a factory environment which does not require clean rooms, or very stringent temperature and lighting controls, in order to perform successfully.

14.2 CASE STUDY I: BODY GAUGING

14.2.1 Body gauging: an application for machine vision?

The accuracy with which the numerous sheet metal panel components are assembled in the construction of a body shell is critical to the subsequent assembly operations and the fit, functionality and cosmetic finish of other components of the completed vehicle.

The requirement for rigorous control of this process is therefore of paramount importance. The major obstacle with traditional contact gauging techniques is that they are slow to operate and hence must be undertaken 'off line', with sample sizes of only one or two per shift being feasible. With further bodies being produced during this checking process, and with such low sample sizes, the results are essentially retrospective and at worst unrespresentative.

Rover Group has overcome these difficulties via the installation of an 'on-line' 100% validation, automatic gauging system, for the framed body shell of the Rover 800 vehicle. In employing automatic visual sensing and computer-based analysis of the data, it encompasses the entire solution by integrating the measurement, analysis control and reporting function.

14.2.2 General requirements

When Rover Group first considered using a non-contact full-body gauging system, a list of requirements was drawn up to identify those features which were considered mandatory and those which were optional. Among the mandatory features of the proposed system were (the list is not exhaustive):

- The need to examine all framed bodies by non-contact means of measurement superseding the means and practices of traditional gauging techniques.
- The gauging system to undertake 100% on-line measurements at line speed for every body shell.
- Real-time speed of response to analysis and report production.
- The need to check four-door and five-door model derivatives and the ability to upgrade the system to accommodate future model variants.
- The need to gauge model styles as imposed by production build schedules.
- Measurements are required of key critical parts of the body geometry, including all aperture conditions.
- Integration with the automated build framing line.
- Essential inclusion of statistical process control tools and techniques.
- Providing a closed-loop feedback to the framing line for process control.
- Ease of use and mantainability.

14.2.3 Manufacturing process: overview

The Rover 800 body framing line is fully automated, with the bodies being robot 'framed' and 'finished welded' through a sequence of stations based on a powered transfer line. The body gauging system is sited at the end of the Sciaky build line on the exit side at station 8. Car bodies are clamped to carriages which index through the series of workstations.

The body is presented to the gauging station, accurately located and clamped to a tolerance of ± 0.2 mm with reference to a jig carriage master location point. Variations in body presentation on this jig carriage can result, however, in potential body deviations of up to ± 3 mm. The line PLC signals the station to measure the body; when the gauging is complete, the system signals the PLC that the body can be transferred out.

The flexibility of the system is demonstrated by the ability to gauge Rover 800 four-door and five-door derivatives in any sequence. Furthermore, the system can be upgraded to accommodate future model variants by simply adjusting the system software and hardware.

14.2.4 Gauging station: concept and construction

Construction

The gauging station consists of a calibrated steel gantry structure (Figure 14.1) known as the 'tunnel', onto which are fixed a total of 62 three-dimensional gauging modules. The gantry structure upon which the modules are mounted is entirely contained within a polycarbonate shielded enclosure, to suppress ambient lighting and electromagnetic interference, and to contain the poten-

Figure 14.1 Body gauging station.

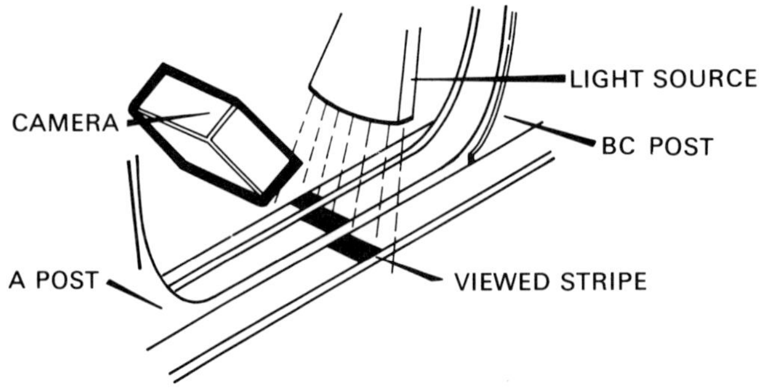

Figure 14.2 Structured lighting concept applied to body gauging.

tially dangerous laser light. The modules are arranged to measure about 70 critical features of a body shell style.

The dimensional checks will gauge front and rear screen aperture, engine and trunk compartment widths, sizes of door apertures and seal conditions, and the general geometric integrity of the side panels.

Machine vision concept

Each module consists of a light source (HeNe laser) fixed to a Panasonic solid-state video camera. The module employs the 'structured lighting' technique for deriving the three-dimensional information needed in visual gauging.

The technique of structured lighting utilizes trigonometry to determine how far a feature or object is away from a camera. As shown in Figure 14.2 the approach is to project a line of light (usually a laser line) onto an object in question. This bright line of light provides excellent contrast and an easily recognized feature.

By projecting the laser at an angle, the displacement of each point on the line is made proportional to the depth of the object at that point. Furthermore, the shape of the line provides additional data about the curvature and edges of the object cut by the line. The versatility of this technique can be enhanced, since if there is a spatial relationship between the gauging modules, then it is possible to compute, for example, door and window apertures.

Structured lighting resolves the distance ambiguity, but at the cost of not illuminating the whole scene; only those parts of the feature struck by the laser beam can be measured.

Hardware architecture

At the heart of the gauging system (refer to Figure 14.3) are three Automatix Autovision 3 (AV3) machine vision controllers, which are located in the

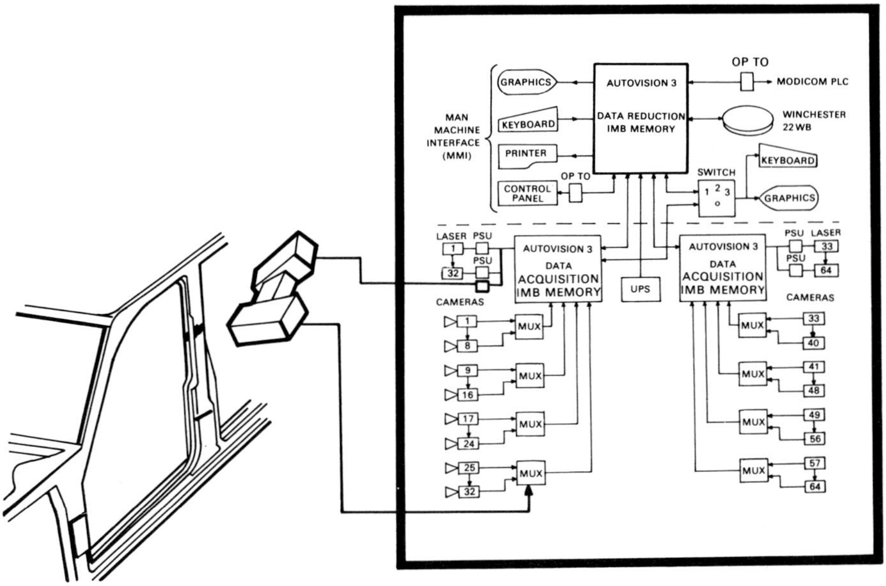

Figure 14.3 System hardware architecture.

operator console. One AV3 controller acts in supervisory mode, sequencing operations and translating information received about either side of the body into real body data.

The remaining two AV3 controllers (known as slaves), each with extended camera port capability via a series of add-on multiplexers, accommodate up to 32 camera modules to undertake the vision process for the two sides of the body respectively.

The operator interface is via keyboard and VDU and an advanced statistical process control (SPC) package incorporated with graphics with hard printouts obtained by menu selection. The Supervisory AV3 processor is also interfaced with the body framing master PLC, to receive the information on the body derivative and also to inhibit the main line function in the event that the bodies are produced out of tolerance or there is a major fault on the body.

14.2.5 Gauging operation

During the automatic gauging cycle, the welded body is automatically shuttled into the station and clamped. Upon receiving the body and model style data from the Sciaky PLC line controller, the supervisory AV3 will initiate the gauging cycle by instructing the two Slave AV3 controllers to commence with acquiring the body data.

Before the main gauging process can begin, the actual position of the body presented to the cell must be established, as body-to-carriage location can vary. This determines the extent to which the laser line will have to be offset from its expected position. This information is obtained by using three 'prime' cameras on two sides of the car to determine coordinate information on the extremities of each side panel. The data processed from these cameras are used in a 'rigid body rotation' software routine which involves the transposition and rotation of measurements relative to the axis of measurement.

Once the body orientation and translation has been established by this 'software fixturing' technique, all the camera/laser modules are initiated sequentially in opposite pairs and the information on the gauging points is stored. The average time taken for image capturing and processing is approximately 1.5 s for each module.

The technique of three-dimensional analysis using structured light is utilized, although due to the rigid body rotation technique being employed, the x, y, z coordinate determined for each pixel is shifted in real space, to take account of the new impinged laser plane.

The 'new' gauge point then has to be related to those expected values stored in the CAD database, to determine whether they fall within the set build tolerance. The stored values are derived by taking measurements of numerous bodies via both the vision system and the CMM, to determine a statistically acceptable mean offset from the original gauge point. This

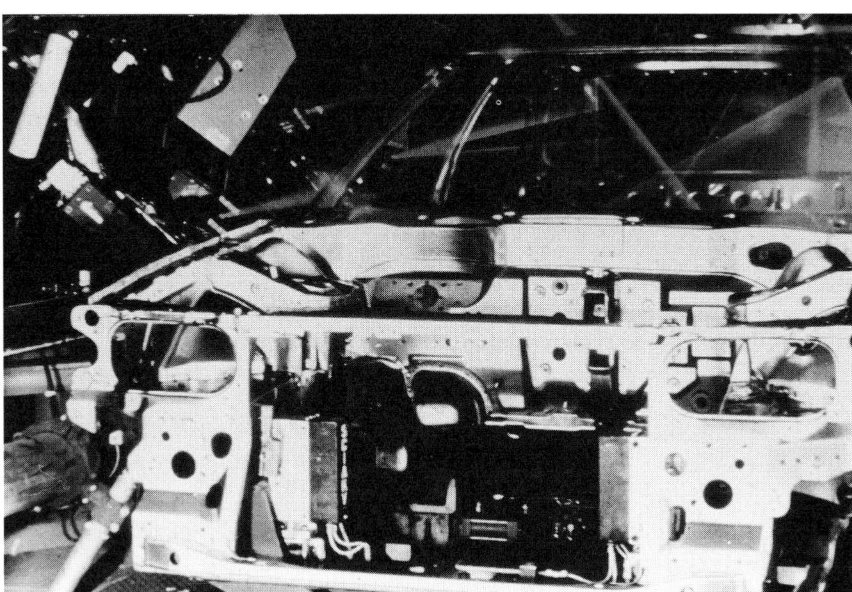

Figure 14.4 Critical features of the body being measured.

constant distance for each of the camera modules is then automatically compensated for in the calculation for all subsequent bodies.

Once the dimensional position of all the gauged points has been determined, the supervisory controller also computes additional information on the distance between gauge points, to provide data on a total of 280 body features. It then directly compares these with the CAD database values to determine whether they fall within the set build tolerances. If any feature is outside this tolerance (typically $\pm$ 3 mm), the main line is immediately stopped with the body still in the gauging cell, with a flashing red beacon indicating the violation to production management. A graphic display of the body is automatically generated on the VDU, indicating the violated feature together with a printout of a full part report. This assists production personnel to ascertain whether the fault was caused by, for example, an incoming poor subassembly or via a fault on the body framing line itself.

The main line can only be restarted by input of a 'release-body' password by authorized personnel, and the offending body shell is directed to a quarantine area for further analysis and subsequent re-work. Figure 14.4 illustrates critical features of the Rover 800 body shell being automatically gauged at the station.

14.2.6 Statistical process control

Of the 280 body feature checks, 125 of the most critical features for each derivative are utilized in an extensive SPC package (Figure 14.5), this being the maximum amount of data that can be handled within the available cycle time.

One of the main features of the package is its trend analysis capability. In essence this provides an advanced problem detection system which can prompt remedial action before deterioration to a line stop condition occurs.

14.2.7 The way forward

The body gauging system for the Rover 800 series model has been in production since 1987, and can gauge up to 70 different dimensions on the body (some relying on intercalibration of separate camera/laser pairs) to an accuracy of 0.2 mm within a cycle time of 96 seconds.

The gauging cell, which represents the state of the art in the field of non-contact gauging has provided a major breakthrough in improved product quality, by ensuring the integrity of the effective 'backbone' upon which all subsequent processes are performed. It also serves to highlight the power of statistical packages as a tool for in-process control and analysis of system performance.

The success of this installation is reinforced by the introduction of a second such measuring rig, recently commissioned for the new Rover 200/400 series model at Rover Group's Longbridge plant, as shown in Figure 14.6.

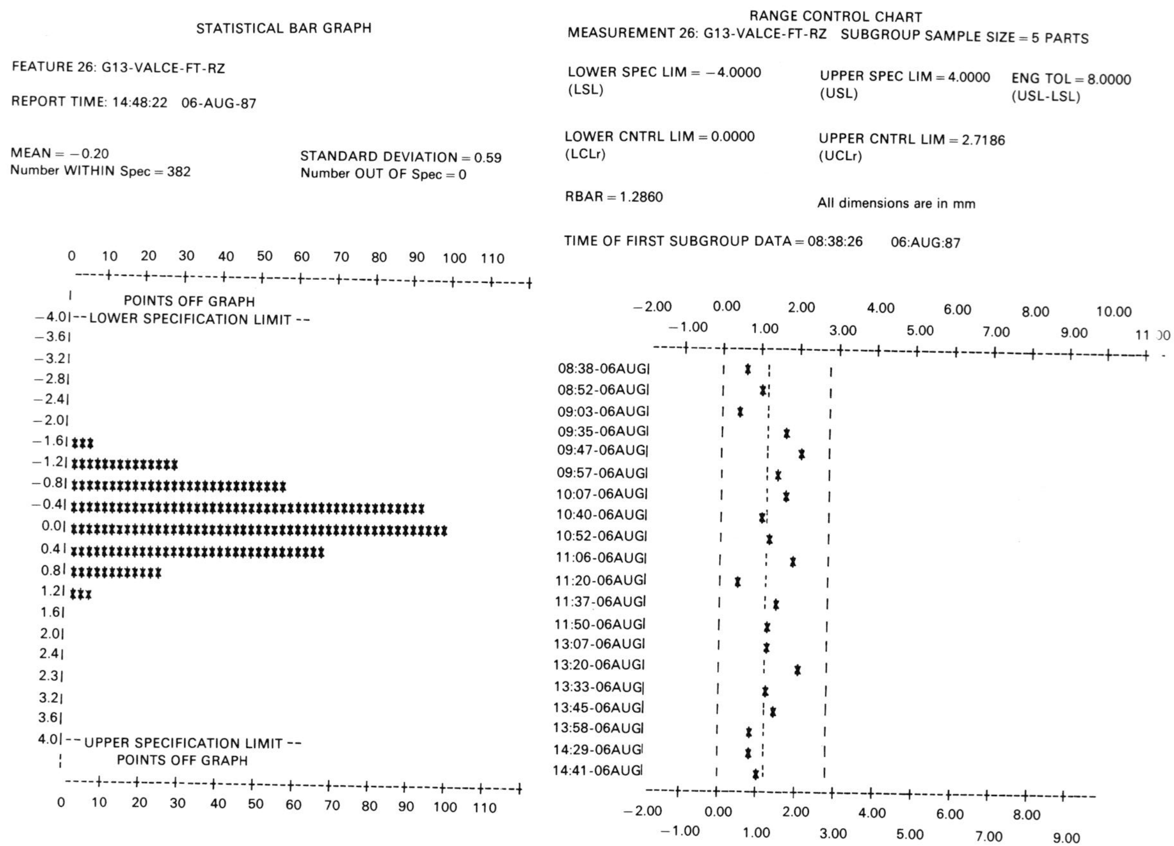

Figure 14.5 Statistical process control–outputs.

Figure 14.6 Rover 200/400 series body gauging station.

14.3 CASE STUDY II: WINDSHIELD FITMENT

14.3.1 Windshield fitment: an application for machine vision?

The potential automation of many vehicle assembly processes is often restricted because the local area of the car body itself cannot be related with sufficient accuracy to the master location features, due to the presence of build tolerances and the requirement for expensive mechanical fixturing.

A stringent body presentation system, achieved either by conventional hard tooling or vision techniques, is therefore not necessarily sufficient to accommodate an automatic assembly fit process that is subject to tight production tolerances.

Automatic windshield insertion is one such example, where an accurate three-dimensional location of the aperture itself is required to ensure a correct fit by automation, for example an articulated robot. This fact, together with the manufacturing tolerances which are imposed on the glass itself, makes this highly complex process an ideal candidate for potential automation. However, these factors are not alone in promoting the application of industrial vision; other criteria need to be taken into consideration.

The inherent variability of the manually performed task renders automation of the process essential in order to guarantee a higher but more consistent standard of product quality. In terms of functionality the glass must be

Figure 14.7 Montego glazing cell.

accurately fitted to the receiving aperture flange, to provide a uniform band with sufficient strength to retain the glass in situ in the event of a vehicle collision. Furthermore, the polyurethane adhesive must form a continuous bead to guarantee a totally watertight fit. The cosmetic appearance is also vitally important since the polyurethane adhesive should be invisible from either inside or outside the vehicle, since any misplacement or excess adhesive will cause local unsightly soiling of the trim items.

Background

Some seven years ago Rover Group unveiled its revolutionary fully automated glazing system, installed on the Montego build lines (Figure 14.7) which is still in operation today. Here the robot gripper assembly not only consists of a glass retention system but also holds the vision system hardware and the drives for the process involving the glass adjustment and insertion. This creates a high payload on the robot wrist, accelerating the mechanical wear of the robot components, which leads to inaccuracies in the vision analysis, resulting in periodic vision failure.

The experience gained from this pioneering system for the Montego series of cars contributed to the evolution of a 'second generation' fully automated glazing system for the Rover 800 series model.

14.3.2 Manufacturing process: overview

The system incorporates a strategy of 'follow on' from the robotic glazing system for the Montego model, but with a significant difference. The advanced vision-guided assembly system is no longer an integral part of the gripper system; instead, the new system employs rigidly mounted cameras, which automatically compensate for variations in aperture presentation.

The glazing cell is fully integrated into the production trim and final assembly system, buffered between two continuously moving assembly tracks by an overhead monorail delivery system. Car bodies are transferred into the cell by individually powered slings and then lowered onto carriages moving around a recirculating floor level monorail system for transferring the car through the glazing cell. The bodies are located by four 'V' blocks welded onto the carriages.

On entry into the cell, the body style is identified by an array of infrared proximity sensors, and this information is matched against the glasses being manually loaded to the preparation system to ensure that this loading sequence is correct.

The glass preparation system consists of two parallel lines for the front and rear screens respectively. Each line comprises a load station, an application station and an unload station with automatic station transfer to the glazing robots. Once within the application station, the glass is clamped by vacuum suckers, centralized and presented beneath a single, purpose-built CNC-controlled five-axis gantry adhesive applicator system. This, in turn, manipulates a mastic applicator along a preprogrammed path, around the periphery of the screen normal to its surface, to apply a triangular cross-sectional bead, with the program controlling both nozzle motion and the sealant-dispensing system. The glass is then shuttled through to the unload station where it is raised above the transfer, providing a suitable access position for the robot to pick up.

The two robots – six-axis Kuka IR662 versions – are mounted on the same side of the track to perform simultaneous windscreen and rear screen pick-up and insertion. Each robot is fitted with a gripper assembly mounted to the wrist, and enhanced capabilities include a collision detection device and pneumatic suction cups integral with light-emitting diodes.

14.3.3 Windshield fitment station: concept and construction

Construction

The optical equipment consists of four Panasonic CCD camera pairs/5 mW HeNe lasers to process the common front screen aperture, and a further 12 camera/laser pairs, four for each of the three derivatives, to process the rear screen. The camera/laser modules are mounted on a gantry (as shown in Figure 14.8) which spans the track with positional adjustment already being

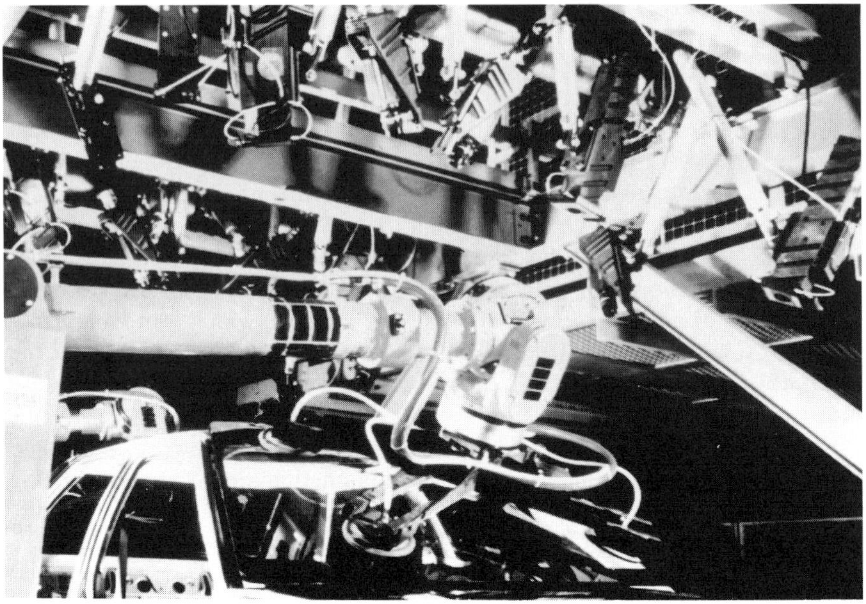

Figure 14.8 Camera/laser measuring module – set-up.

catered for. All of these camera/laser modules are linked to an Automatix-based AV3 vision controller(s).

Machine vision concept

In this application as with the previous case study, the vision controllers utilize the 'structured light' analysis technique, where the plane of the laser stripe impinges on the object feature and the diffuse reflection is captured as an image via the CCD-type camera.

On arrival into the cell, one of the roof camera/laser pairs is operated to determine into which colour category the body falls, i.e. light, medium or dark, which dictates the type of software algorithm to be executed for the main vision analysis. This is simply determined by measuring the relative reflectivity level by counting the number of pixels in the pixel buffer (with millivolt values within preset threshold ranges), the higher millivolt values being generated by the greater reflectivity of the lighter colours.

The cameras at the front and the appropriate derivative cameras at the rear scan the apertures to determine information as to their relative position and orientation. This information is derived by striking each of the laser stripes across both 'A' posts and two targets at the roof panel, which stretch across to and over the receiving aperture flange for the front screen, and an identical arrangement is employed for the rear light aperture as shown in Figure 14.9.

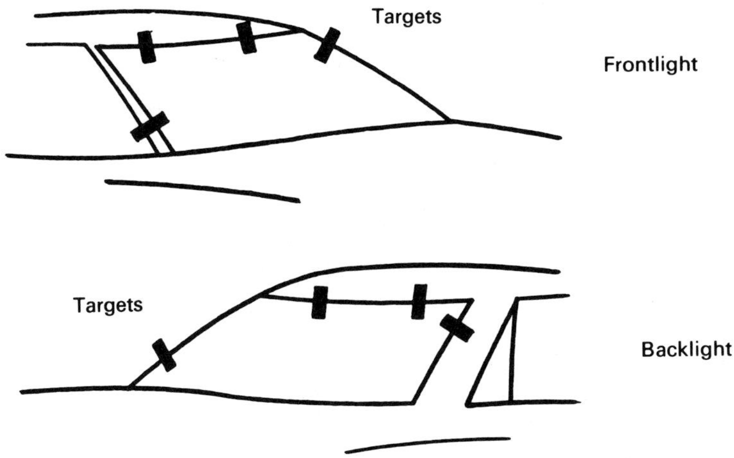

Figure 14.9 Laser targets for front and backlight areas.

The higher reflectivity of the lighter colours enables distinct laser line breaks to be determined by applying simple thresholding techniques. Employing the connectivity analysis algorithm on this resultant image enables the extreme edge of the relevant aperture edge to be determined. However, with the lower reflectivity images resulting from those cars with medium to dark colours, the breaks in the laser line are not readily distinguished. For these cars, edge detection techniques have to be employed, scanning the image from top to bottom in order to determine the centre of the laser line. Using dedicated software routines, the sequence of change in the gradient of the laser line centre is calculated and compared with expected changes to define the position consistent with that representing the aperture edge.

14.3.4 Windshield fitment operation

During this vision analysis of the apertures, the Kuka robots have collected their respective screens and are initiated to manipulate them to a position 50 mm above but in normal plane to the car aperture as shown in Figure 14.10. To ensure that both robots have initially moved to the correct screen gauge position, the two light-emitting diodes which are incorporated into each gripper assembly are then viewed by two of the cameras and their actual spatial location centroid is determined by connectivity analysis and the results compared with their expected position. Any discrepancy in position is then detected at this stage before any attempt is made to fit the screens. Though at this screen gauge position, the glass appears within the field of view of the same set of cameras, the glass is still not considered to be in the correct spatial position because of body presentation, and therefore some form of real-time adjustment, is needed.

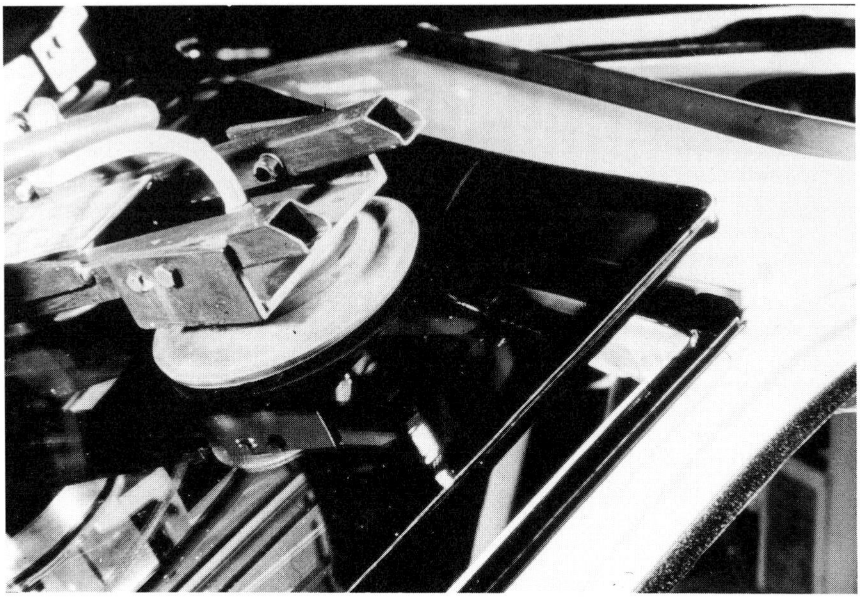

Figure 14.10 Close-up view of robotic fitment of front glass to car.

By using the same set of cameras and lasers, the difference between the glass and the aperture can now be calculated. These data are combined to define uniquely the robot motion which is required to take the robot from its current position to the fitting position. Since the current robot location is known, the final fit position can be calculated and therefore the robot is instructed to move accordingly, to compensate for the necessary adjustment.

Once the vision system has analysed the aperture and the glass information, the gap expected around the edge of the glass is computed and if this is found to be less than zero, then the attempt to fit the glass is aborted. On completion of the fit process and withdrawal of the Kuka robots, the actual post-fit gap is then analysed for condition. This information, together with the data relating to aperture and glass size, is fed directly into a statistical control package which automatically warns of any adverse trends.

14.4 CASE STUDY III: VALVE TIMING GEAR VERIFICATION

14.4.1 Valve timing gear verification: an application for machine vision?

This section describes a machine vision system which is being used to verify the timing gear alignment of the K Series engine as part of its build procedure. This is a typical example of an application of machine vision which is replacing the inspection or verification task which would have been undertaken by a human. The system is designed to validate if the timing belt has been correctly fitted, prior to executing a series of computer-controlled automatic 'cold-test' programs. If the timing belt is found to be incorrectly fitted in relation to the engine timing marks, then the 'cold-test' program is aborted.

14.4.2 Process requirements

Following the assembly of the crankshaft, camshaft pullies and the timing belt, the part-assembled engine is required to undergo a series of validation tests to ensure 'perfect' quality build at that stage of manufacture. The solution sought would comprise an automatic test facility, consisting of two functional test stations, the former checking for leaks in the water cooling and lubrication systems, whilst the latter station would undertake a cold rotational test, measuring the 'torque to turn' and oil pressure values. The main provision is that the cold rotational test could only commence when the relative angular position of the timing marks had been validated.

The overriding objective of the project, therefore, was to find an *automated* means of being able to verify the radial orientation and relationship of the valve timing drive pullies relative to each other. The scope of the project for the K 16 (4v) TBI engine would require examining two camshaft pullies and one crankshaft pulley.

Of the various methods that were analysed, machine vision was considered, on paper, to be the most suitable candidate for selection, even though it meant that the technology would have to be developed for this application.

14.4.3 Test facility layout

The test facility consists of two stations. The first station checks for water and oil leaks. This is achieved by plugging all the apertures that would closed off on a finished engine, and monitoring first the oil gallery, then the waterways for leaks.

Six automatic operations are carried out at the second station, of which the machine vision system is an integral part. The engine is primed with oil and five tests completed. The vision-based timing check is then pencilled in. Following the successful outcome of this check, the oil is forced under pressure through the oil galleries and bearings to prime the lubrication system. If all these are correct, a torque is applied to turn the engine. During rotation comparative compression and oil pressures are checked.

14.4.4 Vision system description

Figure 14.11 is a simple schematic diagram of the timing gear verification system that was finally installed, the front end of the engine with the exposed timing pullies being presented to three fixed CCD cameras. Camera A views

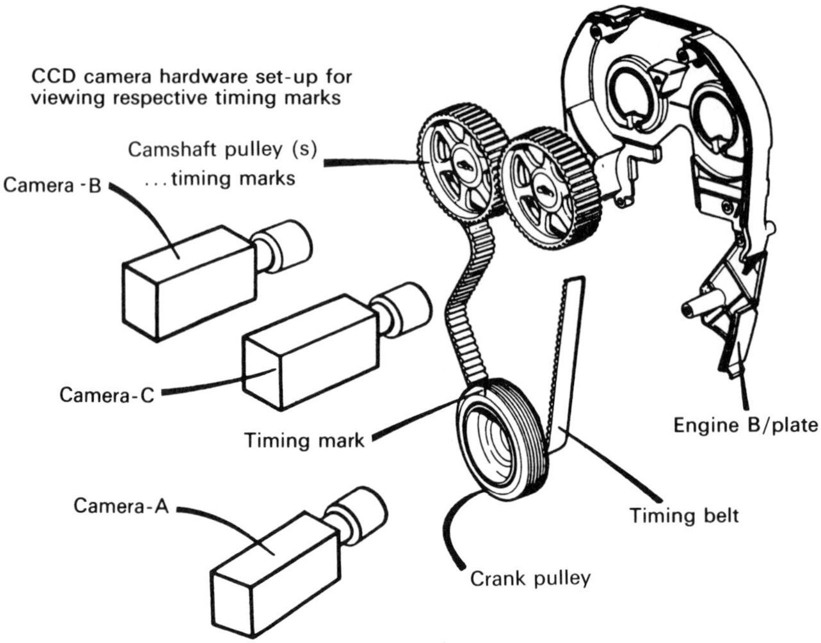

Figure 14.11 Hardware set-up for viewing timing marks.

the timing mark of the crankshaft pulley whilst cameras B and C handle each one of the timing marks of the camshaft pullies.

To maintain the flexibility of the system to accommodate future engine variants, both the field of view and camera on/off can be easily achieved via a software switch. Floodlight sources are used to illuminate the specified areas of the engine part. Correct placement of the light sources ensures that the correct amount of light provides the necessary contrast in order to pick up the respective timing marks relatively simply.

14.4.5 Test sequence

The engine platen automatically transfers to the second test station. Before the tests take place, the conveyor section complete with platen and engine is lowered onto a substantial base. The assembly line controller reads the information from the transponder and identifies the engine variant to the AV5 controller so that it can select the correct software algorithm. When the appropriate communication lines have been established between the controllers, the vision system will automatically begin the valve timing checking procedure (Figure 14.12).

The vision system begins the process by taking a picture of the crankshaft pulley, the digitized image being stored in pixel buffer no. 1. The image in

Figure 14.12 Valve timing check facility and procedure: (a) actual station detail with no engine; (b) timing gear verification process being executed for K8 2v engine.

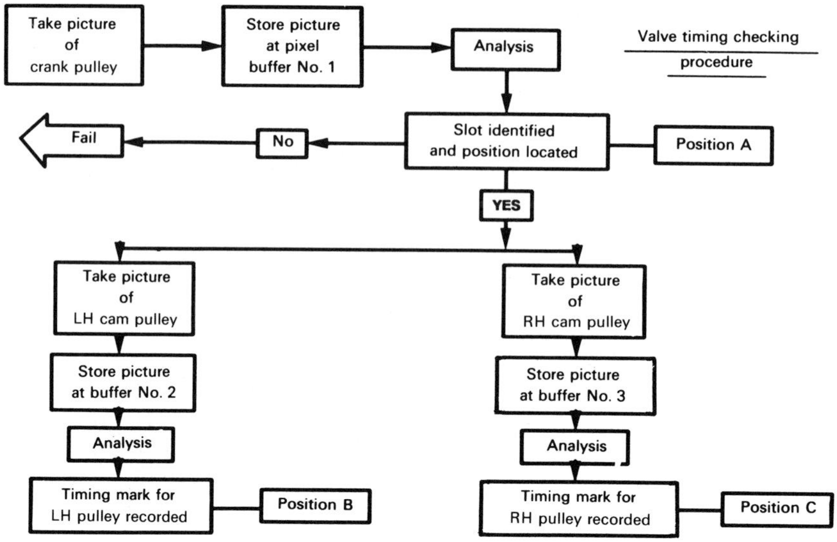

Figure 14.13 Valve timing checking process methodology.

pixel buffer no. 1 is analysed and the position of the timing slot is recorded (Figure 14.13). Once the system has identified the crankshaft timing mark, the system then takes a picture of the left-hand camshaft pulley and the digitized image is stored in pixel buffer no. 2. The image in pixel buffer no. 2 is analysed and the position of the timing mark for the left-hand pulley recorded. Subsequently, the position of the timing mark on the right-hand camshaft pulley is determined in the same manner as the left-hand camshaft pulley with the data being stored to pixel buffer no. 3. Figure 14.14 illustrates a typical digitized image of the right-hand camshaft pulley timing mark.

Having found the position of all the timing marks, the 'mark-matching' software algorithm is executed. This software module will compare the relative position of each timing mark with those stored in the look-up table. If the relationship holds, the vision system will send a 'PASS' signal to the assembly line controller, confirming that the rotational test can begin. If the relationship between the marks is confirmed as being incorrect (i.e. due to engine not timed correctly), then a 'FAIL' signal will be sent to the assembly line controller. The test program is absorted and the engine is auto-routed into a re-work station.

Corrective action is then undertaken and on completion the engine will be diverted back into the automated test facility a second time and subsequently the valve timing checked again; this process is repeated until the engine is validated by the automation, from where it can then continue through the assembly process.

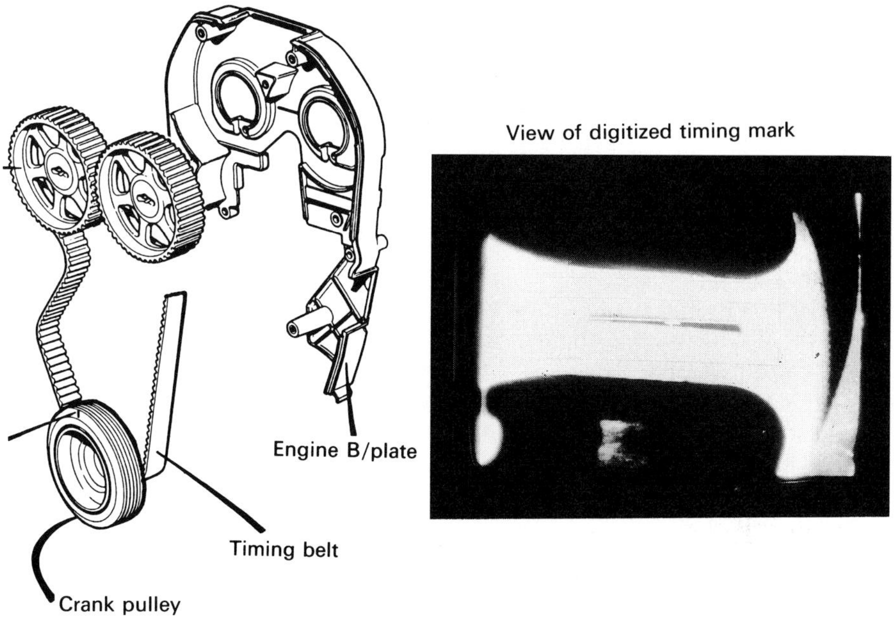

Figure 14.14 Digitized image of camshaft pulley mark.

14.5 CASE STUDY IV: SELECTIVE BEARING SHELL ASSEMBLY

14.5.1 Selective bearing shell assembly: an application for machine vision?

Printed characters (not handwritten) on labels or directly marked on bottles, cans, cartons, etc. can be read by machine vision systems. Systems which can do this effectively and are easily programmed are more specialized units known as optical character recognition devices (OCR). The latest systems use training and recognition techniques to learn the particular type of character font and then enable it to read the label. Grey scale processing is essential to enhance the images prior to the recognition stage to cope with poor printing, variable lighting conditions and variable locations.

This section describes an alphanumeric OCR system which is central to an automated assembly operation involving the fitting of bearing shells to the new K Series engine which is manufactured at Rover Group's Longbridge engine assembly plant, near Birmingham.

14.5.2 Process requirements

The start of the build process covers the requirement for the main crankshaft bearing shells to be assembled automatically to the bearing ladder and the engine block. Subsequently the scope of the specification was extended to cover the selective assembly operation of the con-rod big end. Principally the solution sought would comprise the following three discrete elements:

1. Read the bearing codes/grades inscribed on the various engine components during their manufacture.
2. Identify the grades and calculate the sizes of bearing shells to be selected.
3. Select the correct graded shells and assemble to the block and ladder by robot.

The primary objective of the project was to find an *automated* means of reading and identifying the bearing codes which were to be inscribed on the engine components.

The code arrangement which was finally devised is as follows. The main crankshaft journals were to be identified by five characters marked on the crankshaft and five characters marked on the bearing ladder respectively, whilst for the con-rod big end journals, four characters would be marked on the crankshaft.

Of the various technologies that were evaluated, machine vision was, on paper, considered to be the most suitable candidate for selection, even though the technology had not been developed for this application.

14.5.3 Assembly cell layout

The assembly cell consists of three stations (Stations No. 3, 4 and 5). The first station houses the vision system, which comprises three fixed CCD cameras. Station No. 4, which follows, acts as a standby station to provide the ability to manually load bearing shells and thrust washers whilst the other two stations undergo preventative maintenance work. The final station consists of a Pragma V3000 robot, which automatically assembles the bearing shells and centre thrust shells to the block and ladder.

14.5.4 Station No. 3: vision system description

Figure 14.15 is a simple schematic diagram of the character recognition system. The crankshaft and the bearing ladder to be analysed are placed beneath three CCD cameras. Camera X views the journal grades marked on the bearing ladder whilst Figure 14.16 illustrates the hardware set-up for cameras Y and Z to handle the crankshaft journal and pin grades respectively. There are two variations of the crankshaft, so both cameras Y and Z are fixtured in such a manner that the area under recognition is within the field of view of cameras Y and Z.

Figure 14.16 CCD camera hardware set-up for reading both journal and pin grades marked on the crankshaft.

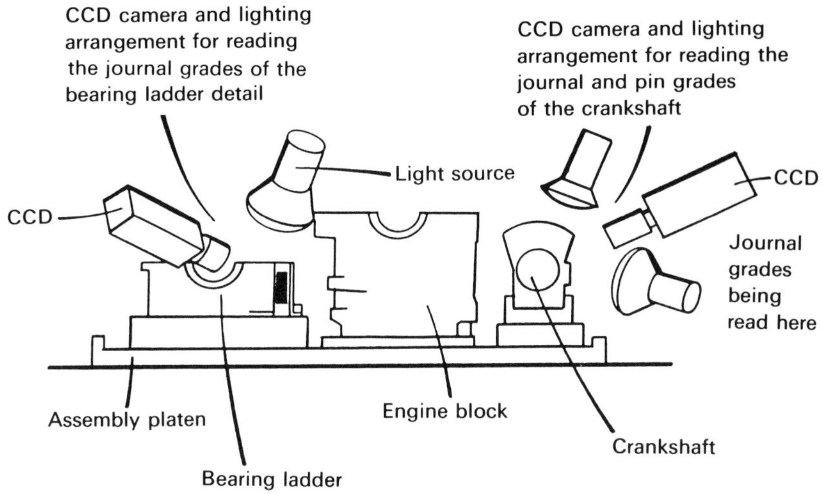

Figure 14.15 Character recognition station – H/W set-up.

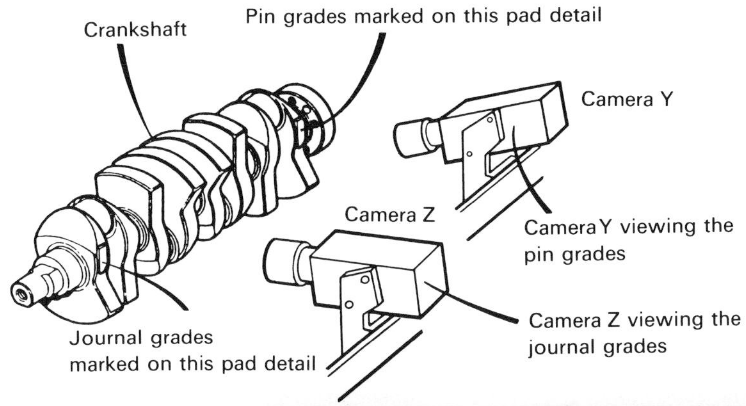

Floodlight sources are used to illuminate the specified areas of the engine part. Correct placement of the light sources ensures that the correct amount of light is reflected from the dot-matrix type of the characters towards the lens of the camera. An Automatix Autovision 5 (AV5) controller is at the heart of the vision system arrangement. The controller is used to take pictures of the parts, digitize images, and carry out the various image-processing tasks as necessary, in order to yield the identification of all the characters prior to robotic assembly. As the production cycle time is at a premium, the architecture of the AV5 controller is configured to process the images captured by all three cameras simultaneously.

14.5.5 Station No. 5: robotic assembly system description

The bearing shell load station (Figure 14.17) comprises a Pragma V3000 series robot which is suspended over both the main assembly conveyor which supports the platen and the feed system. The robot automatically selects the shells from conveyors comprising nine lanes of accurately graded components. The feed direction is terminated with the presence of nine escapements which accurately locate and secure each shell ready for collection by the robot. The Pragma robot traverses a bridge across the assembly track to pick up the parts from these feeders on both sides. Vacuum heads on the robot arms pick up and load the components to the engine. The bearing shells are manually loaded from behind the robot into their designated sized linear feed conveyors. Any low component levels are immediately flagged up, for correction.

The robot is equipped with a double-handed vacuum gripper, in order to maximize its efficiency and also to meet the very short cycle time required

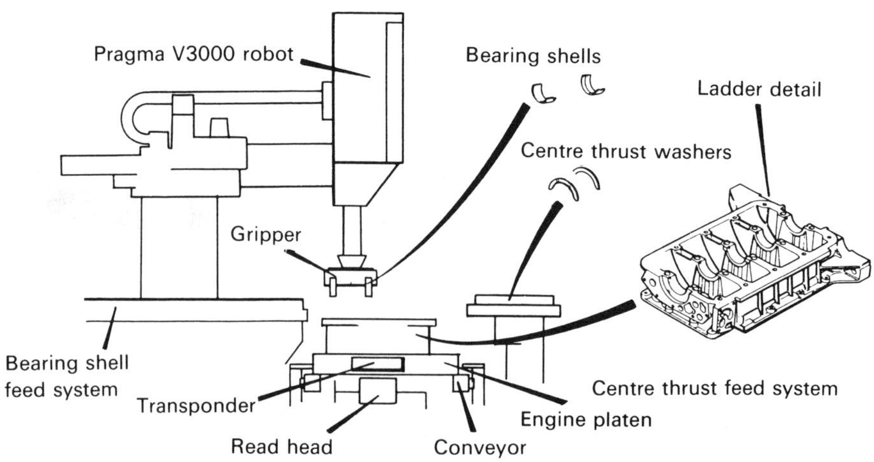

Figure 14.17 Overview of the bearing shell assembly load station.

to complete the load sequence. The vacuum-based gripper has inbuilt sensory features which allow it to intelligently perform the assembly task to specification, such as the ability to check for part presence, and locating the machined surfaces of the ladder and block prior to depositing the shells.

14.5.6 Assembly process overview

The 'K' Series engine assembly process begins with the main engine block and ladder, which are machined as matched pairs, being delivered to the assembly line as a unit. Though these two parts remain together throughout, they are, however, separated, with the block and ladder together with the crankshaft being located accurately to a platen which moves along a non-synchronous assembly track.

The variant required to be built (e.g. K16 (4v) TBi) is electronically coded into a transponder fixed to the underside of the platen from the main production control computer. The transponder also receives and stores coded information from transceivers at each workstation so that the assembly process can be updated with the build status.

As soon as the platen moves into the vision station as shown in Figure 14.18, the assembly line controller reads the information from the transponder and identifies the engine variant to the AV5 controller so it can select the correct software.

Figure 14.18 Character recognition process being executed.

Figure 14.19 Operation showing robotic assembly of bearing shells.

When the appropriate communication lines have been established between the controllers, the vision system will automatically begin the bearing grade data acquisition routine. The vision system takes two separate pictures of the crankshaft and one of the ladder, with the resultant digitized images being stored in the pixel buffers. As the operation time is at a premium, the vision system then starts the process of recognizing the characters from the various captured images, with all three algorithms being executed simultaneously.

When the characters have been identified, the grade codes are read in by the assembly line controller and downloaded to the Pragma robot controller. The robot controller uses this information about the bearing codes to calculate the correct grade of shell to match each crankshaft journal diameter. The pin grades are written into the platen transponder and read out at a subsequent station to tell an operator the size of the con-rod big end shell diameter to select.

The platen at the vision station is released (passing through Station No. 4) and enters the robot shell load station as shown in Figure 14.19. At this station, a robot automatically selects the appropriate-sized bearing shell from one of the nine conveyor lanes of graded shells. The robot assembly cycle commences with the loading of two centre thrust washers and terminates with the five sets of bearing shells to the block and ladder. When the assembly operation has been completed satisfactorily the platen is automatically released and allowed to enter the next robotic assembly station.

14.6 CONCLUDING DISCUSSION

This chapter has considered the issue of machine vision and, in particular, its deployment within the automotive industry. The chapter has presented work on machine vision for the prospective end-users and not the designers of such systems.

The content of each case study should have provided sufficient background about the subject to separate machine vision promises from reality and permit intelligent decisions regarding machine vision application to be made.

The main application areas for machine vision within the automotive industry can be classified into two main categories. The first of these relates to inspection-based systems for the critical dimension measurement and the integrity of component parts. The second classification has been identified for robot guidance systems, which form an integral part of manufacturing processes.

The advantage realized through the use of machine vision for gauging is primarily that, because of its speed of operation, it is now possible to gauge every body rather than periodic samples. Being a non-contact device, it also contains no moving parts and, because of its inherent built-in intelligence, variations in body presentation can be accommodated.

Furthermore, due to the speed and nature of the data capture process, it provides feedback in 'real time' to the robotic welding stations on developing errors, thereby taking early preventative action on the process.

The glazing system is a successful working example of position sensing by using vision technology, highlighting the opportunity it now presents for the potential automation of numerous applications, previously deemed impractical due to problems experienced with inconsistent body/feature presentation.

The last two case studies, covering the deployment of machine vision for the assembly process of engineers, have shown the following benefits, namely:

- Providing reduction in manning requirements, thereby leading to cost savings.
- It has given consistent production over multi-shift working.
- Removing the repetitive decision-making task from the operator, by using achievable technology.
- Total quality both within the product and assembly process can be assured.
- The speed and accuracy of operation cannot be matched by its human counterpart.
- Gives total confidence of a consistent process.

In conclusion, despite the limited availability of industrially based machine vision systems, the success of implementation is not always guaranteed, as the technology both imposes technical limitations and introduces new human engineering considerations.

By demonstrating the application and the implications of the technical requirements for both the 'staging' and the 'image-processing' power required of the machine vision system, this chapter has shown that the most significant elements of a successful application are indeed the lighting optics, component design, etc. – the 'staging'. From the case studies investigated, optimized 'staging' has resulted in the need for less computing power in the machine vision system. Inevitably, greater computing power not only requires more time but is generally more expensive.

The experience gained from these projects has demonstrated that machine vision technology is a realistic alternative means of capturing data in real-time, since the current limitations of the technology are well suited to the delivery process of the quality function within the manufacturing process. Relevant literature on the implementation of vision systems for automotive manufactures is listed [1–15].

REFERENCES

1. Bailey, G. (1985) Machine vision systems, *Design Engineering*, 53–6.
2. Gill, K. (1989) Machine vision technology for engine assembly, at Rover Group, in *Proceedings of the 21st ISATA Conference*, Wiesbaden, West Germany 1989, ISATA.
3. Moffatt, J. (1987) Machine vision. *Management Services*, 20–3.
4. Gill, K. (1990) The application of machine vision technology as used within the Rover Group, in *Image Processing Systems Conference*, London, England 1990.
5. Lapidus, S. (1985) The unblinking link to inspection and positioning systems. *Plant Engineering*, 78–83.
6. Summerfield, P. (1987) Electro-optics manufacturing systems. *Production Engineer*, December.
7. DeVicq, A. (1985) Set your sights on vision. *Metalworking Production*, 97–102.
8. Jawaid, A. and Gill, K. (1990) Implications of robotic assembly for power train components, in *Proceedings of NAMRC XVIII Conference*, Pennsylvania, USA 1990.
9. Oshima, M. (1983) Object recognition using three-dimensional information. *IEEE*, **5**, No. 4, 353–61.
10. Gill, K. (1990) Body beautiful. *Image Processing Journal*, 27–30.
11. Rosenfeld, A. (1987) Survey, picture processing: 1986. *Computer Vision*, 171–4.
12. Mantas, J. (1986) An overview of character recognition methods. *Pattern Recognition*, 425–30.
13. Kahan, S. (1987) Recognition of printed characters. *IEEE*, **PAMI-9**, 274–88.
14. Lam, L. (1988) Structural classification and matching of handwritten code numbers. *Pattern Recognition.*
15. Mahdavieh, Y. (1988) Automatic recognition of characters on machined parts. *Sensor Review*, July.

PART EIGHT

Smart Sensors and Systems

CHAPTER 15

Intelligent sensors for vehicles

15.1 INTRODUCTION

The substitution of mechanical control units by electronic systems offers attractive benefits and improvements in the further development of the automobile. In some case, the realization of new and additional functionality has already been possible. In systems related to safety, however, the mechanical redundancy was usually retained. In general, initial development and optimization was performed on separate, individual systems.

In a second phase it was discovered that it is of advantage when these electronic systems communicate with one another. On the one hand, this makes it possible to reduce the number of components, e.g. through multiple use of a single sensor. On the other hand, the proper operation of individual systems can be checked with lateral information exchange and plausibility comparisons. In case of failure many functions can be overtaken by other systems for emergency operation. In this phase, in which we presently find ourselves, new multipurpose connections between the units are also being established: bus systems. It is particularly their flexible and transparent structure that makes them superior to conventional wiring in complex vehicle systems.

At this stage, however, it was recognized that the individual systems are related to each other by much more than multiple use of sensor signals and emergency functions. They are also closely connected in their basic operation. A great advantage lies in their possible co-operation to support a single concept. For this, however, the systems must – with reference to total vehicle operation – be rearranged; expressed more simply, who tells whom what, in the sense of issuing commands must be defined.

It appears that a hierarchical modular arrangement of vehicle operations is superior to a complicated structure composed exclusively of equal systems. Such a hierarchical arrangement concept is shown in Figure 15.1.

This hierarchical structure should be considered separately from the attendant hardware structure in which it is actually realized. Modular hardware design corresponds to the modular nature of the control functions. The goal is to get individual, independent, separately developed and testable peripheral units. The availability of intelligent sensors and actuators is a precondition for the realization of such a hardware concept as shown in Figure 15.2.

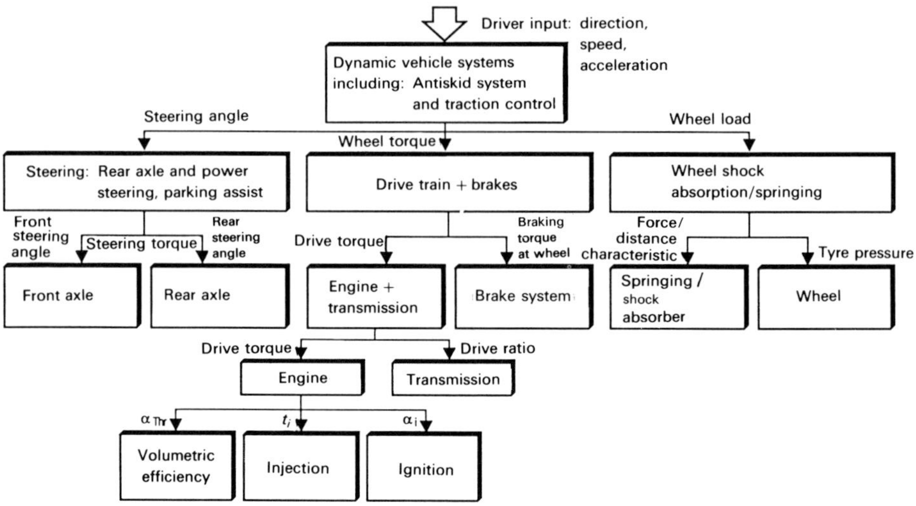

Figure 15.1 Hierarchical arrangement concept.

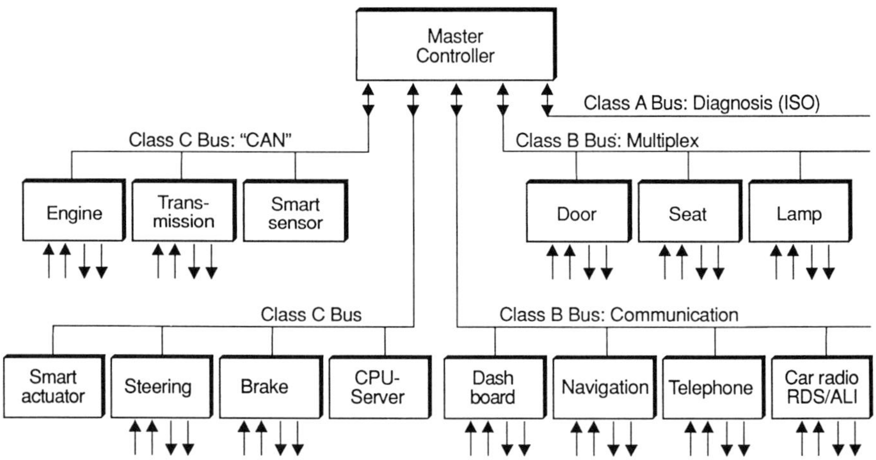

Figure 15.2 Decentralized hardware concept.

The peripheral units, also called part electronics, are connected to one another and to the master controller by bus systems of different transmission speeds. Bus systems suitable for the car are currently being created for information distribution. A well-known bus system is called controller area network (CAN). As may be inferred from Kiencke, Dais and Litschel [1] and Kiencke and Dais [2], it is flexible and has special facilities ensuring interference-suppressed operation.

The master controller in Figure 15.2 functions as a gateway connecting the different bus systems. In case of failure it is responsible for the organization of emergency operation. There is, however, no knowledge about the physical processes which are observed and controlled. This intelligence is completely moved to the part electronics at the periphery.

15.2 CONCEPTS FOR INTELLIGENT SENSORS

15.2.1 Definition

There has always been a discussion about the suitability or otherwise of using the attribute 'smart' or 'intelligent' for sensors. Anyway, the technical meaning of this now well-established term has become quite clear during the course of many discussions and may be regarded as being defined according to Tränkler [3].

Sensors with sensor-specific measuring signal processing serving the following purposes are described as smart or intelligent:

- bringing the utilizable information content of a sensor to a higher level;
- gaining the interesting and useful information from a variety of less informative single pieces of information.

Intelligent sensors may be identified in concrete terms by means of the features listed in Table 15.1. Static and dynamic corrections of the measuring signal may indeed be implemented, in principle by analog or digital means; however, the digital storage of specimen-specific sensor data, as well as its digital

Table 15.1 Features of smart sensors

Limitation of the tolerances using electronics:
Acquisition and consideration of disruptive influencing variables on the spot (static correction)
Improvement of the dynamic behaviour (dynamic correlation)
Consideration of specimen-specific characteristics (calibration)
Standardized interface:
Bidirectional
Adaptable
Communicating with digital systems (bus capable)
Possibility of:
Inherent monitoring
Multivariable acquisition
Recalibration
Additional ouput of influencing variables

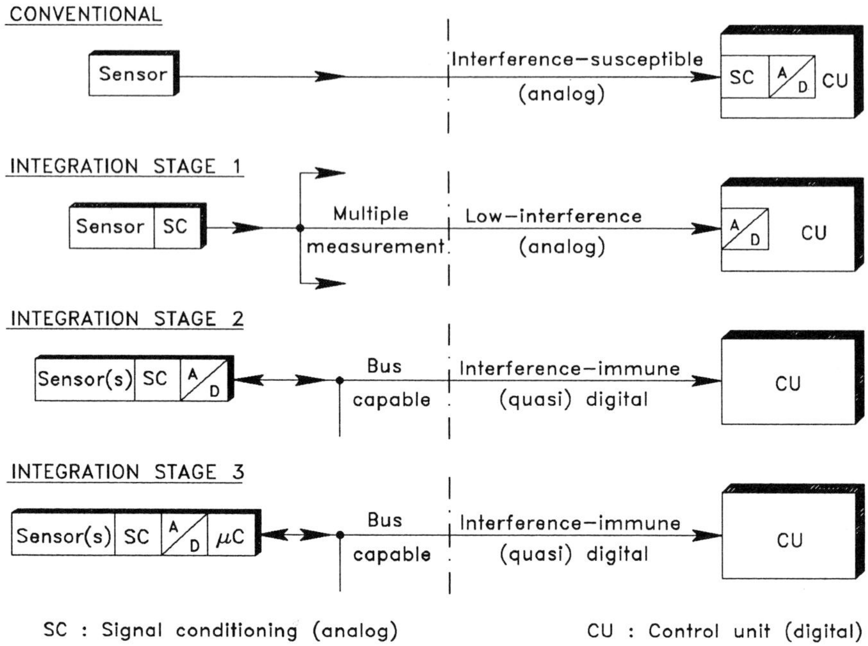

Figure 15.3 Integration stages for sensors.

processing and transfer, are playing an ever greater role. The prerequisite is always the validity and familiarity of a specific sensor model as well as the constancy of the specimen-specific model parameters once acquired.

Figure 15.3 shows possible stage between conventional sensor techniques with no electronics on site up to the highest integration stage. In this stage (integration stage 3), electronics on site include the circuitry for the analog signal conditioning, the AD conversion and the error correction unit, realized for instance by a microcontroller.

15.2.2 Sensor model

As shown in Figure 15.4, a real sensor generates a disturbed measuring signal x_d that depends not only on the analog quantity x but also on different influencing variables, denoted y_1 up to y_n. The most important among these is the temperature ϑ. Also note that the sensor transfer function

$$x_d = f_t(x, y_1, \ldots, y_n) \tag{15.1}$$

is not necessarily linear even if all disturbing variables are elminated. Thus there are two main effects disturbing the measuring signal: the non-linearity of the sensor element and the influence of the environment temperature.

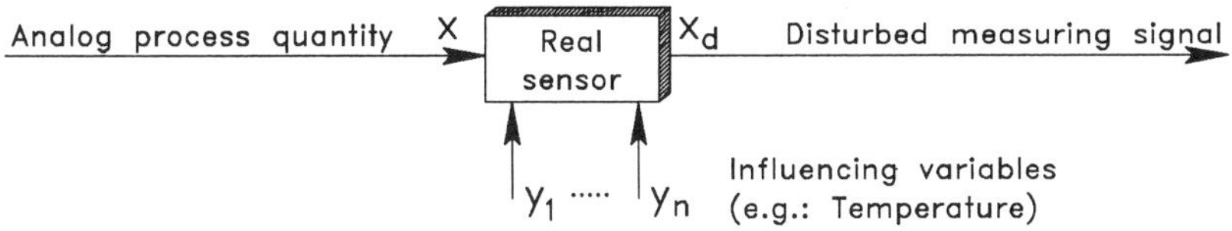

Figure 15.4 General sensor model.

The sensor transfer function is unit specific. The influence of the unit variation can be described by a set of m parameters $p_{1,i},\ldots,p_{m,i}$, where i denotes the unit number. Thus Equation (15.1) can be written at full length:

$$x_d = f_t(x, y_1, \ldots y_n, p_{1,i} \ldots, p_{m,i}) \tag{15.2}$$

The idea behind the intelligent sensor concept is that a fault-free measurement signal x_f can be derived from the disturbed signal x_d. For this reason, a sensor model

$$x_f = f_m(x_d, y_1, \ldots, y_n, p_{1,i} \ldots, p_{m,i}) \tag{15.3}$$

must be known. It can be obtained by inverting the sensor transfer function f_t.

The signal x_f which is a fault-free representation of the process quantity x can be calculated with the sensor model f_m if the influencing variables $y_1, y_2, \ldots, y_n$ are also measured. The basic smart sensor circuit using this principle is shown in Figure 15.5.

The sensor model is a function with free parameters. It is applicable to every sensor unit in the same way. By applying the specimen-specific set of model parameters $p_{1,i}, \ldots, p_{m,i}$ to the sensor model f_m, the unit-specific sensor correction function $f_{c,i}$ is obtained. Thus Equation 15.3 can also be written as follows:

$$x_f = f_{c,i}(x_d, y_1, \ldots, y_n) \tag{15.4}$$

In this way the individual fluctuations in production variation are covered

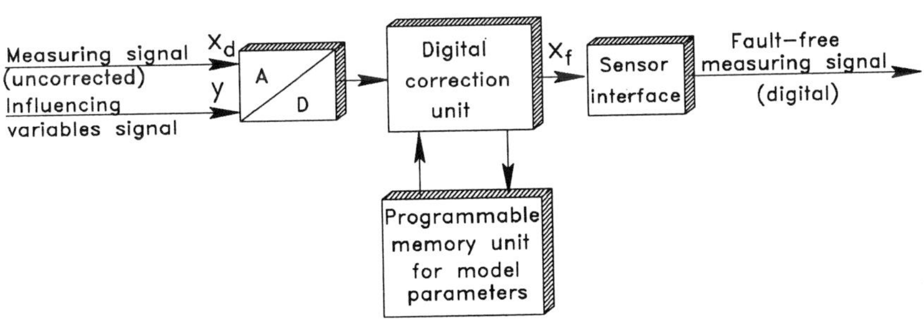

Figure 15.5 Basic smart sensor circuit.

by the smart sensor principle. It is thus a constant attribute of intelligent sensors that the electronics integrated therein include a programmable memory unit (PROM or EPROM) for the storage of the specimen-specific model parameters $p_{1,i}, \ldots, p_{m,i}$.

15.2.3 Structures of intelligent sensors

The basic smart sensor circuit as shown in Figure 15.5 is suitable for most applications of intelligent sensors. The digital correction unit can be a microcontroller or a dedicated processor for the specific correction alogrithm. The AD converter must, however, fully satisfy the corresponding measuring signal requirements regarding converter speed and accuracy. If the analog circuit generates a signal with an analog period length, the AD conversion can be done easily by using digital counter structures.

In special applications a simpler sensor structure may be sufficient. In order to utilize the advantages of digital correction with the least possible expense whilst simultaneously maintaining the high-frequency dynamics of analog sensors, a structure like that in Figure 15.6 is of advantage; it may be used by all sensors with approximately linear characteristic curves.

It leaves the uncorrected sensor measuring signal x_d in analog form and only converts the (generally slowly changing) influencing variables y into digital form. As the influencing variables generally affect only a small part of the analog process quantity's nominal range (typically 5–20%), no high resolution or high working speed of the AD converter is necessary. The digital converter result is, at the same time, the address for a PROM, from which values for zero point and slope correction may be read. As the PROM values are also only correction quantities, no high resolution is required here either. They are transformed back in the analog range and are added or multiplied into the analog signal train.

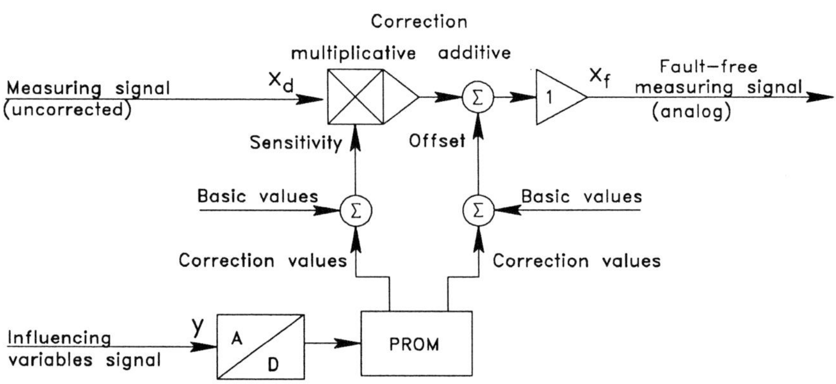

Figure 15.6 Simple correction circuit.

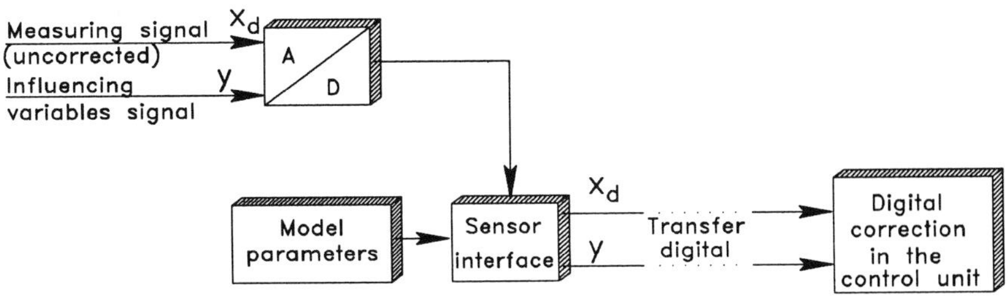

Figure 15.7 Smart sensor with separate correction.

Under especially difficult operating conditions there will be absolutely no desire to increase the utilization of electronics more than it is really necessary. In this case no correction algorithm is carried out at the measuring point. There are, as shown in Figure 15.7, only circuits for the digitalization of the unprocessed signal, a PROM for the specimen-specific sensor parameters as well as interface electronics for the information transfer to the central correction computer. The contents of the PROM are transmitted during the initialization phase. Such a structure is advantageous if an electronic vehicle system contains several sensors of the same model whose correction algorithm may be carried out in the same way at the central point under less critical operating conditions for all sensors, for example during tyre pressure measurement.

Film techniques and micromechanics as described by Heintz and Zabler [4] and Csepregi [5] also make it possible to virtually simultaneously acquire the analog process quantities using several identical, integrated sensors without considerably increasing the volume of a packaged sensor. Such arrangements mean that by statistical methods even uncontrolled drifts of unlimited extent may be reduced and outliers or failures may be recognized.

15.2.4 Summary of the advantages

The step of introducing electronics – especially digital – to the measuring point may be used to considerably improve utilization of the information contents of sensor structures. This is possible by using individual, digital correction to a greater level than has been technically possible until now. There remains the demand for high stability and reproducibility of the sensor properties over time.

The signal preprocessing and information condensation carried out on site also satisfies the need to relieve the central control units. By now, it is to a great extent the electronics belonging to peripheral components which make the control units bigger and bigger. The available space, however, is limited by the given housing as well as by the maximum possible heat dissipation.

As intelligent sensors are bus capable, they are multiply usable. In addition, such comprehensive and comfortable information distribution results in additional operating safety. If a sensor malfunctions, replacement signals may always be resorted to and emergency operation may be maintained. Moreover, sensors may be more easily monitored by virtue of the cross and plausibility comparison with the signals of other sensors.

The transfer of voltage or amplitude analog signals is subject to natural limits regarding transferable accuracy by interference voltage influences. Transfer operations for analog signals increase greatly as measuring accuracy increases; the levels of accuracy achievable using smart sensors may be more easily transferred in digital form in which they are already mostly present on location thanks to the special correction procedure. Quasi-digital signal forms such as frequency or time analog representations are also transferable interference-free, but are mostly subject to other limitations. Moreover, electronics on site facilitate the realization of a very flexible, e.g. programmable, interface via which communication of the sensor may take place in both directions. This means, for example, that changing the operating mode (working speed, diagnosis, interface, logging, calibration, emission of additional signals, emergency operation, etc.) is also possible.

Shielded instrument leads to prevent interference radiation or irradiation are in general not desired in motor vehicles for cost reasons. Electronics on site make it possible to demodulate carrier signals at the measuring point or to considerably decrease the arising signal frequency by means of dividers so that there is no necessity to have screened instrument leads. For example, interesting sensors may be used in accordance with the HF damping principle, magnetoelastic sensor films, saturation core probes as well as temperature- and power-dependent quartz crystal units for oscillation and ultrasonic measurement up to operating frequencies of several megahertz. Multi-sensor systems have until now for reasons of pin and connection economy, as well as the associated operating safety, not been possible. Integrated electronics on site can use the connection and bonding techniques of hybrid and monolithic circuit technology so that only a few instrument leads are required despite the complex sensor structure. Complex sensor structures permit either the utilization of statistical methods to increase measuring safety or the gaining of superior analog process quantities of great expression and/or precision. Information compression of any type does, however, greatly relieve the transfer system used and the following processing units.

For reasons of transfer safety, only sensors with great measurement effects have been used until now in conventional sensor techniques, e.g. sintered ceramic semiconductor resistors for temperature measurement (NTC thermistors) and potentiometers for displacement/angle measurement. Electronics on site allow the use of advantageous sensor techniques (film and semiconductor sensors) with smaller measurement effects.

Finally, digital electronics on the spot provide special possibilities of self-diagnosis and the emission of detailed diagnosis information at the sensor output. The possibilities of intelligent sensors compared to conventional

Table 15.2 Pressure sensors in the motor vehicle

Analog process quantity	*Measuring range (bar)*
Intake manifold or load pressure	1–5
Tyre pressure	5
Brake pressure (pneumatic)	10
Pneumatic spring pressure	16
Combustion chamber pressure	100
Hydraulic pressure	200

sensor techniques are discussed in the next section by introducing several interesting examples.

15.3 POSSIBILITIES FOR INTELLIGENT SENSORS IN CURRENT APPLICATIONS

15.3.1 Pressure sensors

In motor vehicles pressure sensors are at the top of the list concerning acquisition of dynamic effects. As regards the measurement of static pressures, accuracies of 1.5–2% FSO currently form the lowest possible limit with low-cost sensors. However, only deviations far under 1% FSO are permissible for various applications. Table 15.2 shows the most important applications for pressure sensors.

The methods and structures of smart sensors seem specially suited to overcome existing limits. The prerequisite for this is, however, a sensor technology of sufficient long-term stability as well as housing and connection technology preventing the introduction of uncontrollable mechanical interference forces.

High accuracy as well as the high-frequency dynamics of an analog signal path with simultaneously very low utilization of electronics are offered by the sensor structure shown in Figure 15.6. It was applied for semiconductor pressure sensors known by the designation PROMCOMP, described in more detail by Ritter, Bomholt and Keller [6]. A separate temperature sensor in addition to the strain gauge pressure measureing bridge is not required if the strain gauge resistors already possess a sufficiently high temperature coefficient. This circuit is shown in Figure 15.8. Pressure and temperature measurements are implemented at the same place, and the mostly positive temperature compensation of the strain gauge resistors already effects, in conjunction with the injected supply current, most of the temperature compensation required for measuring sensitivity. Figure 15.9 shows a commercially available digitally corrected semiconductor pressure sensor of high accuracy, and Figure 15.10 the characteristic curves of the sensor before and after compensation.

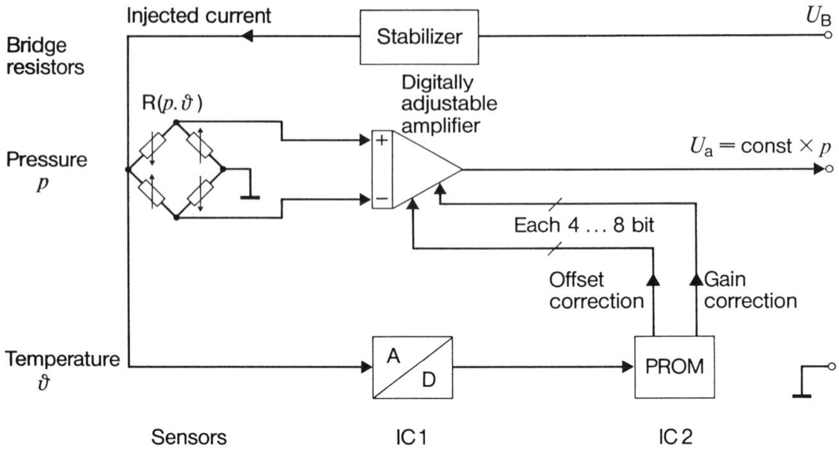

Figure 15.8 Principle of a smart pressure sensor.

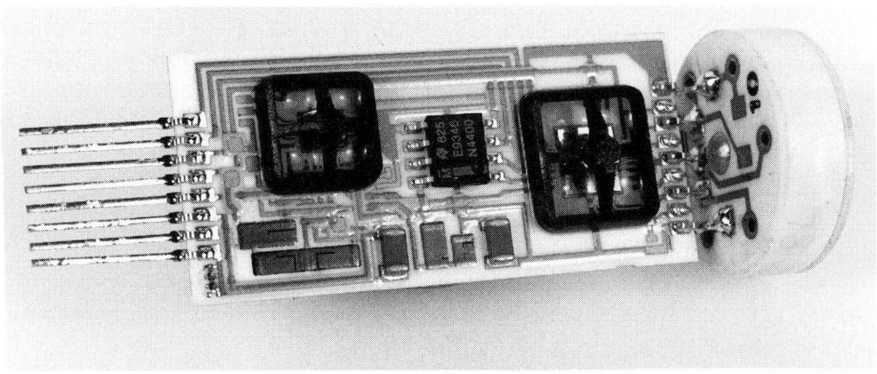

Figure 15.9 Semiconductor pressure sensor with digital correction.

A further reduction in the price of such a sensor – direct transition from the sensor to signal conditioning – would seem possible with a somewhat lower level of accuracy, using the very simply structured thick film pressure sensor described by Glas [7]. As the strain gauge resistors used in this case are temperature-independent, there is the need for a separate temperature sensor which may be realized by using the thick film technique. The usage of metal substrates, in conjunction with thick film wire strain gauge sensors, is also plausible for higher pressure ranges (> 10 bar).

Such smart pressure sensors are certainly also necessary for contactless tyre pressure measurement during travel. This must take place irrespective of the rotation position of the wheels at the time and without pneumatic

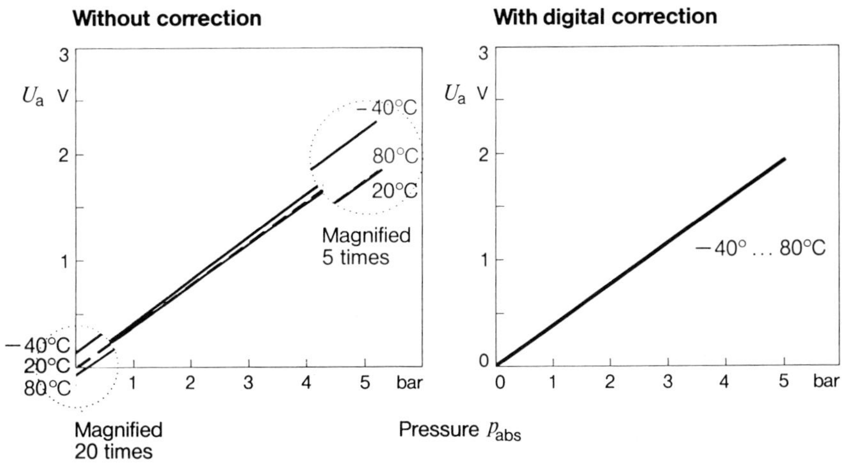

Figure 15.10 Characteristic curve of a semiconductor pressure sensor before and after digital correction.

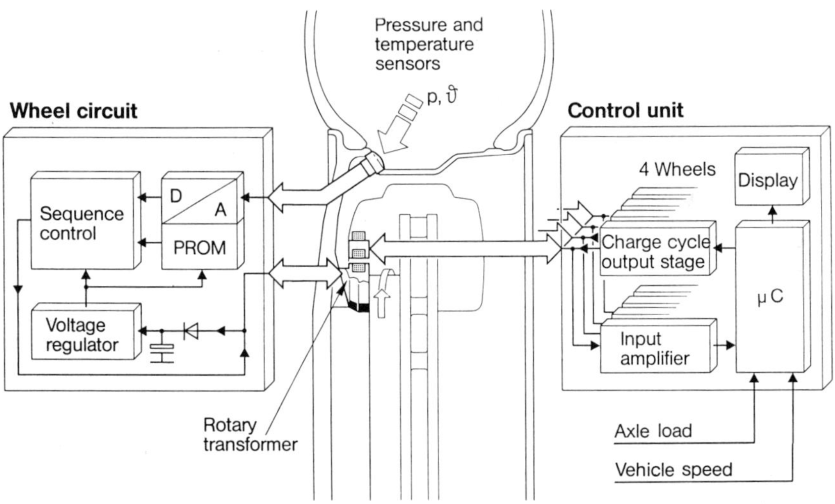

Figure 15.11 Principle of contactless tyre pressure measurement.

connection to the chassis as shown in Figure 15.11. This measurement is necessary in order to provide – at least in superior vehicles – maximum driving safety and comfort dependent on driving speed and load conditions and to minimize the trye wear and fuel consumption. In addition to tyre pressure the influencing variable tyre temperature is also to be included in the monitoring; the latter is also especially important for commercial vehicles in which the avoidance of tyre and vehicle fires is of paramount importance.

15.3.2 Acceleration sensors

The example of acceleration measurement clearly illustrates how electronics on site may be used to improve the time response of sensors and to expand their frequency-measuring range.

For instance, spring-mass systems with thin film wire strain gauge tapping, as used to a great extent to trigger passenger protection systems (e.g. airbag), have proved to be very successful.

Adaptation of this sensor principle to lower measuring ranges would unjustifiably worsen the time response for many purposes (e.g. control of shock absorbers), if the previous form of signal conditioning, mainly only composed of proportional gain of the bridge signal, were to be maintained. However, if the first and second time derivations of this signal are generated from the bridge signal using electronics on site, they can be superimposed on the source signal with the weighting specified by the valid differential equation. Thus the frequency range may be theoretically extended without any change in the amplitude and phase values previously obtained, i.e. even far beyond the resonance frequency of the system, as shown in Figure 15.12. In practice the bandwidth may, for technical reasons, be increased by approximately a factor of 10.

15.3.3 Displacement and angle sensors

Displacement and angle measurement sensors are the most widely used sensors in motor vehicles. Without using any electronics on site, sliding action potentiometer taps are mainly used in this case. As potentiometer sensors have serious reliability problems, they are replaced by advantageous contactless and wear-free sensors in ever more applications.

A new principle for contactless displacement and angle measurement is based on magnetoresistive thin film elements, the so-called Barber poles. Electronics on site permit the utilization of this completely new sensor structure which seems to be suited for highly precise applications in the motor vehicle.

Figure 15.13 illustrates the function of the sensor elements of very fine, ferromagnetic FeNi or FeCo films. Their resistance only depends on the angle, which is embraced on the one hand by the magnetization M_s of the film resistance, and on the other hand by the sensor current I. The resistance may be varied within certain limits (2–4%) by altering the M_s direction by an external magnetic field H. An especially favourable characteristic curve form is achieved if an angle of 45° is forced by means of very low ohmic surface metallization at rest, $H = 0$. This sensor form, known as a Barber pole, is used to reduce temperature influences mostly as a half or complete bridge.

The S-shaped characteristic curve obtained when a bar magnet is run over such a Barber pole divider is naturally also reflected in the scan results of a multi-sensor structure as shown in Figure 15.14. Whilst the individual scan

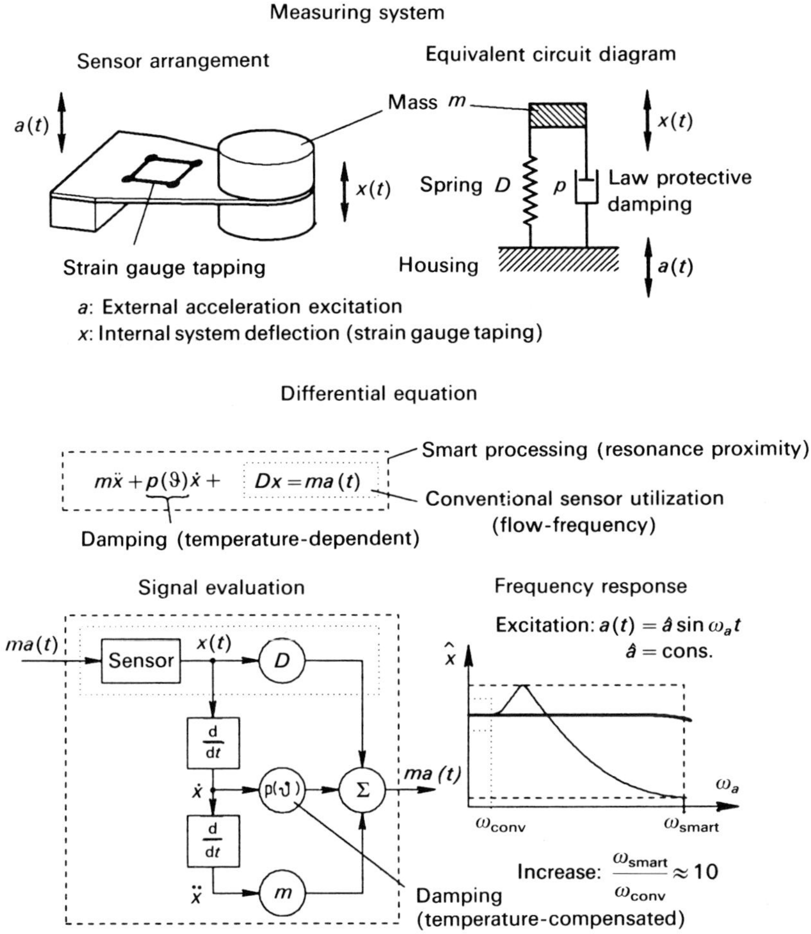

Figure 15.12 Principle of an intelligent acceleration sensor.

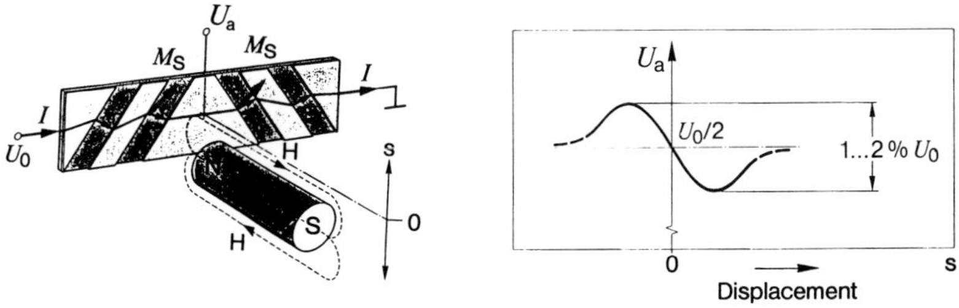

Figure 15.13 Function of magnetoresistive sensor elements (Barber pole).

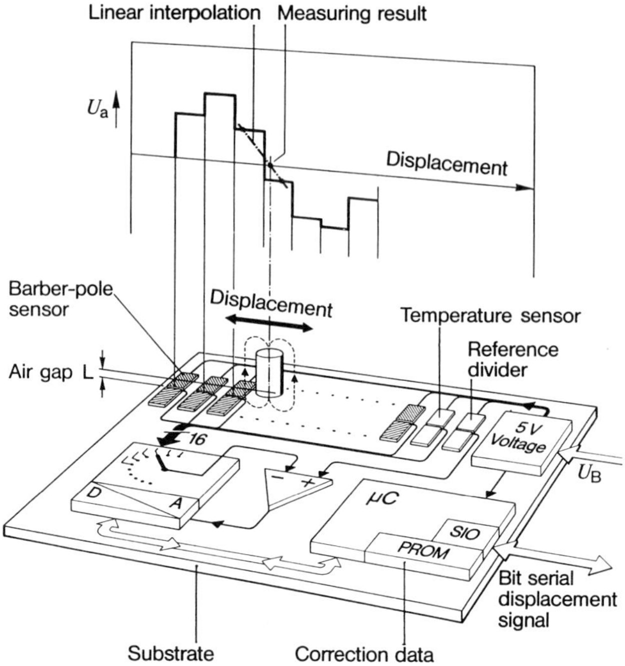

Figure 15.14 Principle of an intelligent Barber pole displacement sensor.

results are to a great extent dependent on the air gap L between the bar magnet and the surface of the sensor array as well as on the temperature, the interpolated measuring result of the scan results next to the central line (field strength $H = 0$) remains independent to a great extent. This point of disappearing field strength H represents with a high level of accuracy the position of the permanent magnet and thus the analog process quantity to be determined. The microcomputer circuit integrated with the sensor array on the same carrier permits the storage of correction data for every individual sensor element.

The first obvious application example is the tapping of a deeply set gravitational pendulum provided with a permanent magnet as a vertical sensor to release passenger protection facilities (e.g. overturning protection) depending on the angle position and angle speed around the roll and pitch axis of a vehicle. Here extremely short measuring periods are obviously not required. Otherwise, with corresponding further development the measuring principle can ideally replace conventional sensors for large displacement and angles, such as the sliding action potentiometer, in all cases where a high level of accuracy and freedom from wear is necessary.

In the next section, another principle of contactless displacement and angle measurement is described in detail. For the example of the short-circuit ring

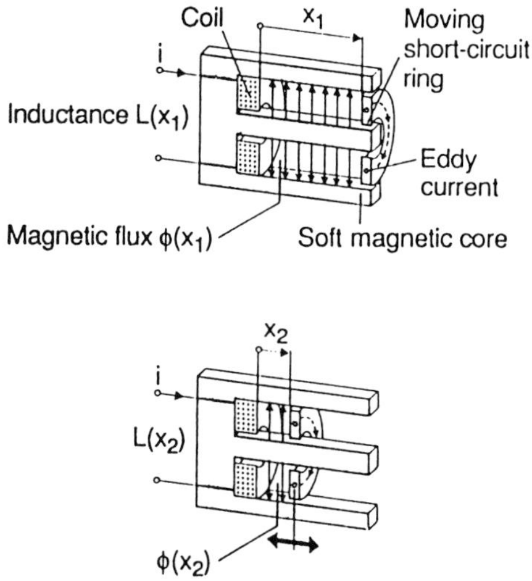

Figure 15.15 Short-circuit ring displacement sensor.

displacement sensor, both conventional and intelligent sensor versions are introduced and compared to one another.

15.4 CONCEPT OF AN INTELLIGENT SHORT-CIRCUIT RING DISPLACEMENT SENSOR

15.4.1 Sensor principle and conventional solutions

Sensors constructed according to the short-circuit ring (SCR) and short-circuit plate (SCP) principle are an important alternative for contactless and wear-free displacement and angle sensors. They offer increased operational reliability and interference resistance. These principles, which have already been discussed in the literature [8, 9], are shown in Figures 15.15 and 15.16 for a SCR and SCP displacement sensor, respectively. Figure 15.7 shows the application of these principles to angle measurement.

The magnetic flux of a soft iron core, such as a laminated plated E-core, is limited according to the position of a moving SCR or a SCP made of copper or aluminium, thus varying the inductance measurable at the excitation coil terminals. These sensor principles are extremely flexible, suitable for adaptation to various measurement tasks, and are characterized by robustness and high measurement effect.

Figure 15.18 shows the SCR displacement sensor for an electronically controlled diesel pump. The shape of the iron core has changed from the simple E-form to a rather complicated shape with two systems and a non-

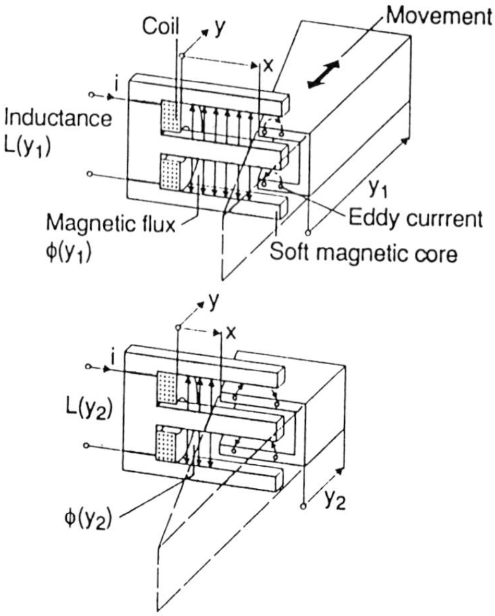

Figure 15.16 Short-circuit plate displacement sensor.

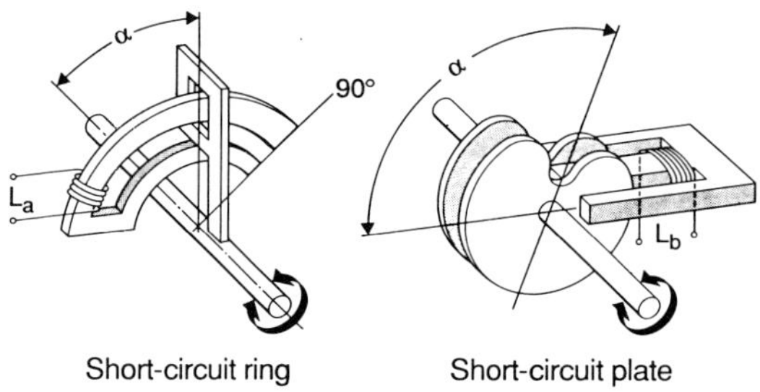

Figure 15.17 Angle measurement.

constant air gap between the core limbs. This can also be seen in Figure 15.19, where the conventional sensor structure is shown.

The reason for this tapering is to achieve better linearity. As the magnetic flux decreases with the coil distance, this effect can be compensated by a smaller air gap. In order to minimize temperature errors, a reference system has been established which is permanently fixed during adjustment. By evaluating the inductance relationship $L_1(s)/L_2$, the temperature influence is strongly suppressed, particularly in place of measured distance s where $L_1(s) = L_2$.

Figure 15.18 SCR sensor for electronic diesel control.

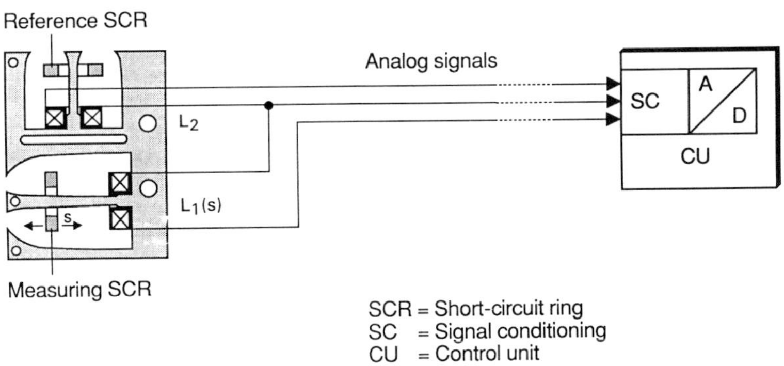

Figure 15.19 Conventional sensor structure.

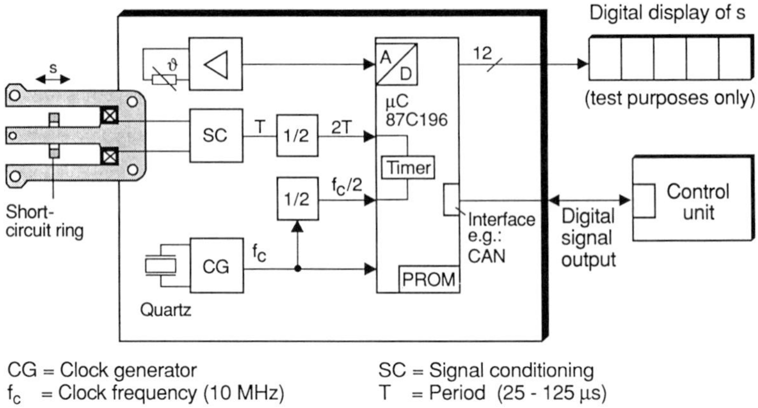

Figure 15.20 Structure of an intelligent SCR displacement sensor.

As shown in Figure 15.19, sensor and analysis electronics are separate; the sensor is at the measurement location, while the circuitry is part of the control unit. Each of the two must be interchangeable and individually compensated for error. Thus it is inevitable that individual variations add up. Still other factors are the interference influences resulting from the long wiring between sensor and circuitry.

For this sensor structure, the measured deviation can be kept in the range 85–170 μm, depending on the measured displacement. Relative to a measurement range of 21 mm, this corresponds to approximately 0.4–0.8% in FSO.

15.4.2 An intelligent sensor version with microcontroller

However if – according to the rules of intelligent sensors – the sensor elements and the inherent evaluation circuitry are integrated, the deviations of both the sensor and the electronics might be adjusted and compensated. A SCR sensor with electronics on site can be realized as shown in Figure 15.20. A reference inductance is no longer necessary. No special contour of the E-core legs must be provided for linearizing, because all those deviations are precisely eliminated on site by a correction algorithm. There is only an additional need for a very simple temperature sensor (e.g. Ni thin sheet technology). It must be mounted on the sensor's core to achieve a good temperature contact.

Using all the advantages of a sensor provided with electronics on the spot, it was the target to essentially increase the accuracy of this sensor shown in Table 15.3.

The core of the electronics on site is a 16-bit microcontroller with integrated AD converter, 16-bit counter and EPROM memory. The EPROM is used for storing the processor's instruction code and the unit-specific model

Table 15.3 Specifications of the intelligent SCR displacement sensor

Measurement range	22 mm
Measurement time	500 μs
Accuracy	0.1% of measurement range (corresponding to 10 bit or 22 μm)
Temperature range	-40 to $+120$ °C
Resolution (for 2 ms measurement time)	12 bit

parameters. The AD converter transforms the voltage analog signal from the temperature sensor into an 8-bit digital signal. The conditioning circuit for the inductance uses the principle of a LR relaxation oscillator. This circuit consists of a Schmitt-trigger with a LR-low pass in the feedback path as shown in Figure 15.21. It generates a square wave oscillation with a variable period length. The shortest period is 25 μs ($s = 0$), the longest 125 μs ($s = 22$ mm), corresponding to a frequency variation from 8 kHz to 40 kHz. This frequency is divided by 2, and thus the counters must cover a measurement range from 50 μs to 250 μs. They are driven with the maximum clock frequency of 5 MHz. In this way, a 10-bit resolution for the period measurement can be obtained.

Now the digital values for both the inductance and the temperature are available. Note that in the following the period length T always corresponds to inductance value L. The temperature is denoted by ϑ.

15.4.3 Representation of the sensor correction function

The specimen-specific sensor correction function $s(T, \vartheta)$ must contain all information necessary to correct errors caused by unit variation, non-linearity and temperature influence.

There are various possibilities of describing $s(T, \vartheta)$ with a set of model parameters. One method with less memory requirement uses, for example, a fifth-order multinomial:

$$s(T, \vartheta) = a_0(\vartheta) + a_1(\vartheta)T + a_2(\vartheta)T^2 + a_3(\vartheta)T^3 + a_4(\vartheta)T^4 + a_5(\vartheta)T^5 \quad (15.5)$$

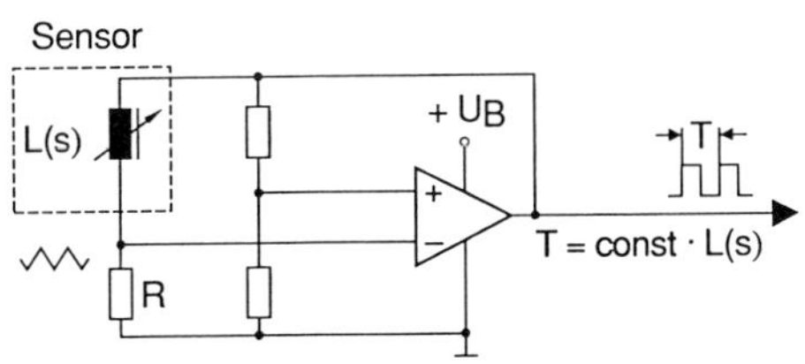

Figure 15.21 Basic oscillation circuit.

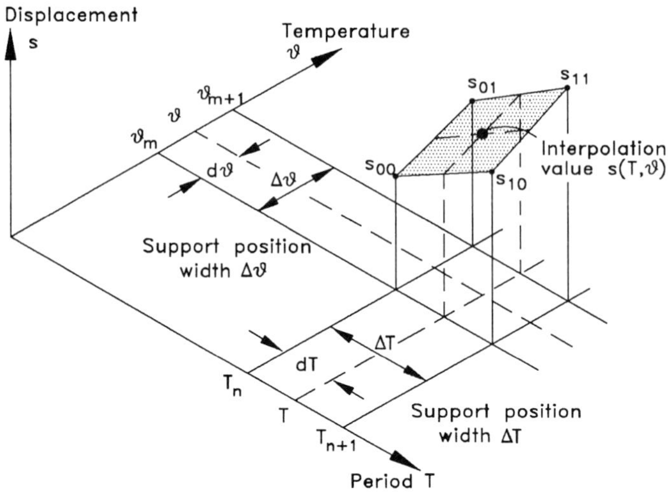

Figure 15.22 Employed interpolation principle.

The coefficients a_i ($i = 0$–5) depend in a similar manner on the temperature ϑ:

$$a_i(\vartheta) = b_{i0} + b_{i1}\vartheta + b_{i2}\vartheta^2 + b_{i3}\vartheta^3 + b_{i4}\vartheta^4 + b_{i5}\vartheta^5 \tag{15.6}$$

Only the 36 individual coefficients b_{ij} have to be stored in the PROM. However, as this method demands a lot of calculation time, it is not suitable if short measurement cycles are required.

Another method is the representation of the two-dimensional sensor correction function by a certain number of support points. In this case, the model parameters are given by the true measurement values at the support points, which form a two-dimensional characteristic diagram. The remaining function values are calculated by an interpolation procedure, for instance as shown in Figure 15.22. The four surrounding support points of a measured point (T, ϑ) define a bilinear area that approximates the sensor correction function at the point (T, ϑ). This can be expressed by the following formula:

$$\begin{aligned} s(T,\vartheta) = {} & s_{00} + (s_{10} - s_{00})\mathrm{d}T/\Delta T + (s_{01} - s_{00})\mathrm{d}\vartheta/\Delta\vartheta \\ & + (s_{00} + s_{11} - s_{10} - s_{01})\mathrm{d}T/\Delta T\,\mathrm{d}\vartheta/\Delta\vartheta \end{aligned} \tag{15.7}$$

ΔT and $\Delta\vartheta$ are the distances between two support points. They can be chosen equidistant in order that the addresses of s_{00}, s_{10}, s_{01} and s_{11} in the memory unit can be easily calculated.

As shown in Figure 15.23, a density of 64 support points for T and 16 support points for ϑ is sufficient to obtain an error not greater than 1 bit. This is a quantization error which may always occur in principle if the number of support points is smaller than the number of possible (T, ϑ) values. As the intended resolution for the displacement values is 12 bits, a total

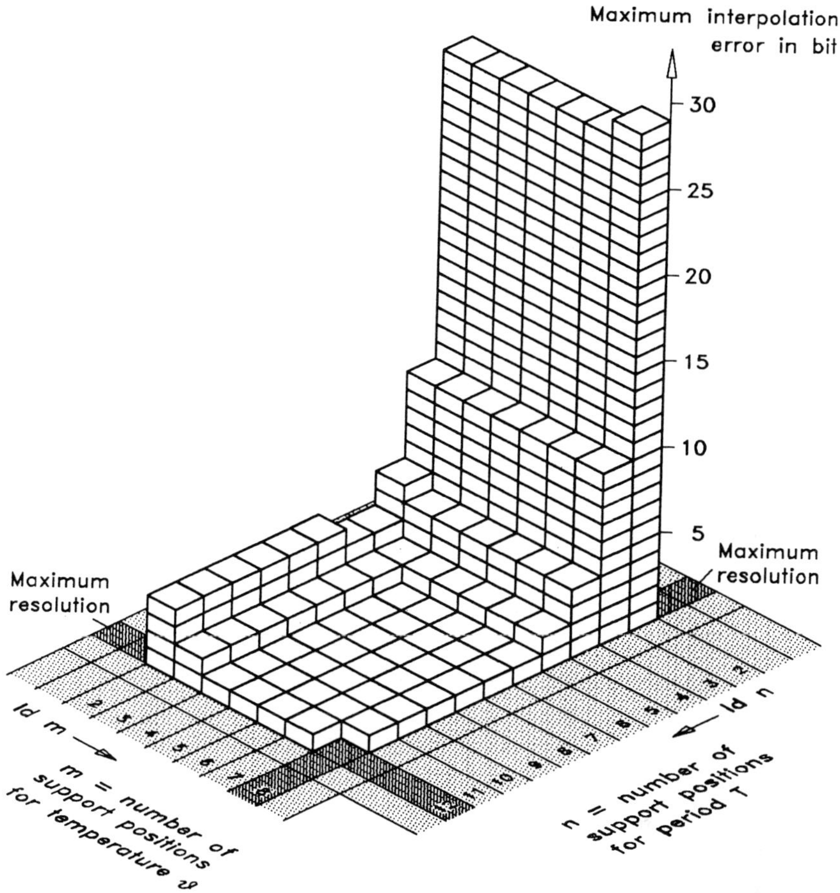

Figure 15.23 Interpolation errors depending on support point density.

memory space of 1 k × 12 is required to store the parameters for the sensor correction function.

The characteristic diagram is an excellent representation of the sensor correction function, which has the great advantage that the corrected measurement values can be calculated in a simple manner. As shown by Zabler *et al.* [10], the desired measurement time could not be achieved with the first method. Therefore, the intelligent SCR displacement sensor was realized by using a two-dimensional characteristic diagram.

15.4.4 Measuring results

The test arrangement for the intelligent SCR sensor based on a microcontroller is shown in Figure 15.24. It is a PC-controlled measurement system

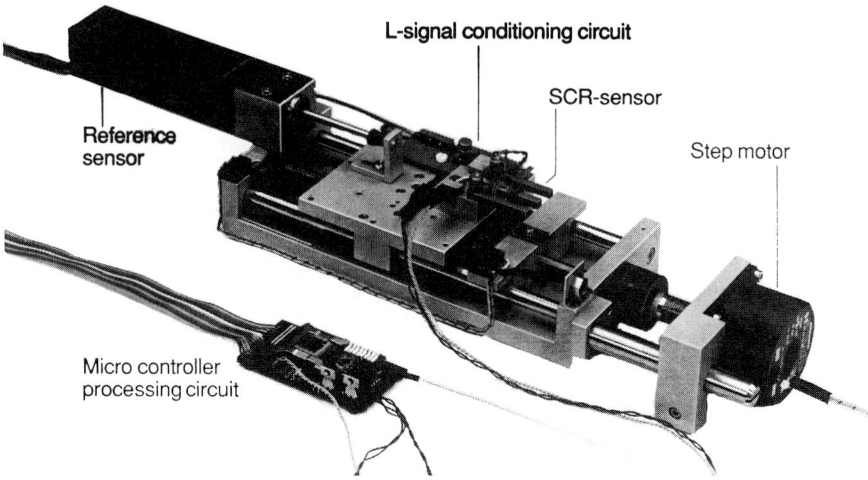

Figure 15.24 Employed test arrangement.

using automatic advance to adjust different displacements. An incremental glass scale with a precision of 0.5 μm serves as reference sensor.

With the selected 16-bit microcontroller the calculation of Equation 15.7 can be carried out in 250 μs. As the longest time for the L measurement is also 250 μs, the desired measurement cycle time of 500 μs can be obtained. It is ensured that a new temperature measurement value is present for at least each third measurement cycle, and thus after a maximum of 1.5 ms.

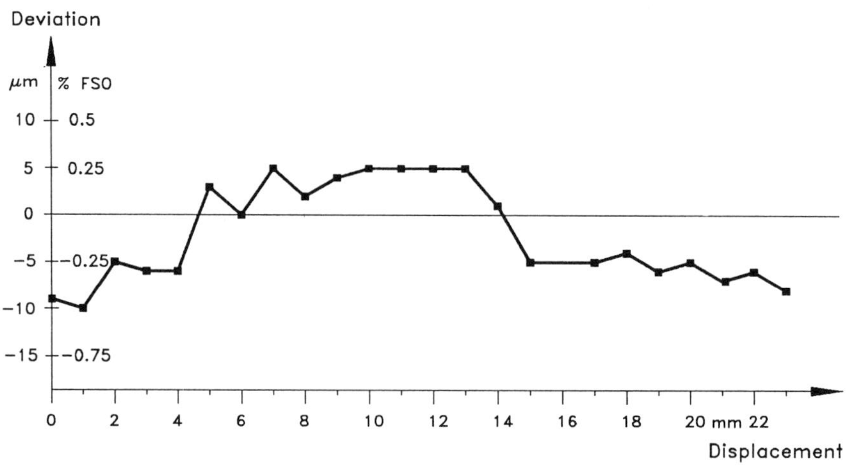

Figure 15.25 Smart displacement sensor deviation at room temperature.

Table 15.4 Measurement of temperature (20–70 °C)

Displacement	*Deviation with temperature correction*	*Deviation without temperature correction*
6 mm	− 10 μm (0.05%)	+ 120 μm (0.6%)
16 mm	+ 10 μm (0.05%)	+ 180 μm (0.9%)

This is completely sufficient, as the thermal time constant of the sensor is several magnitudes higher.

Figure 15.25 shows the systematic deviations of the sensor as measured at room temperature. In comparison with a sensor of conventional type, the accuracy was increased at least by the factor 4, and at some operating points even by the factor 7. The intended accuracy of 22 μm has been achieved, partially even exceeded. The characteristic shown in Figure 15.25 assumes that quantization errors are eliminated by evaluating a statistical mean value. Since the sensor signal and the pulse generator output are totally asynchronous, a quantization error can occur with a certain probability. Therefore, an additional error of 10 μm is possible.

The temperature response achieved by the intelligent sensor appears similarly positive. Table 15.4 shows the measurement of temperature response for two displacement points ($s = 6$ mm and $s = 16$ mm). The deviation determined without temperature measurement shows the magnitude of the temperature correction and corresponds to the deviation of a non-compensated SCR inductance.

15.4.5 Using a dedicated interpolation processor

A disadvantage of the presented microprocessor solution is the great hardware overhead that is not necessary for the given interpolation task. Furthermore, it would be desirable to integrate the processor hardware in a signal-processing chip together with the other parts, i.e. the measurement circuits for the process quantity and the temperature, the memory unit for the model parameters and the sensor interface (e.g. CAN). Thus a dedicated processor with minimum hardware will save both space and costs. In addition, the calculation speed can be increased.

The following considerations on the structure of a signal-processing chip for intelligent sensors are also guided by the example of the SCR displacement sensor. The described processor hardware, however, is in principle applicable for the whole range of sensors that can be characterized by a two-dimensional sensor correction function.

In order to achieve minimum hardware for the measurement unit, the Schmitt trigger circuit of Figure 15.21 can also be used for the temperature measurement. The LR low pass is replaced by a RC low pass. The resulting square wave oscillation is also digitized by means of digital counters. In case of interference between the two frequencies, it might be useful if only one oscillation circuit was active at one time.

The size of the EPROM storage depends generally on the special sensor characteristic, e.g. on the type of temperature influence. As shown in Figure 15.23, a considerable reduction in the number of support points is normally possible without losing any accuracy.

Standard interfaces for digital circuits can be used for the sensor interface. The main emphasis must be put on the development of a well-suited interpolation processor which is guided by the following considerations.

For the different additions in Equation 15.7, the arithmetic unit of the processor must contain an n-bit adder where n denotes the required resolution in bits. The subtractions can be reduced to additions by inverting the subtrahend and by adding 1. If the multiplication is carried out sequentially, no additional hardware is required. A multiplication with an n-bit operand may be executed by n shift operations and n conditional additions. The divisions in Equation 15.7 can be done by shift operations if ΔT and $\Delta\vartheta$ are powers of 2.

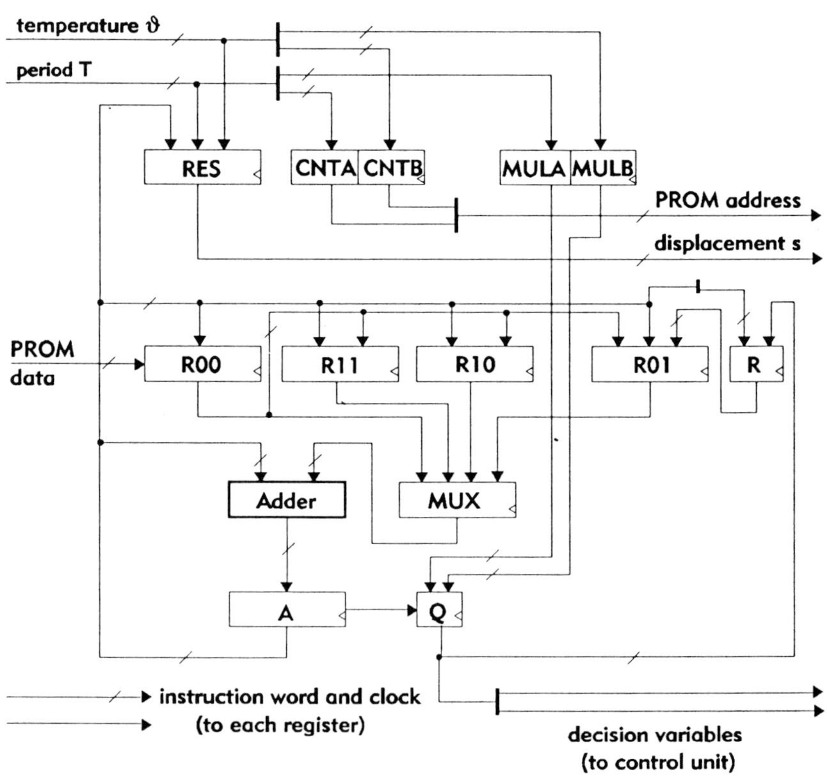

Figure 15.26 Architecture of the arithmetic unit.

Figure 15.26 shows the architecture of the arithmetic unit. The possible move operations are limited as shown by the drawn data paths. Altogether, about 30 basic instructions are sufficient to carry out the calculation given in Equation 15.7. All register functions are selected by an instruction word provided by the control unit. An optimal encoding ensures that the decoding hardware not shown in Figure 15.26 can be minimized.

The simple structure of the control unit is shown in Figure 15.27. By the introduction of conditional add instructions a completely sequential program flow is made possible. Conditional add operations are used to implement the multiplication algorithm and the rounding procedure at the end of the calculation. The program flow can be controlled by a simple counter.

A combinatorial circuit generates the instruction words depending on the status of the program counter and on two decision variables provided by the arithmetic unit. At certain states of the program counter, they determine whether the code for a NOP (no operation) or for an ADD instruction is generated. Another circuit decodes the end of the calculation.

An optimal program flow uses the registers R00 up to R11 for storing the function values at the support points and for storing provisional results. The minimum register set as shown in Figure 15.26 is sufficient if Equation 15.7 is rearranged as follows:

$$s(T,\vartheta) = [(s_{11} - s_{01} + s_{00} - s_{10})\mathrm{d}T/\Delta T] + (s_{01} - s_{00})\mathrm{d}\vartheta/\Delta\vartheta \\ + (s_{10} - s_{00})\mathrm{d}T/\Delta T + s_{00} \qquad (15.8)$$

For instance, s_{11} appears only once, the register R11 is immediately available for provisional results as soon as the first subtraction has been carried out.

Altogether, about 150 program steps are necessary to complete this calculation. The floating point operations are exactly carried out by simple means; division remainders with influence on the integral result are not neglected. Assuming, for instance, a clock frequency of 5 MHz, the calculation

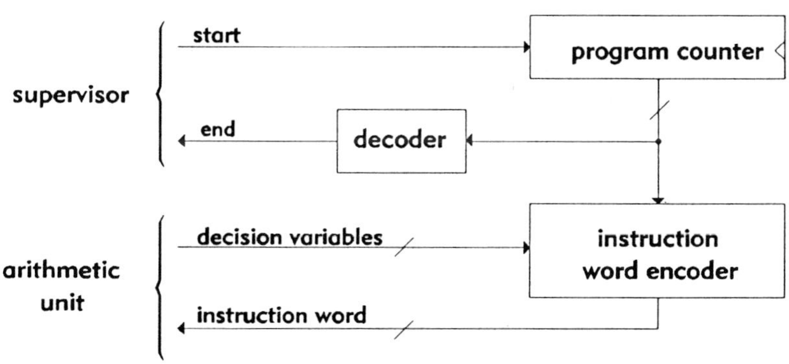

Figure 15.27 Structure of the control unit.

time will take only 30 μs. Thus a dedicated processor might be about 10 times faster than a standard microcontroller although all multiplications are carried out sequentially.

The dedicated processor concept will satisfy all requirements for measurement correction in a two-dimensional characteristic diagram. As it can be easily integrated with the other parts of the signal-processing chip, this processor can facilitate many future applications of electronics on site. Nevertheless, the chip costs compared to a microprocessor solution might be reduced by approximately a factor of 3.

15.5 CALIBRATION PROCESS FOR INTELLIGENT SENSORS

For obtaining the specimen-specific sensor parameters, an intelligent sensor must be measured at several representative operating points with regard to its actual state given by a reference sensor. Therefore, the intelligent sensor must be capable of a calibration mode in which the uncorrected values for the measured process quantity and the measured values for the influencing variables are directly led to the sensor output. These values are picked up by an external computer which calculates the sensor model parameters and writes them back to the sensor's programmable memory unit. These two calibration phases are shown in Figure 15.28.

After the first phase, a calibration program is executed at the external computer which calculates the sensor's model parameters. It is a problem of great importance to find a minimum number of calibration points that represent the sensor's behaviour with the desired accuracy. This is essential for an effective manufacturing process.

The calibration procedure is now discussed using the well-known example

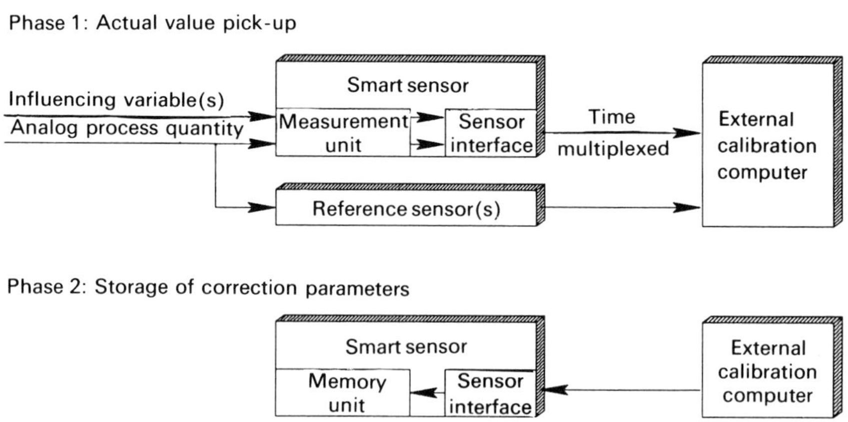

Figure 15.28 Sensor calibration process.

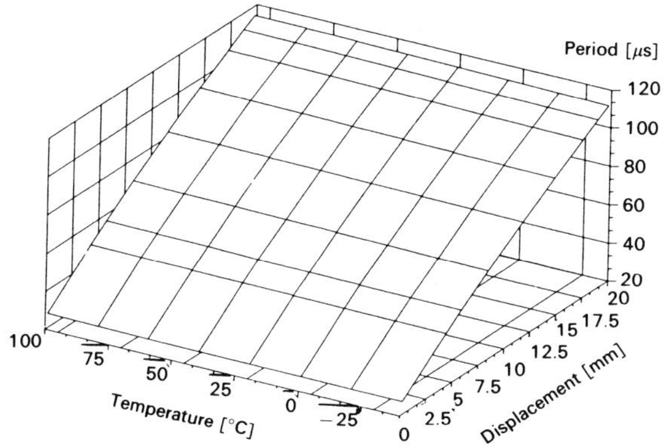

Figure 15.29 Measurement data.

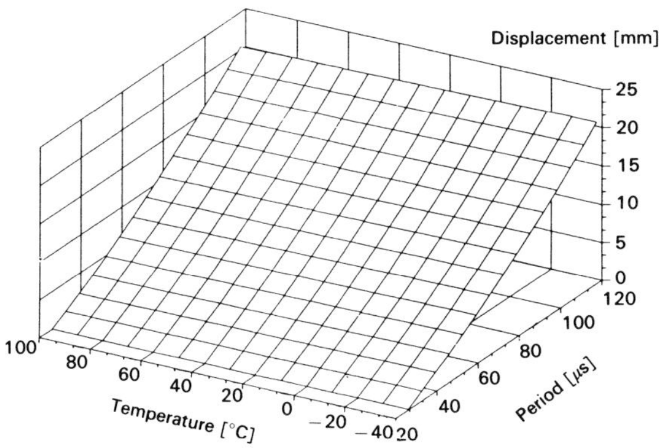

Figure 15.30 Sensor correction function.

of the SCR displacement sensor, again using a representation of the sensor correction function in a two-dimensional characteristic diagram. Thus the model parameters to be calculated are the true measurement values at the selected support points.

First, the room temperature characteristic curve is very precisely determined with the test arrangement already shown in Figure 15.24. The measured performance curve can be approximated by a regression polynom, for instance with a fifth-order multinomial. The maximum deviation would be 2 μm in this case. It can be avoided by using a spline interpolation method as is well known in numerical mathematics [11, 12].

In the next step, nine SCR positions are chosen and the inductance is measured for eight different temperatures. For this, no reference sensor or automatic advance is required. The displacements corresponding to the nine SCR positions are calculated using the induction value at room temperature and the approximation of the room temperature characteristic.

Thus we get a two-dimensional measurement diagram with 72 points as shown in Figure 15.29. By executing an interpolation procedure and by calculating the inverse function, the characteristic diagram as shown in Figure 15.30 is obtained.

The last step is to digitize the ranges and to select the appropriate number of support points. In our example, 64×16 support points are sufficient, as has already been shown in Figure 15.24. The 1024 function values can be digitized and stored in the sensor's memory unit.

The number of representative operating points for the pick-up phase might be reduced dramatically if a systematic part of the temperature influence could be found. In our example, a linear function might be sufficient to describe the temperature influence at a given SCR position. It can be assumed that the slopes of these temperature characteristics also increase linearly with the displacement. This must still be confirmed by examining an adequate number of specimens. Thus a further improvement of the calibration process seems to be practicable. If, in addition to room temperature, only a few calibration points must be measured, there will be decisive savings in costs and time during the manufacturing process.

15.6 CONCLUSIONS

Electronics on site facilitate new concepts of information distribution which are more efficient and transparent than the standards used until now. Electronics on site also make intelligent sensors possible, utilizing various measurement effects which are today very little applied in conventional sensor techniques. This is possible thanks to new structures and methods. The accuracy limits regarded as being fixed for large-scale production are thus broken through and the central control units are relieved to an ever greater extent by the higher information level of these sensors.

The list of possible applications of intelligent sensors is almost limitless. In this chapter, however, basic considerations have been discussed by using the instructive example of the SCR displacement sensor. Intelligent SCR sensors with both microcontroller and with a dedicated interpolation processor have been described in detail. The calibration procedure has been discussed in so far as results from current investigations are already available.

The concept of the intelligent sensor has been proven to overcome existing limits caused by bad properties of the sensor elements such as non-linearity and temperature dependence. The only precondition is that the sensor element does not change its characteristics during operation time. On the whole, there should be no hesitation in using such sensors in order to enjoy their advantages as soon as possible.

REFERENCES

1. Kiencke, U., Dais, S. and Litschel, M. (1986) *Automotive Serial Controller Area Network*. SAE Technical Paper No. 860 391.
2. Kienckc, U. and Dais, S. (1987) *Application Specific Microcontroller for Multiplex Wiring*. SAE Technical Paper No. 870 515.
3. Tränkler, H.-R. (1989) *Sensorspezifische Meßsignalverarbeitung*. NTG Technical Report No. 93, 3rd Professional Convention, Bad Nauheim, Germany, March.
4. Heintz, F. and Zabler, E. (1986) *Motor Vehicle Sensors Based on Film-Technology: An Interesting Alternative to Semiconductor Sensors*. SAE Technical Paper No. 870 477.
5. Csepregi, L. (1983) *Technologie dünngeätzter Siliziumfolien in Hinblick auf monolithisch integrierbare Sensoren*. Final report on BMFT Study NT 2604, Munich, Germany, December 1983.
6. Ritter, M.v., Bomholt, J. and Keller, H.W. (1986) A new approach to nonlinear sensor signal correction, in *Conference Capteur 86*, Paris, France, June 1986.
7. Glas, J. (1987) *Integrated Hyorid Pressure Sensor Using a Piezoresistive Thickfilm Sensor Element*. SAE Technical Paper No. 870 286.
8. Weckemann, A. (1973) Ein neuer Weggeber zum elekrischen Messen großer Verschiebungen. *Archiv für technisches Messen*, **14**, 1121–31.
9. Zabler, E. and Heintz, F. (1983) Shading-ring sensors as versatile position and angle sensors in motor vehicles. *Sensors and Actuators*, **3**, 315–26.
10. Zabler, E., Heintz, F., Dietz, R. and Gerlach, G. (1991) Mechatronic sensors in integrated vehicle architecture, in *Eurosensors V*, Rome 1991.
11. Stoer, J. (1989) *Numerische Mathematik*, Vol. 1, Springer–Verlag, Berlin, Heidelberg.
12. Hoschek, J. and Laser, D. (1989) *Grundlagen der Geometrischen Datenverarbeitung*, B.G. Teubner, Stuttgart.

CHAPTER 16

Development of intelligent sensor technology in Japan*

16.1 INTRODUCTION

Various sorts of sensor are presently employed in the automobile industry. Most of the sensors applied to automobiles are composed of discrete parts or hybrid IC. However, intelligentialization of sensors has rapidly advanced in response to demands such as reduction in size and weight, enhancement of function, cost cutting etc.

This chapter describes intelligent sensors that were recently developed in Japan. They are integrated single or plural sensors and amplification circuits on single chip with LSI manufacturing technology [1, 2].

Figure 16.1 shows the development trends of intelligent sensors. The third step is the established arrangement to be rendered as the final system, one typical embodiment of which is an intelligent image sensor consisting of the three-dimensional IC shown in Figure 16.2 [3, 4]. This sensor has a multi-layer structure; the upper, middle and lower layers are in charge of photo-electric conversion, signal transfer/memory/processing, and power supply/applying action respectively.

It is expected that the practical application of such sensors may greatly contribute not only to development in robots and visual sensing for automatic operation of automobiles but also to system development in various industrial fields.

The intelligentialization statuses sensors are as shown in Table 16.1, from which some concrete examples will be drawn.

16.2 SEMICONDUCTOR PRESSURE SENSOR

The automobile pressure sensor is one of the most important sensors and is used in systems such as fuel injection control, exhaust gas recirculation (EGR) control, and supercharger pressure control for an engine. In fuel-injection control, for instance, a system controlling an air/fuel ratio after

*The content of this chapter was presented at Future Transportation Technology, 1989, Society of Automotive Engineering, Inc.

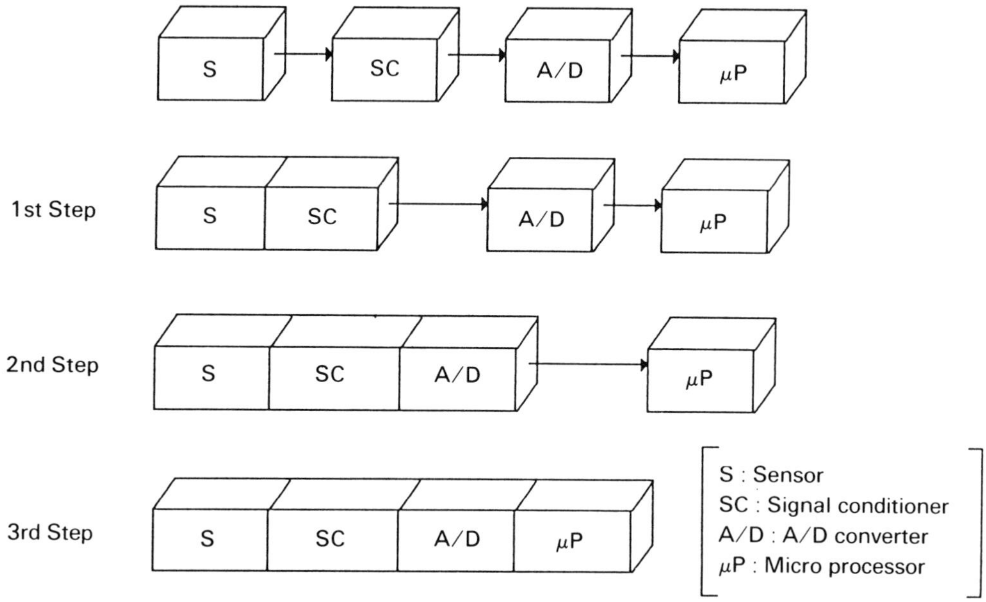

Figure 16.1 Development trends of intelligent sensors. S = sensor; SC = signal conditioner; A/D = A/D converter; μP = microprocessor.

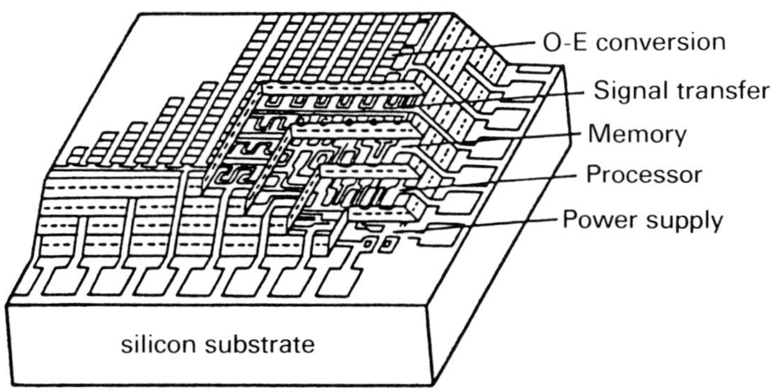

Figure 16.2 Three-dimensional intelligent image sensor.

measuring an intake air quantity in an intake manifold by means of an airflow sensor of through-vane type, hot-wire type or Karman Vortex type has been conventionally used. Recently, however, a speed density system has been applied which measures intake air pressure by means of a pressure sensor to compute an intake air quantity for controlling air/fuel ratio.

Mass production of a semiconductor pressure sensor has been carried out in our company for several years (Figure 16.3). The sensor is formed by

Table 16.1 Intelligentialization status of various sensors

Sensor	*Kind of sensor*	*Configuration of inteligent sensor*
Photosensor	Visual image sensor	Photosensor + signal cumulation/ transfer circuit
	Infrared image sensor	Infrared sensor + Si CCD
	Contact-type image sensor	Photosensor + reading circuit
Pressure sensor	Piezoresistive effect Si pressure sensor	Pressure sensor bridge + amplifying/ temperature compensating circuit
	Capacitive Si pressure sensor	Capacitive sensor + oscillation/low impedance amplification circuit
Magnetic sensor	Hall IC	Hall element + amplification/wave-form shaping/temperature compensation circuit
	IC magnetic sensor	Magnetic resistance element + amplification/waveform shaping circuit
Temperature sensor	Bipolar IC temperature sensor	Tr temperature sensor + bipolar signal-processing circuit
	CMOS IC temperature sensor	Tr temperature sensor + CMOS signal-processing circuit
Ion sensor	ISFET	FET sensor + CMOS multiplexer

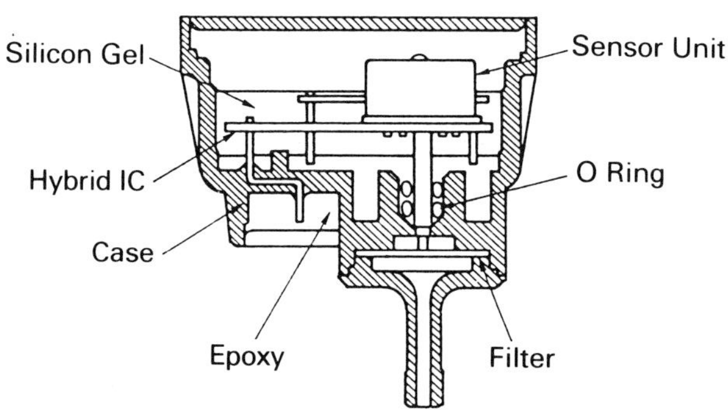

Figure 16.3 Structure of pressure sensor.

putting, in a single module, both a silicon diaphragm type of pressure sensor utilizing piezoresistive effects of the semiconductor and an amplification circuit unit of hybrid IC construction. The uses of this sensor can be further enlarged in future because of merits such as low price, small size, low weight and high reliability.

The following discussion will relate to the intelligent pressure sensors including an intelligentialized article consisting of a sensor unit an amplification circuit unit being currently developed in our company [5, 6].

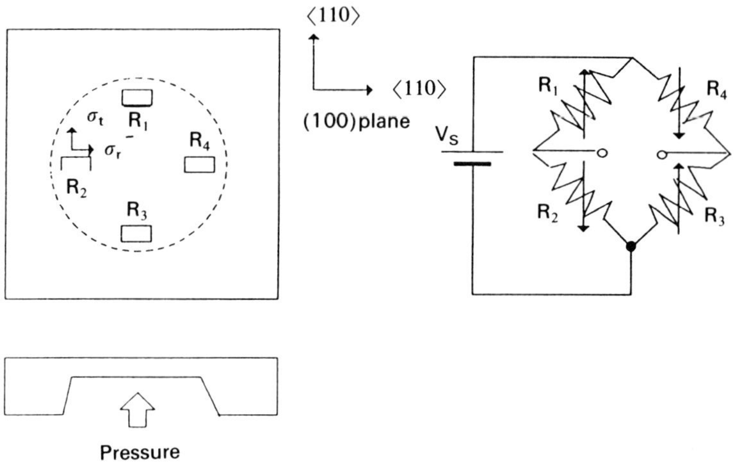

Figure 16.4 Configuration of silicon diaphragm and bridge circuit.

16.2.1 Principle of operation

One preferable embodiment of the construction and detection circuit of a conventional silicon diaphragm type of pressure sensor is shown in Figure 16.4.

P-type piezodiffused gauges (R_1, R_2, R_3, R_4) having all of their longitudinal directions oriented in parallel to the ⟨110⟩ axis are formed in respective positions on the main obverse side of a circular diaphragm obtained by thinning the central area of an n-type silicon substrate with (100) plane into a thin film shape by means of aeolotropic etching. The diaphragm deforms to yield surface stress when any pressure is applied on one side of the diaphragm. Stress in the radial direction and the tangential direction are defined as σ_r and σ_t respectively, as shown in the following formulae:

$$\sigma_r = \frac{3P}{8h^2}[(1+v)a^2 - (3+v)r^2] \tag{16.1}$$

$$\sigma_t = \frac{3P}{8h^2}[(1+v)\mathrm{a}^2 - (1+3v)r^2] \tag{16.2}$$

where a is the radius of the diaphragm, h is thickness, P is applied pressure, r is the distance of each gauge from the centre of the diaphragm and v is Poisson's ratio.

Each gauge on the diaphragm yields a variation in the value of its resistance due to these surface stresses, σ_r and σ_t, and the amount of variation (ΔR) can be expressed in the following formulae:

$$\Delta R_1/R_1 = \Delta R_3/R_3 = \frac{1}{2}\pi_{44}(\sigma_r - \sigma_t) \tag{16.3}$$

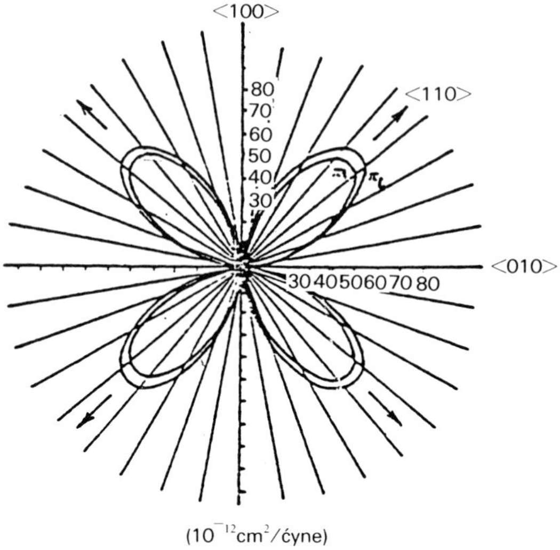

Figure 16.5 Piezoresistive coefficients in the P-Si {100} plane.

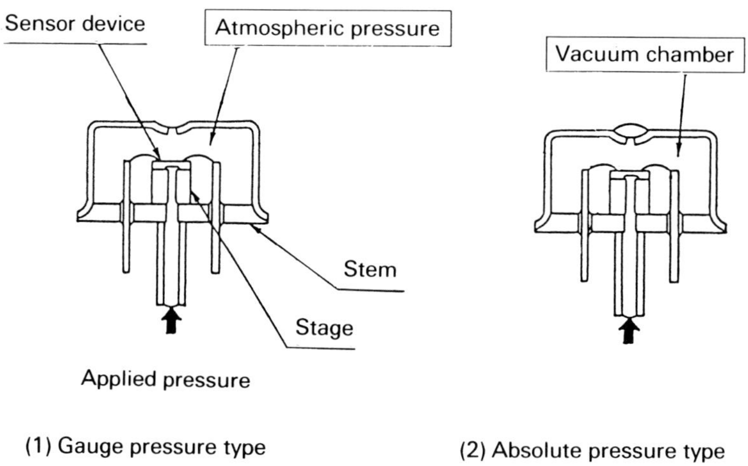

Figure 16.6 Structure of the pressure sensor unit.

$$\Delta R_2/R_2 = \Delta R_4/R_4 = -\frac{1}{2}\pi_{44}(\sigma_r - \sigma_t) \tag{16.4}$$

where π_{44} stands for the piezoresistive coefficient and the magnitude of π_{44} of a P-type gauge in the (100) plane is shown in Figure 16.5. Thereby, a voltage applied to a bridge is assigned to V_s, and output voltage V_0 with each gauge disposed in its position corresponding to $r = a$, is expressed by the following formula:

$$V_0 = \frac{3P}{4h} 2(1 + \nu)a^2 \pi_{44} V_s \tag{16.5}$$

Assuming that radius of diaphragm $a = 1$ mm, thickness $h = 40\,\mu$m, and $V_s = 4$ V, an output voltage of 40 mV is obtained if a pressure of 1 kg/cm^2 is applied to the diaphragm.

This sensor element is packaged to a hermetically sealed terminal etc. via a base of glass or silicon. The object of this base is to absorb and attenuate the thermal stress generated due to a differential linear expansion coefficient between the silicon and case material, and monocrystalline silicon or glass (#7740) is normally used as the material of this package. Figure 16.6 shows one example of our company's package.

16.2.2 Structure and characteristics

The output voltage of a silicon diaphragm type of pressure sensor consists of a signal component proportional to the applied pressure voltage with various sorts of erroneous components superposed thereon. The erroneous components may be classified in such categories as dispersion of sensitivity, temperature dependence and hysteresis. In order to design a good intelligent sensor, it is necessary to determine the elementary structure, circuit composition and processes to improve the synthetic accuracy; the errors should be classified into two groups, one of which should be eliminated by improving the sensor unit, and the other by means of an amplifier circuit. The circuit

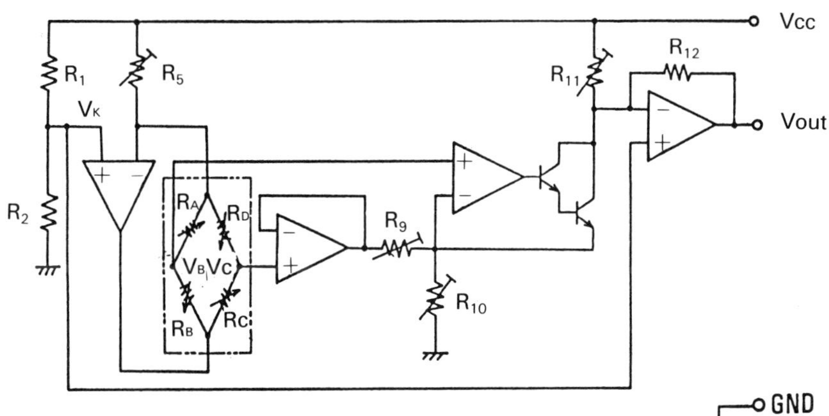

$$\text{Vout} = V_K - \frac{R_{12}}{R_{11}}(V_{CC} - V_K) + \frac{R_{12}}{R_9}(V_B - V_C) + \frac{R_{12}}{R_{10}} V_B$$

Figure 16.7 The circuit configuration of the intelligent pressure sensor.

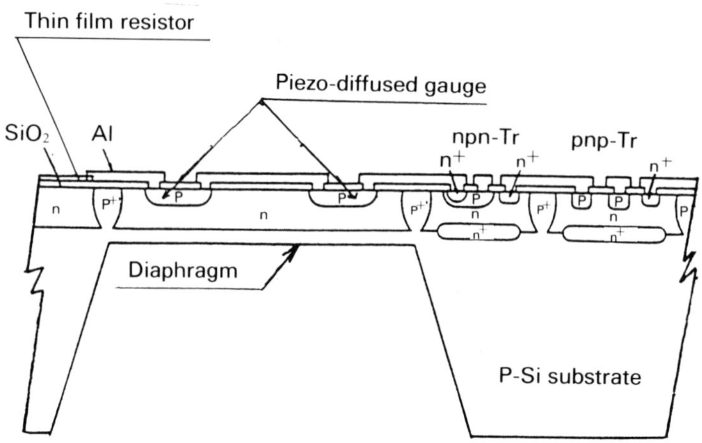

Figure 16.8 Cross-section of intelligent pressure sensor.

Table 16.2 Chip dimension

Sensor	{100} plane, P-type gauge/full bridge
Circuit configuration	Op-Amp + thin film resistor (bipolar IC)
Diaphragm size	$1.8 \times 1.8\,mm^2$
Diaphragm thickness	$35\,\mu m$
Chip size	$3.5 \times 3.5\,mm^2$

Table 16.3 Specification of intelligent pressure sensor

Supply voltage	5 ± 0.25 V DC
Operating pressure range	$150 \sim 750$ torr (abs)
Operating temperature range	$-30 \sim 120\,°C$
Output voltage range	$1.2 \sim 3.6$ V
Accuracy	± 10 torr

constitution and element construction of an intelligent pressure sensor developed by our company are shown in Figure 16.7 and Figure 16.8 respectively. Meanwhile, three items consisting of offset voltage adjustment, sensitivity adjustment and temperature adjustment are being carried out through the on-chip laser trimming of a trimming resistance comprising a thin film resistor in an amplifying circuit provided on the chip. The sources of error other than the above three have been eliminated by improving the sensor unit, and the chip dimension and sensor specification of this intelligent sensor are shown in Table 16.2 and Table 16.3 respectively. In addition, one example of a chip photo is shown in Figure 16.9.

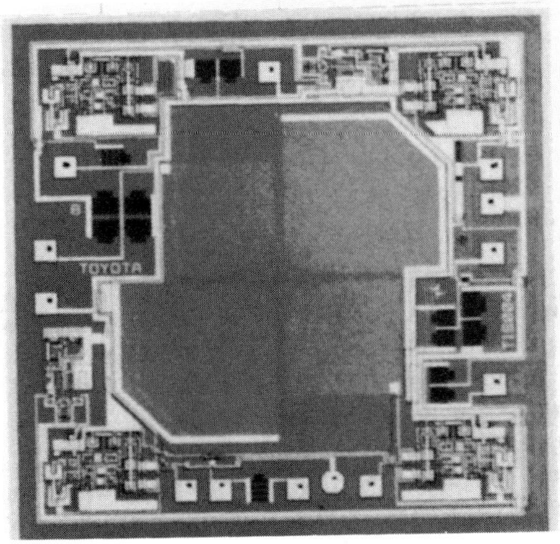

Figure 16.9 Intelligent pressure sensor device (Toyota Research and Development Laboratories Inc.).

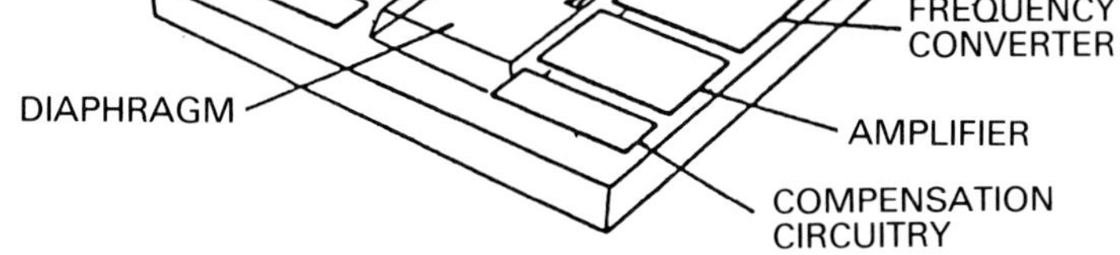

Figure 16.10 Conceptual drawing of intelligent sensor with voltage and frequency output (Toyota Research and Development Laboratories Inc.).

The sensor system shown in Figure 16.10 is an advanced one, provided with a voltage/frequency converter together with both analog and digital outputs, as the result of further promotion of intelligentialization, so that it also has the merit of being capable of being directly connected to a microcomputer with no AD converter. The concept diagram and pressure–frequency characteristics are displayed in Figure 16.11 and Figure 16.12 respectively [7].

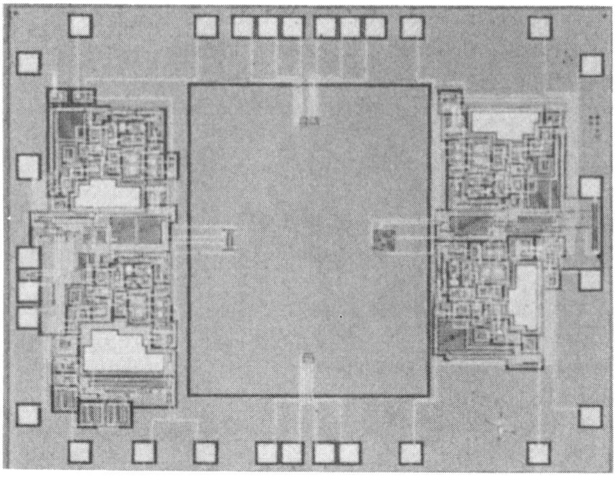

Figure 16.11 Intelligent pressure sensor device with voltage and frequency output.

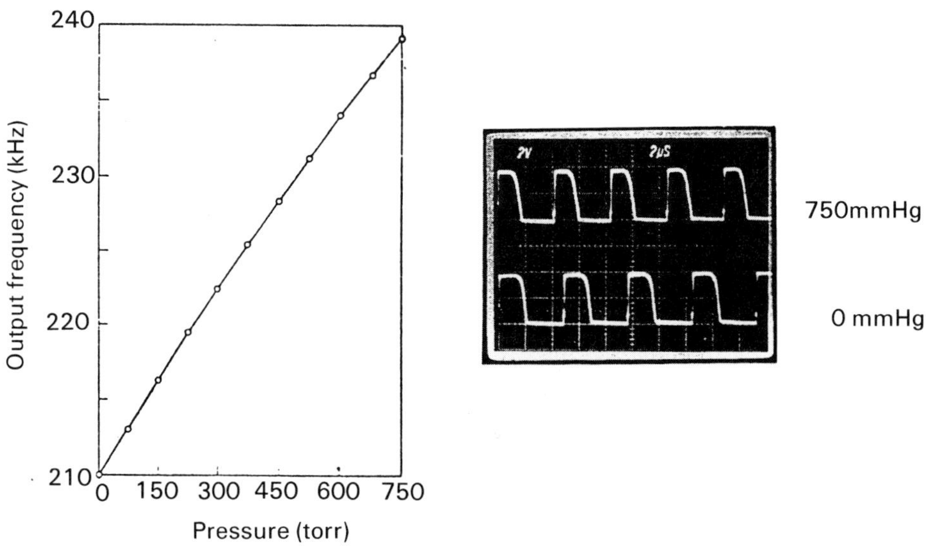

Figure 16.12 Frequency output characteristic.

16.3 MAGNETIC SENSOR

The term 'magnetic sensor' normally stands for an apparatus having the function of magnetoelectric conversion, but the one most frequently used in automobiles is a solenoid pick-up, which is composed of a coil and magnet.

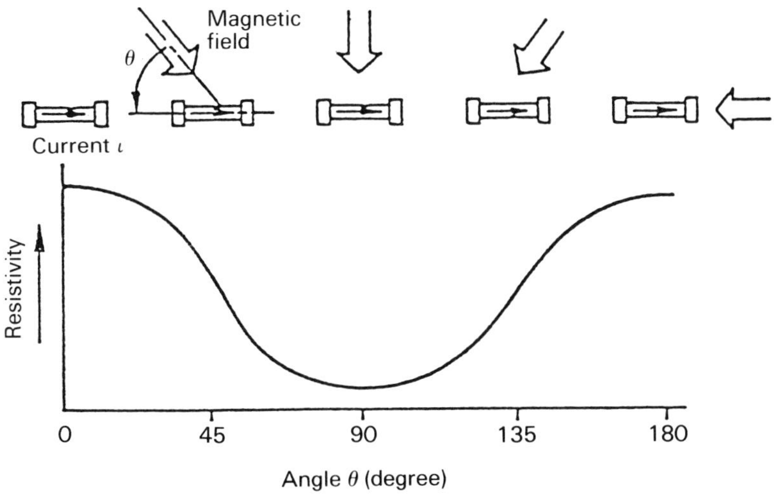

Figure 16.13 MRE's resistivity change by the angle between the current and the magnetic field.

There are many kinds/grades of magnetic sensors, from a simple coil-type sensor to a SQUID-utilizing superconductor available at extremely low temperature. This section will relate to an intelligent magnetic sensor in which a ferromagnetic resistance element developed in our company a rotary sensor for automobiles is employed.

16.3.1 Principle of operation

It is a general characteristic of a ferromagnetic metal that its resistance value rises to the maximum level when the directions of current and magentic field are parallel to each other and drops down to the minimum level when the directions cross each other at right angles (Figure 16.13). This phenomenon is called the anisotropic magnetoresistive effect [8]. The resistance value R under crossing angle θ may be expressed by the following equation:

$$R = R_{\parallel} \cos^2 \theta + R_{\perp} \sin^2 \theta \tag{16.6}$$

where

$R_{\parallel}$ = resistance value when directions of current and magnetic field are parallel to each other
$R_{\perp}$ = resistance value when directions of current and magnetic field cross each other at right angles
θ = angle between directions of current and magnetic field

The resistance change ratio $\Delta R = R_{\parallel} - R_{\perp}$ is determined on the basis of

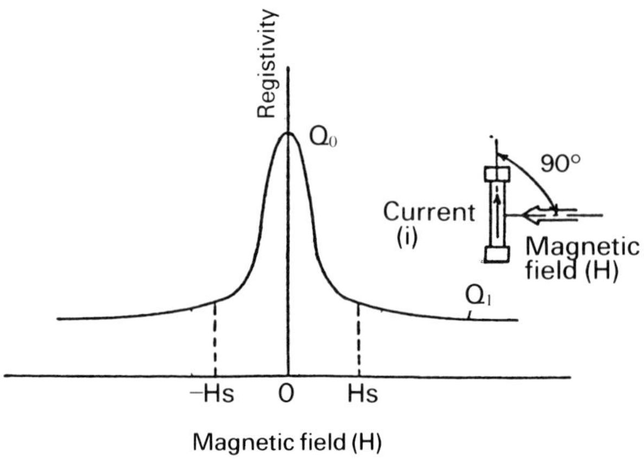

Figure 16.14 MRE's resistivity change by the magnitude of the magnetic field applied.

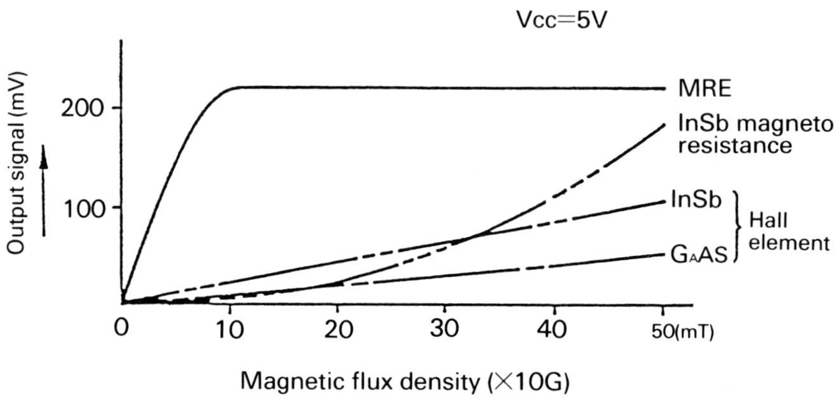

Figure 16.15 Output characteristics of various magnetic sensors.

material composition. Presently, nickel–cobalt alloy and nickel–iron alloy are used as materials having large ΔR.

Figure 16.14 shows variations in a resistance value taking place when the magnetic field intensity is changed by applying an additional magnetic field thereto at an angle perpendicular to the direction. Here, R_0 is a resistance value corresponding to zero magnetic field. It is known from the data in this figure that a magnetic resistance element yields some resistance value with magnetic field intensity at more than a certain level. This level of magentic field intensity is called 'saturated magnetic field intensity (H_s)' and expressed in the following formula:

$$H_s = K \times t/W + H_k \tag{16.7}$$

where

K = constant
W = width of resistor
t = thickness of ferromagnetic film
H_k = anisotropic magnetic field

H_k is inherent in every kind of material and is generally smaller in a nickel–iron system than in a nickel–cobalt system. Figure 16.15 shows the relation between the output and magnetic flux density of various kinds of magnetic sensor element.

The output from a ferromagnetic resistance element increases in response to an increase in the magnetic flux density in the case of a weak magnetic field but stays at an approximately constant level when the magnetic field is H_s or more. This phenomenon means that a metallic material can be used even at a high temperature because of a small variation in its output due to a temperature change.

16.3.2 Structure and characteristics

Since a ferromagnetic resistance element possesses the above characteristics, a combined unit consisting of this element and an amplified wave from a shaping circuit combined therewith is widely used in various fields as a rotation-detecting sensor. In conventionally produced sensors a sensor unit and an implifier unit are composed of respective individual chips. Vehicle speed sensors for automobiles of hybrid IC construction have been manu-

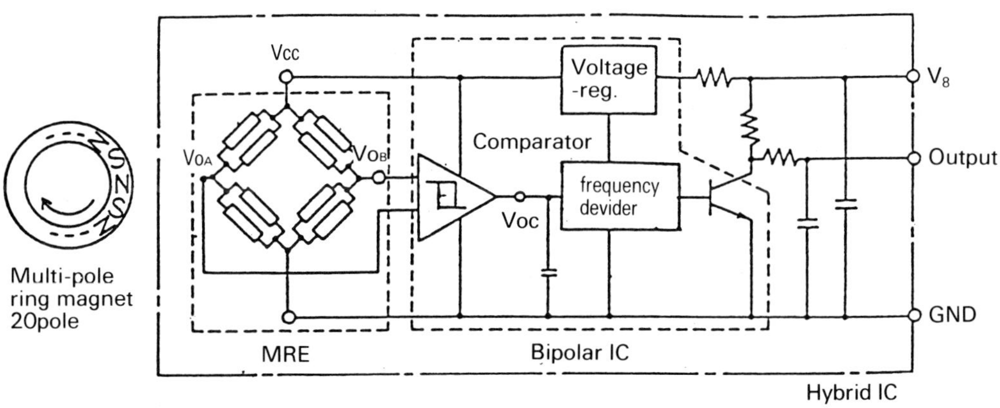

Figure 16.16 The circuit configuration of MRE vehicle speed sensor.

Figure 16.17 The hybrid micro-circuit.

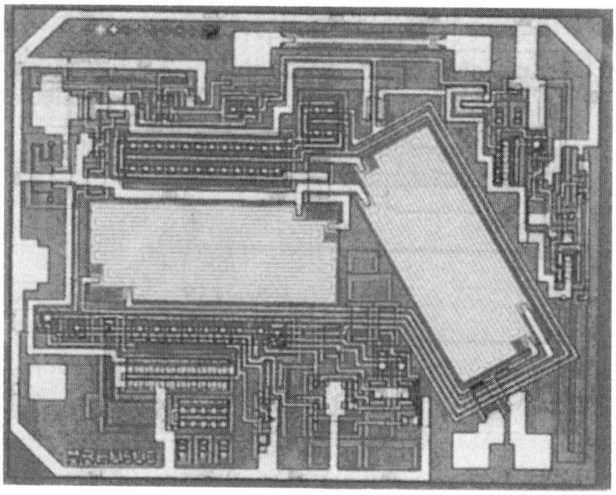

Figure 16.18 Intelligent MRE device.

Figure 16.19 Vehicle speed sensor.

Table 16.4 Specification of vehicle speed sensor

Item	*Spec.*
Supply voltage range	8–16 V DC
Operating temperature range	$-30 \sim 120\,°C$
Measurable transmission rev/min	$5 \sim 3000$ rev/min
Output type	20 pulses per rotation duty ratio; $50 \pm 10\%$ Open collector

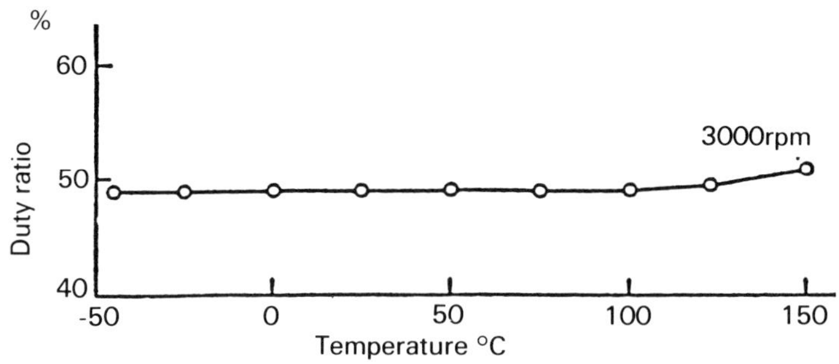

Figure 16.20 Temperature characteristic of vehicle speed sensor.

factured for five years in our company [9]. Figures 16.16 and 16.17 show the circuit configuration and exterior view of the hybrid IC.

Recently, a silicon substrate provided on its surface with a sensor unit and an amplifier unit integrated with each other has been developed [10]. Figure 16.18 displays an intelligent magnetic resistance element developed in our company and having its circuit constitution equalized to that of the above-mentioned hybrid IC. Figure 16.19 shows the exterior view of a vehicle speed sensor with the element mounted thereon. Its product specification and output characteristics are shown in Table 16.4 and Figure 16.20 respectively. The duty ratio of this product is described as less than 2% in a temperature range from $-40°$ C to 150° C.

16.4 IMAGE SENSOR

The function of an optical sensor depends upon the photoelectric transfer action of semiconductor, and a number of these sensors arranged in an array from a solid image sensor for converting image information to electrical signals. In this sensor, the CCD element, MOS element and PCD element etc. are put into practical use.

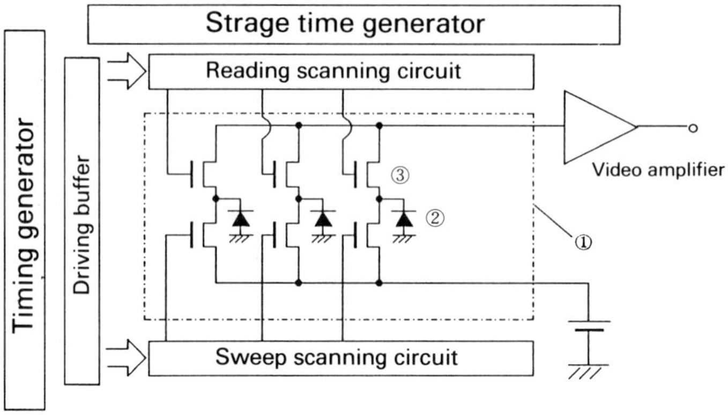

Figure 16.21 Configuration of intelligent image sensor.

Table 16.5 Chip dimension

Sensor type	*C-MOS (one-dimensional line sensor)*
Pixel number	1024 pieces
Pixel size	$14^{(H)} \times 14^{(V)}\ \mu m^2$
Photosensitive area of pixel	$14^{(H)} \times 1.26^{(V)}\ \mu m^2$
Chip size	$16.4^{(H)} \times 1.26^{(V)}\ \mu m^2$

H = horizontal; V = vertical.

The reason why these elements are presently considered desirable is that they have such features as compact size and light weight, easy intelligentialization, low power consumption, high reliability, etc.

Presently, these elements are used for bar-code readers, image scanners, etc. and are also considered useful for back monitors, automatic wiper sensors etc. The following relates to a C-MOS-type image sensor developed in our company [11].

16.4.1 Structure and characteristics

The configuration of an intelligent image sensor comprising 1024 pixels is shown in Figure 16.21 and chip dimensions are shown in Table 16.5. 1024 pieces of photodiode (2) are arranged in an array in a light-receiving region (1); a C-MOS transistor (3) for reading a pixel is connected to each diode and each gate of this transistor is connected to a reading scanning circuit. On the other hand, each drain of (3) is connected to an output unit via a common A1 wire for recognizing an image. Thus, functions of a crystal

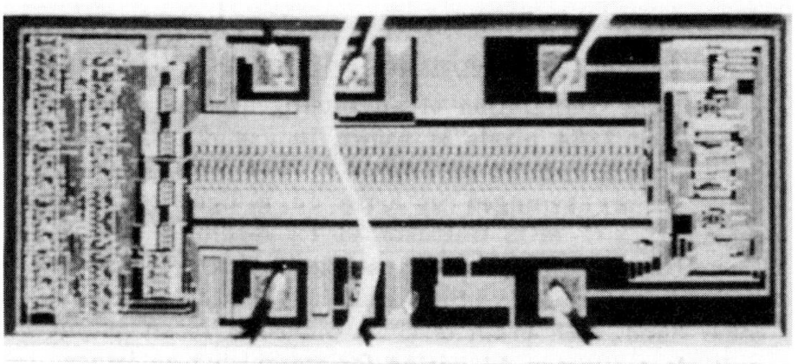

Figure 16.22 Intelligent image sensor device.

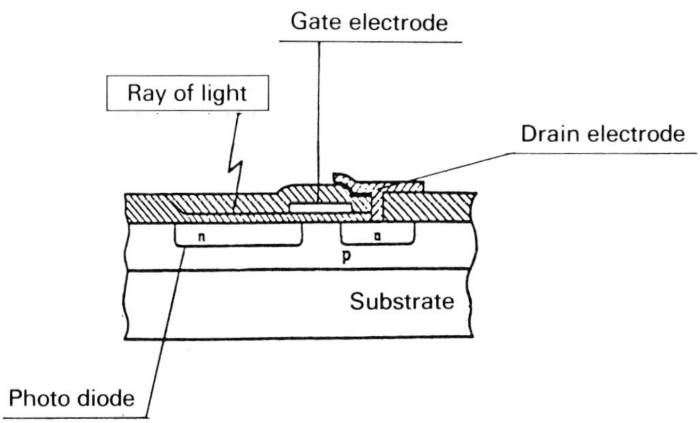

Figure 16.23 Cross-sectional view of pixel.

oscillation circuit, a timing generation circuit, a video amplifier provided with sample-holding function and an electronic shutter are integrated together.

An actual sensor chip is shown in Figure 16.22. The size of one pixel is $14 \times 14\,\mu m^2$, its photosensitive area is $14 \times 10\,\mu m^2$ and its rate of hole area is 79%. The cross-sectional view of a pixel unit is shown in Figure 16.23.

This sensor can operate with a 5 V unit power supply and its power consumption is less than that of a CCD element. The optical characteristics of this image sensor and its wavelength sensitivity characteristics are shown in Table 16.6 and Figure 16.24 respectively. Its sensitivity can cover a range of wavelengths from 400 to 900 nm with its peak at 530 nm and it can be used with a light source of green LED (565 nm), red LED (660 nm), etc. Figure 16.25 shows the exterior view of a product formed in a package provided with a sight glass.

Table 16.6 Optical characteristic of image sensor

Characteristic	*Symbol*	*Min.*	*Typ.*	*Max.*	*Unit*
Saturation output voltage ($R_L = 4.7\,k\Omega$)	V_{sat}		0.8		V
Saturation exposure	SE		0.8		$(\mu W/cm^2)s$
Responsivity	R		1		$(\mu W/cm^2)s$
Photoresponse nonuniformity	PRNU			10	%
Dark signal voltage	V_{dark}		10		mV
Spectral response	SR	400		900	nm
Resolution (pixel pitch)	R		28		μm

Note Operation frequency is 6 MHz; $V_{dd} - V_{ss} = 5$ V, $T_a = 25$ °C; PRNU is measured at 50% of SE (typ.) under uniform illumination; PRNU is defined as follows:

$$\text{PRNU} = \frac{V_{max} - V_{min}}{V_0} \times 100, \quad V_0 = \frac{\sum_{n=1}^{510} V_n}{510}$$

where

V_n = output of n pixels (average of 10 times measurement); V_{max} and V_{min} = maximum and minimum value of V_n.

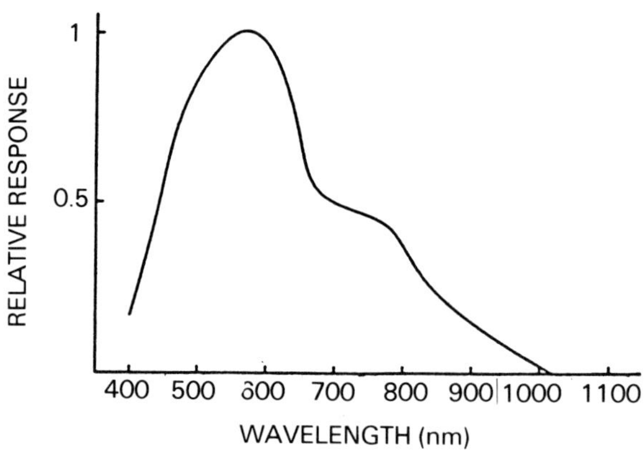

Figure 16.24 Relative spectral response as a function of wavelength.

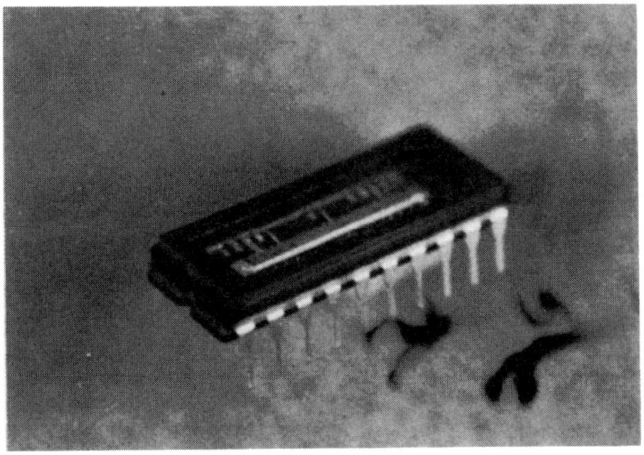

Figure 16.25 Packaged image sensor.

16.5 CONCLUSION

The present intelligentialization arose from demands for compact size, light weight and low cost. It is presently on the first step, but will progress to the third step by fully utilizing LSI technology from now on.

However, the reliability of sensors is liable to be lowered because of the much worse environment to which the sensor will be exposed in comparison with the conventional semiconductor element. Furthermore, the cost of the chip will increase due to the intricate work processes needed for integrating various functions on one chip, and because of a natural drop in yield.

REFERENCES

1. Middelhoek, S. (1985) Digest of Technical Papers, *Transducers '85*, pp. 2–7.
2. Wise, K.D. (1982) *Sensor and Actuators*, **2**(3), 229.
3. Kataoka, S. (1985) Digest of Technical Papers, *Transducers '85*, pp. 440–442.
4. Yamasaki, H. (1984) *Proceedings of 4th Sensor Symposium*, pp. 69–76.
5. Sugiyama, S. (1983) *Sensors and Actuators*, **4**(1), 113.
6. Hirata, M. (1984) *Proceedings of 4th Sensor Symposium*, p. 237.
7. Sugiyama, S. (1983) *Proceedings of 3rd Sensor Symposium*, p. 209.
8. Thomson, W. (1857) *Proceedings of the Royal Society of London*, **8**, 546.
9. Yoshino, Y. (1987) SAE Technical Papers No. 870 470.
10. Konno, H. (1988) *Annual Report IEICE*, C-2-73.
11. Yoshino, Y. (1989) *Internal Combustion Engine*, **28**, No. 351, 22.
12. Ina, O. (1989) SAE Technical Paper No. 891 709.

Index